IEE ELECTROMAGNETIC WAVES SERIES 35

Series Editors: Professor P. J. B. Clarricoats
Professor Y. Rahmat-Samii
Professor J. R. Wait

VHF and UHF ANTENNAS

Other volumes in this series:

Volume 1 **Geometrical theory of diffraction for electromagnetic waves** G. L. James
Volume 2 **Electromagnetic waves and curved structures** L. Lewin, D. C. Chang and E. F. Kuester
Volume 3 **Microwave homodyne systems** R. J. King
Volume 4 **Radio direction-finding** P. J. D. Gething
Volume 5 **ELF communications antennas** M. L. Burrows
Volume 6 **Waveguide tapers, transitions and couplers** F. Sporleder and H. G. Unger
Volume 7 **Reflector antenna analysis and design** P. J. Wood
Volume 8 **Effects of the troposphere on radio communications** M. P. M. Hall
Volume 9 **Schumann resonances in the earth-ionosphere cavity** P. V. Bliokh, A. P. Nikolaenko and Y. F. Flippov
Volume 10 **Aperture antennas and diffraction theory** E. V. Jull
Volume 11 **Adaptive array principles** J. E. Hudson
Volume 12 **Microstrip antenna theory and design** J. R. James, P. S. Hall and C. Wood
Volume 13 **Energy in electromagnetism** H. G. Booker
Volume 14 **Leaky feeders and subsurface radio communications** P. Delogne
Volume 15 **The handbook of antenna design, Volume 1** A. W. Rudge, K. Milne, A. D. Olver, P. Knight (Editors)
Volume 16 **The handbook of antenna design, Volume 2** A. W. Rudge, K. Milne, A. D. Olver, P. Knight (Editors)
Volume 17 **Surveillance radar performance prediction** P. Rohan
Volume 18 **Corrugated horns for microwave antennas** P. J. B. Clarricoats and A. D. Olver
Volume 19 **Microwave antenna theory and design** S. Silver (Editor)
Volume 20 **Advances in radar techniques** J. Clarke (Editor)
Volume 21 **Waveguide handbook** N. Marcuvitz
Volume 22 **Target adaptive matched illumination radar** D. T. Gjessing
Volume 23 **Ferrites at microwave frequencies** A. J. Baden Fuller
Volume 24 **Propagation of short radio waves** D. E. Kerr (Editor)
Volume 25 **Principles of microwave circuits** C. G. Montgomery, R. H. Dicke, E. M. Purcell (Editors)
Volume 26 **Spherical near-field antenna measurements** J. E. Hansen (Editor)
Volume 27 **Electromagnetic radiation from cylindrical structures** J. R. Wait
Volume 28 **Handbook of microstrip antennas** J. R. James and P. S. Hall (Editors)
Volume 29 **Satellite-to-ground radiowave propagation** J. E. Allnutt
Volume 30 **Radiowave propagation** M. P. M. Hall and L. W. Barclay (Editors)
Volume 31 **Ionospheric radio** K. Davies
Volume 32 **Electromagnetic waveguides: theory and application** S. F. Mahmoud
Volume 33 **Radio direction finding and superresolution** P. J. D. Gething
Volume 34 **Electrodynamic theory of superconductors** S.-A. Zhou

VHF and UHF ANTENNAS

R. A. Burberry

Peter Peregrinus Ltd. on behalf of the Institution of Electrical Engineers

Published by: Peter Peregrinus Ltd., London, United Kingdom

Peter Peregrinus Ltd.,
Michael Faraday House,
Six Hills Way, Stevenage,
Herts. SG1 2AY, United Kingdom

British Library Cataloguing in Publication Data

A CIP catalogue record for this book
is available from the British Library

ISBN 0 86341 269 6

Printed in England by The Redwood Press, Wiltshire

Contents

Attributions

Where illustrations have been taken from source documents not published by the IEE, this is shown on the relevant diagram by the words 'Reference X'. The reference number refers to the particular chapter to which the diagram belongs: thus Reference 7 on Fig. 2.3 implies the 7th reference of Chapter 2, in this instance Schelkunoff. The exceptions are for illustrations taken from two papers by the author, not referenced, and published by the Royal Aeronautical Society. These are identified by the words 'Courtesy RAeS'.

Textual references are in the form 'according to Y [Ref X]'.

Chapter 1
Introduction

Antennas do not become obsolete since they are based on unvarying physical principles. Sometimes the applications for which specific antennas were developed themselves disappear and the antennas fall out of use. They may nevertheless be admirably suited for new applications. One purpose of this book is to present to the antenna engineer as wide a range of antennas as possible, to indicate their performance characteristics and to comment on any limitations. In the course of 46 years of antenna design, the author has on many occasions been able to adapt old designs to new applications simply because he was aware of the original design. If this book gives the reader new ideas for antennas it will have served that purpose.

It is a book primarily for the antenna designer and user. Mathematics has been kept to the minimum necessary to provide guidance, particularly where the number of parameters involved would require a large number of graphs to give the information. In some instances a number of different formulae have been evolved for one particular antenna. Where possible, the merits and demerits of competing theories are discussed.

Internationally the VHF and UHF bands cover the frequency range 30 to 3000 MHz. At the upper end of this range the small wavelengths permit the use of antenna techniques which are impractical at lower frequencies. Some of these techniques are more appropriate to a book on microwave antennas and will not be discussed here. In general, the antennas described are usable over most of the frequency range with appropriate attention to manufacturing techniques.

Five basic types of antenna are described in detail. These are the dipole, monopole, loop, slot and notch and they are the building blocks for most antennas and antenna systems. Although these types would appear to be distinct, this distinction may not be so obvious in some of their derivatives. For example, when does a folded dipole become a loop? Is a transmission line antenna a top-loaded monopole or a vestigial notch? These reminders that there may be more than one way of analysing an antenna may help the reader to a better appreciation of a specific antenna or, indeed, of antennas in general.

Few antennas are used in situations where the influence of local surroundings can be neglected so considerable space is devoted to examining the impact of those surroundings. These may be a fixed structure such as a mast or a building, or they may be a vehicle or even a person. Methods of predicting performance may be theoretical or practical: advantages and disadvantages of different methods are discussed in terms of accuracy, time and cost.

It has sometimes been claimed that an antenna system extends as far as the terminals of the radio equipment. If this is true, then the antenna system performance may be made or marred by the performance of the feed system.

Components of the system — baluns, combiners, phase shifters — are described but space is also given to analysis of the complete system including cables and connectors. This is particularly important when the complete cable run has to accommodate a number of breaks either for ease of installation or for servicing.

Few of the books on antenna measurement tackle the problem of measuring 'installed performance', that is, the performance of an antenna in its working environment. The author was involved for some 15 years in attempts by the International Electrotechnical Committee (IEC) to produce specifications for methods of measurement of antennas in the field, on motor vehicles, ships and aircraft. This work involved not only the definition of measurement methods but also the design and approval of gain standards. Some of the IEC's work has been published but some appears to have foundered on the very real difficulties of measuring installed performance. The general principles and problems are discussed and details are given of some of the appropriate gain standards. What became obvious in the course of this work was that some of the methods used by Government agencies in Britain and abroad to assess antennas submitted to them were irrelevant to the intended use of the antennas. How many antennas have been rejected in consequence is not known but this is one outcome of the failure of the world-wide antenna community to get some agreed standards.

Chapter 2

The dipole

The theory of the dipole has been so extensively covered in print that it is not proposed to do more than summarise the main characteristics.

(i) For a straight thin dipole the current will be assumed to be sinusoidal falling to zero at the ends. If the dipole is not thin the zero current point is at the centre of each end; i.e. there is current on the end surface.

(ii) If the dipole is fed on its axis the radiation pattern in the equatorial plane will be circular. If the diameter is sufficiently large for the surface currents to be considered as a ring of thin dipoles fed in phase, the radiation pattern then becomes dependent on $J_0(kR)$ where R is the radius of the dipole cross-section and can be zero for some values of R.

2.1 The centre-fed dipole

2.1.1 *Radiation patterns*

Examination of the radiation pattern in the E plane shows that when the total length of the dipole exceeds about 1·25λ the maximum radiation ceases to be in the equatorial plane. Some typical patterns are shown in Fig. 2.1 for a thin dipole. As the ratio of length/diameter is decreased, subsidiary nulls are blurred. The axial null appears to be unaltered but pattern changes, particularly for overall lengths of 1·25λ, 2·5λ etc., are very much dependent on diameter. These longer dipoles are of little practical importance and in fact for most purposes dipoles of half-wave and full-wave length are used. The elevation pattern for a thin dipole of length $2L$ is given by

$$E=\left[\frac{\cos(kL\cos\theta)-\cos kL}{\sin\theta}\right]$$

where θ is the angle from the dipole axis.

2.1.2 *Impedance*

2.1.2.1 *The cylindrical dipole*

Various formulae have been proposed for the impedance of a cylindrical dipole of length $2l$ and radius a. Fig. 2.2 due to Schelkunoff [7] illustrates several important points:

(i) The electrical length of the dipole increases as the ratio l/a decreases. This is shown by the shorter lengths at which resonance occurs in the case of the fatter dipole.

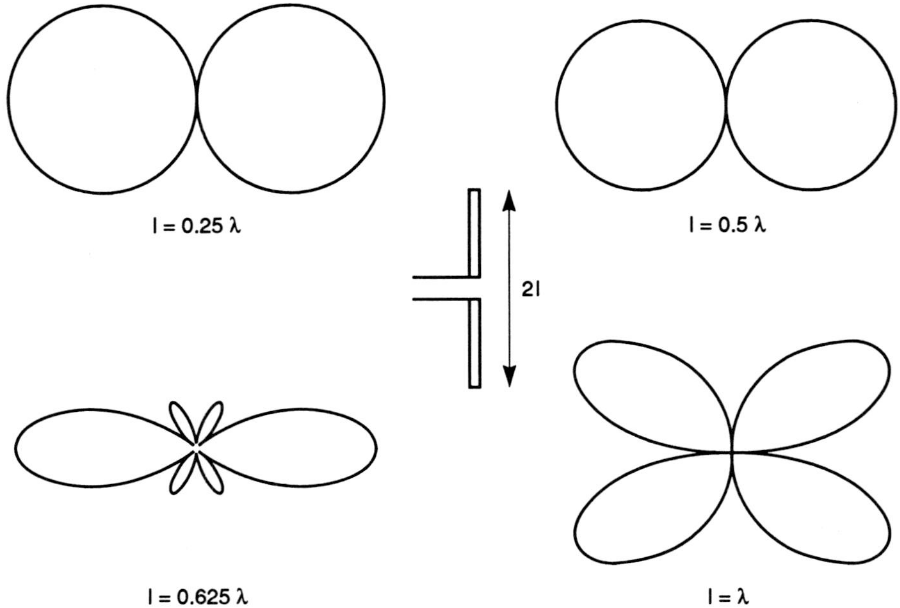

Figure 2.1 *Elevation patterns of a vertical centre-fed dipole as a function of length*

(ii) The radiation resistance at the 'half-wave' resonance changes only slowly with thickness compared with the marked reduction at the full-wave resonance.

(iii) The percentage shortening for resonance is greater at the full-wave than the half-wave case.

In Schelkunoff and in several other publications the parameter used is K_a, the mean characteristic impedance:

$$K_a = 120\left(\ln \frac{2l}{a} - 1\right)$$

It is worthwhile considering the range of l/a which is likely to be used in the VHF–UHF range. At the lower frequencies dipoles (as opposed to monopoles) are likely to be used on static installations. One main concern is mechanical strength so it is probable that at 30 MHz l/a will be of the order of 100. At 100 MHz l/a might be 150 for a radio and television receiving antenna but as low as 30 for a transmitting antenna. Similar conclusions can be drawn for higher frequencies: at 1 GHz the thinnest antenna could have $l/a = 37{\cdot}5$ and the fattest might be $l/a = 7{\cdot}5$. Fig. 2.3 shows the slow change of resistance at the half-wave resonance as a function of K_a, representing a range of l/a from 87·5 to 30 000. The reduction in length as a function of diameter is shown by Fig. 2.4. Smith [9] shows that the centre spacing affects the resistance: the smaller the spacing the lower the resistance at resonance. Interestingly the resonance for a 16:1 range of centre spacing occurred for the same lengths of dipole arms, the spacing having no apparent effect on the reactance.

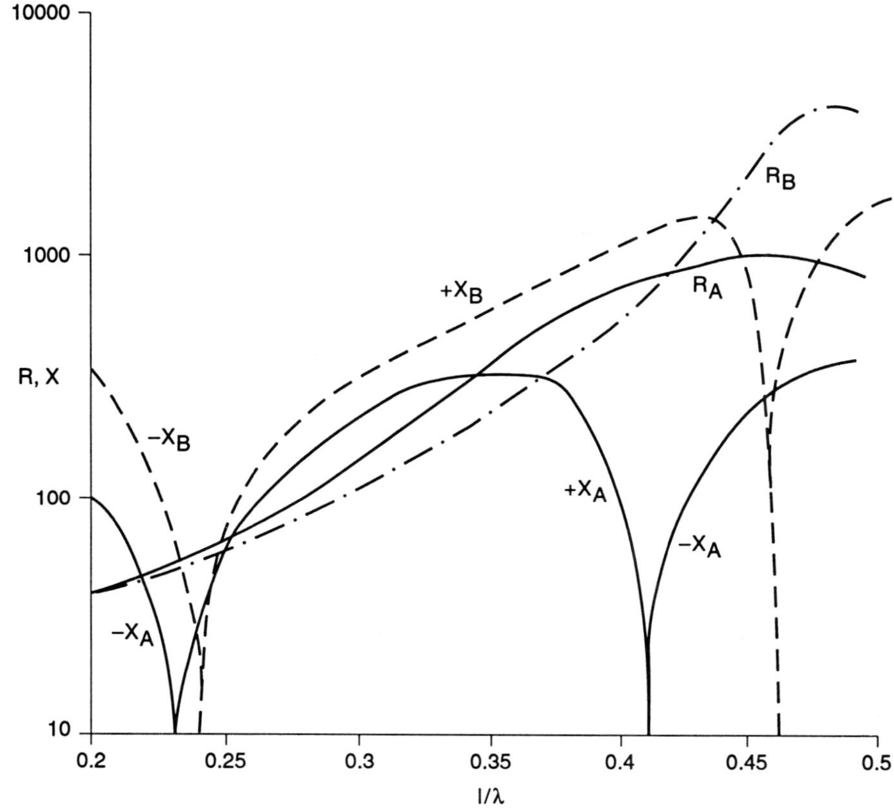

Reference 7

Figure 2.2 *Resistance and reactance of cylindrical dipoles as a function of length and diameter*

A: $l/a = 87$
B: $l/a = 30000$
a = radius

The dependence of resistance on diameter at anti-resonance is much more marked than in the half-wave case. Unfortunately much of the published work on cylindrical antennas with small l/a was done with monopoles and there is not a direct read-across to dipoles because of the different base capacitance effects. Of the genuine dipoles, Fig. 2.5 is a good example. The ratio, l/a, is 6·76 and the length at 'full-wave resonance' is about 0·60λ. This example is due to Brown and Stanier [2]. Fig. 2.6 shows resistance and reactance for a full-wave dipole reported by Buschbeck [3]. Here $l/a = 9{\cdot}7$ and the length at resonance is 0·69λ. The curves shown are for the dipole without the series reactance provided by the open-circuit line shown.

The impedance measured for any real dipole will be affected by the capacitance between the inner ends of the arms, the inductance between the arms and the feed system and the characteristics of the feed system itself. For

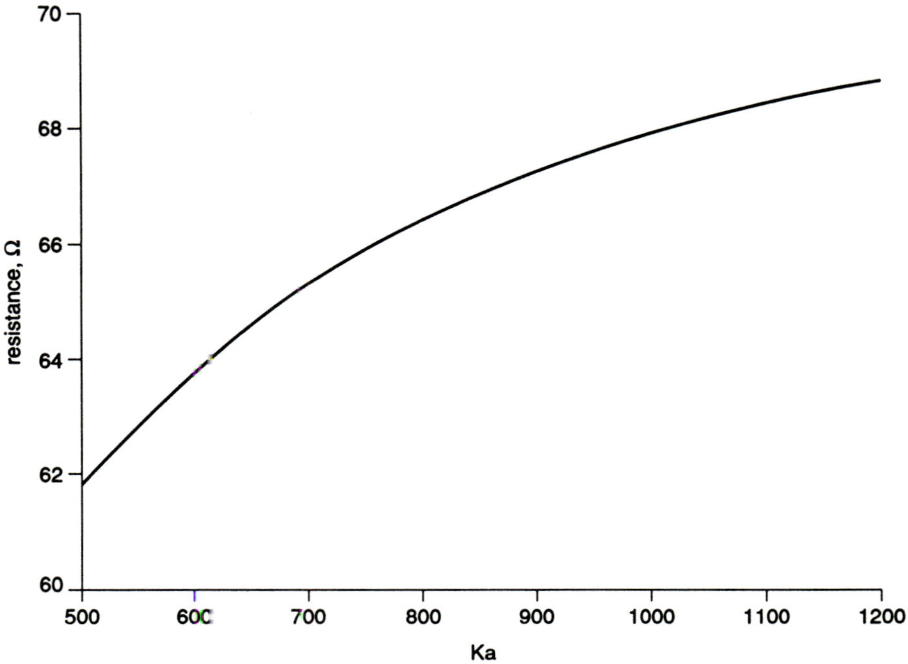

Reference 7

Figure 2.3 *Resistance of cylindrical dipoles at half-wave resonance as a function of diameter*

the radiation patterns to be as described in Section 2.1.1 the dipole has to be fed in a balanced manner. The most natural way of doing so is by twin transmission line. This is excellent for the higher impedances of full-wave dipoles but is unsatisfactory for half-wave dipoles. Some form of 'balun' or balance-to-unbalance transformer is then required.

2.1.2.2 *Feed system*

(*a*) *Quarter-wave sleeve*

One of the simplest arrangements is the 'quarter-wave can' shown in Fig. 2.7. In this arrangement the coaxial line formed by the outer of the feed line and the inside of the sleeve provides a high impedance at AB thus minimising current flow on the outside of the feed line along its length. The Z_0 of the coaxial line needs to be as high as possible to maintain a high impedance over an adequate frequency band; for practical reasons a Z_0 of much more than 100 ohms is improbable. This feed arrangement is not very good at very high frequencies because of the imbalance inherent in the method of connecting to the coaxial line: the dipole arm connected to the line inner conductor will have some capacitance to the outer line as well as some additional series reactance. The sleeve itself does not contribute to the dipole impedance.

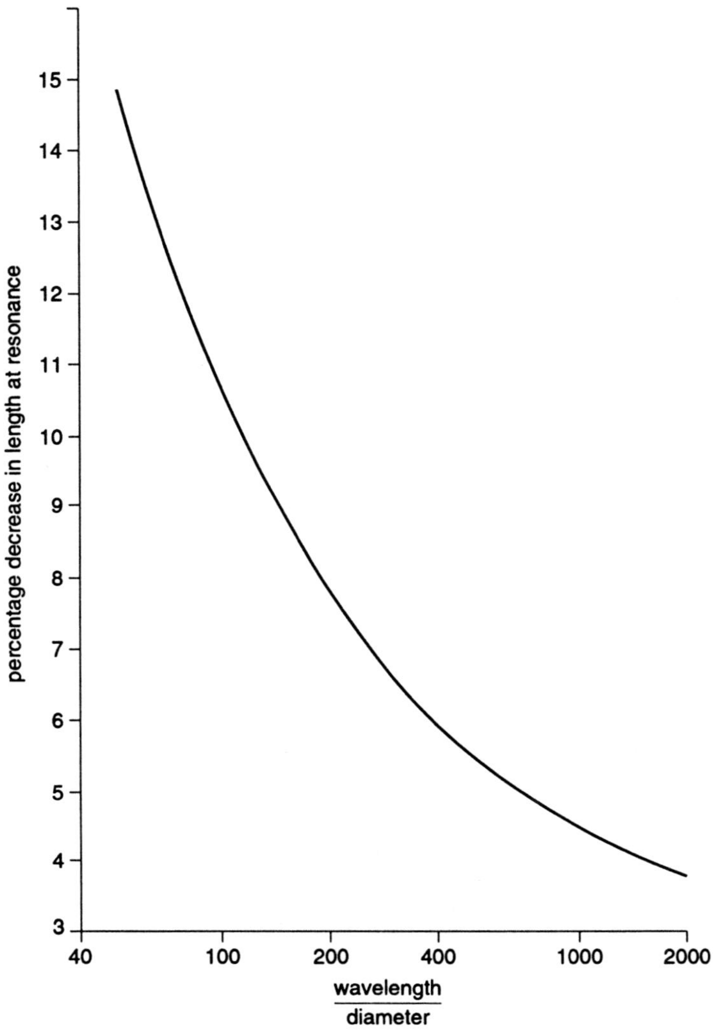

Figure 2.4 *Percentage reduction in length of half-wave dipole at resonance as a function of diameter*

(b) Twin line balun

The twin line balun of Fig. 2.8 has the great merit of maintaining balance whatever the line length. The form shown in (*a*) is sometimes known as the Pawsey stub after its British inventor. This version modifies the dipole impedance by the shunt reactance due to the short-circuited twin line plus the series inductance due to the connecting link. It is often desirable to sheath the balun in a metal tube for ease of mounting the antenna, Fig. 2.8*b*. This reduces the Z_0 of the twin line. The balun of Fig. 2.8*a* has the disadvantage of presenting a DC short-circuit across the feed line thus making it impossible to check for

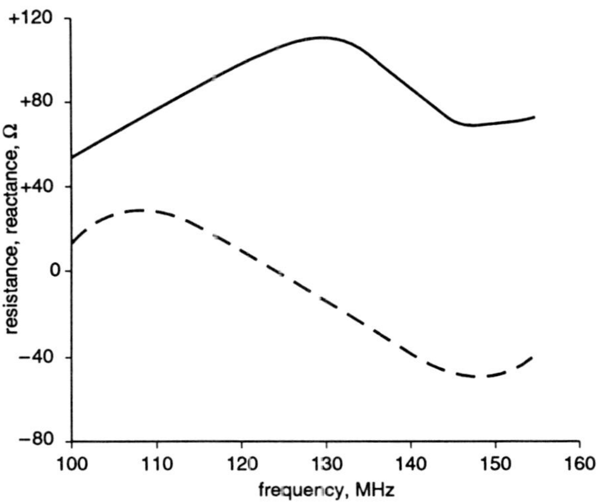

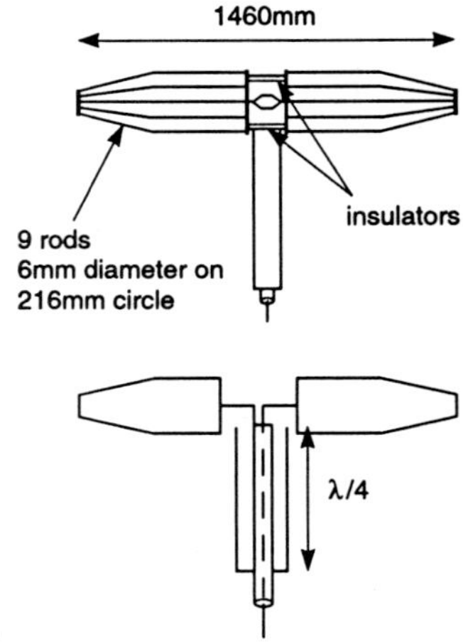

Figure 2.5 *Impedance of a fat cage dipole*

—— resistance
– – – reactance

inadvertent short-circuits by continuity testing. The version of Fig. 2.8*c* overcomes this problem. Because the reactance due to the short-circuited twin line is of the form $jZ_0 \tan kl$ and is in shunt with the dipole impedance it can be used to improve the impedance bandwidth of a half-wave dipole. Shnitkin and Levy [8] give details. For full-wave dipoles the compensation method shown in Fig. 2.6 is appropriate. This type of balun is useful for frequencies between about 80 MHz and 1500 MHz: below this band the line becomes inconveniently long and above it the diameter of suitable solid-jacketed cables may make the gap at the centre of the dipole proportionately so large that the series inductance is too large. This type of balun is not suitable for construction in microstrip as the limited width of the 'ground-plane' allows too much radiation from the 'inner conductor'.

(*c*) *Split balun*

The split balun or slotted feed is suitable for frequencies above about 300 MHz and is more compact than the twin line. Fig. 2.9 shows a typical arrangement. It should be noted that this system provides a transformation. If the unsplit line has a characteristic impedance Z_0 then the left hand side of the split section has an impedance $2Z_0$. This produces a transformed impedance Z_A at AA which is shunted by the short-circuited section at the right side of the split, a reactance $2jZ_0 \tan kl$. It may therefore be desirable to alter the diameter of the inner of the coaxial line within the split section to obtain the desired output impedance. The top view in Fig. 2.9 shows a wide strap from inner to outer; this is to reduce the series inductance at this point.

One problem that may arise in the presence of water across the split: putting a dielectric over the slots may cause the unsupported left-hand side to move inward thus altering the Z_0. Dielectric in the slots will modify the electrical

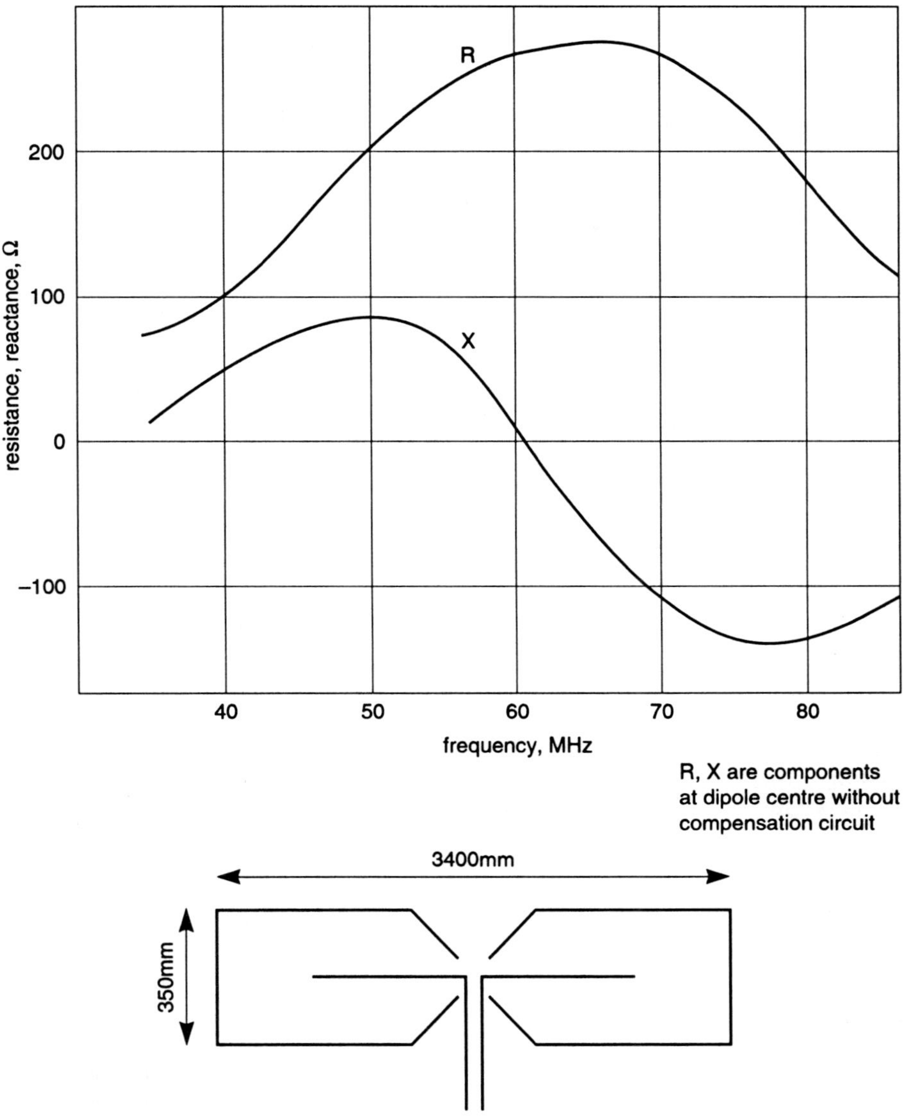

Figure 2.6 *Impedance of a cylindrical 'full-wave' dipole*

length so some care is needed in the mechanical construction. It may be concluded that this type of balun should not be used for permanent outdoor installations; certainly one should be aware of the possible problems. The slots should not be made too wide or the characteristic impedance will be altered. Attempts have been made to use flat strips for the slotted section. Blackband [1] does show in his Fig. 18.22 a formula for such an arrangement but it is probable

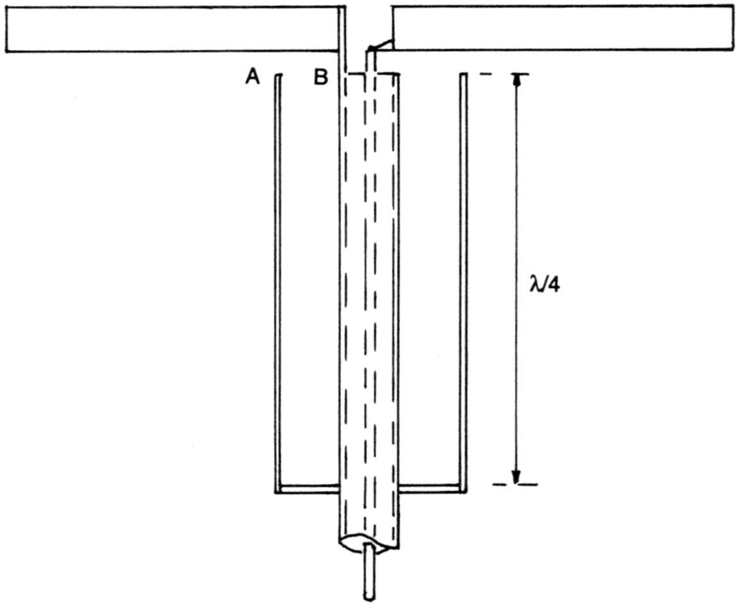

Figure 2.7 *Quarter-wave sleeve feed*

that the strip widths would have to be much wider than is practical for the split balun.

This type of balun does lend itself to construction in strip line. One such arrangement using a non-contacting system is shown in Fig. 2.10 with a fat full-wave dipole. There are two advantages to be gained:

(*a*) Isolation between inner and outer is provided by the dielectric.
(*b*) The line impedance can readily be altered by adjusting the strip width.

The slot still remains vulnerable to moisture but the whole assembly could be surrounded by low density closed-cell foam with a more robust cover outside it.

(*d*) *Lumped-circuit network*

At the lower frequencies and where space is at a premium a lumped-circuit balun may be appropriate. If jX is the reactance of the components in Fig. 2.11 then an antenna impedance Z_A is transformed to X^2/Z_A. This type of balun is narrow band, depending as it does on the reactances all remaining equal. It has been used at frequencies up to nearly 200 MHz; with printed circuit coils and close tolerance capacitors, the range could be extended to higher frequencies. A neat test rig was devised for checking that balance has been achieved. The dipole incorporating the balun was mounted in front of a plane reflector at a spacing of about 0·1 λ. A radiation pattern was measured in the plane of the dipole normal to the reflector. The dipole was then rotated through 180° and the pattern repeated. If the dipole is balanced then the two patterns will be identical. By changing the frequency it is easy to determine which component is in error.

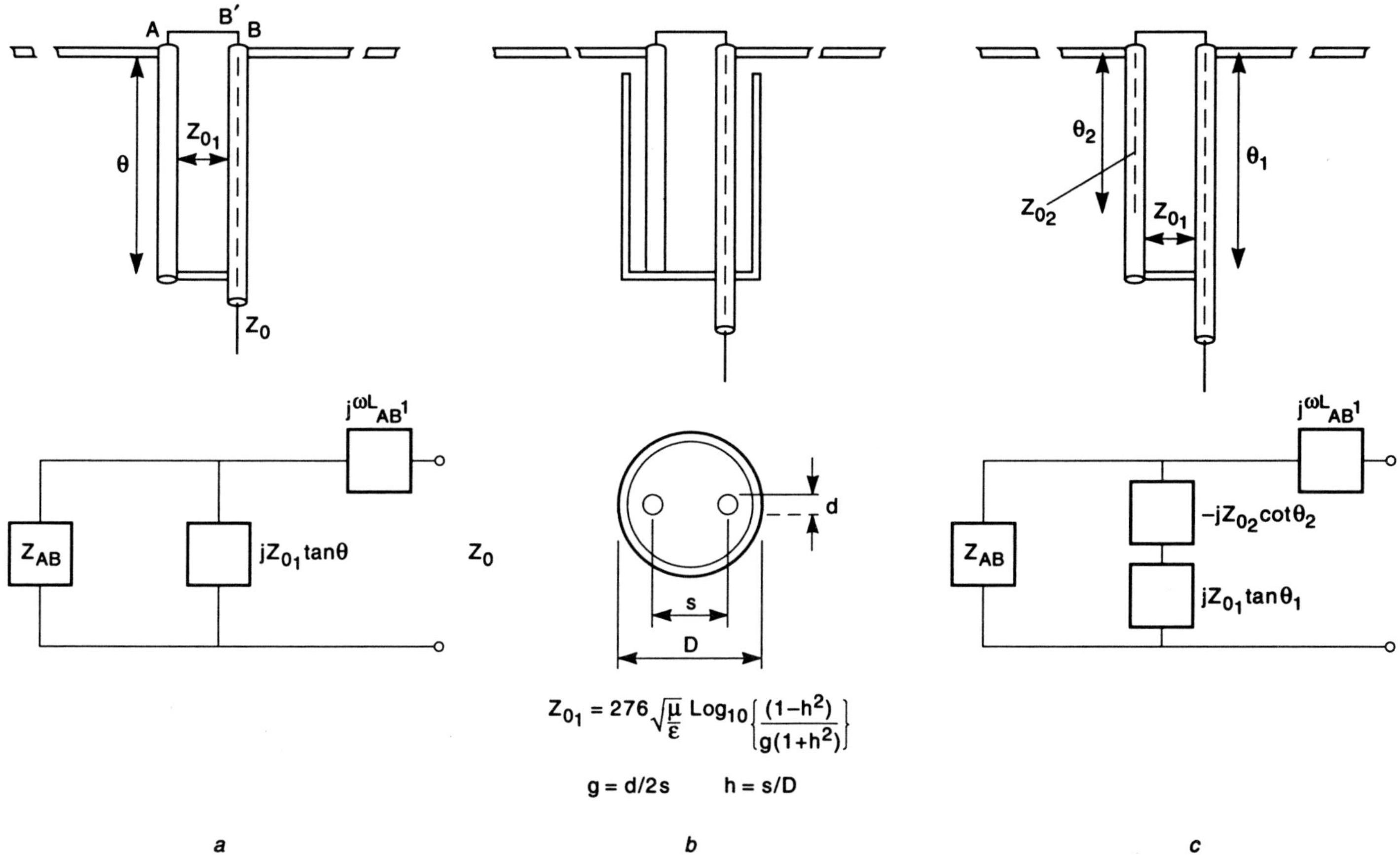

Figure 2.8 *Twin line baluns* *a* Twin line *b* Shielded twin line *c* Twin line with DC open circuit

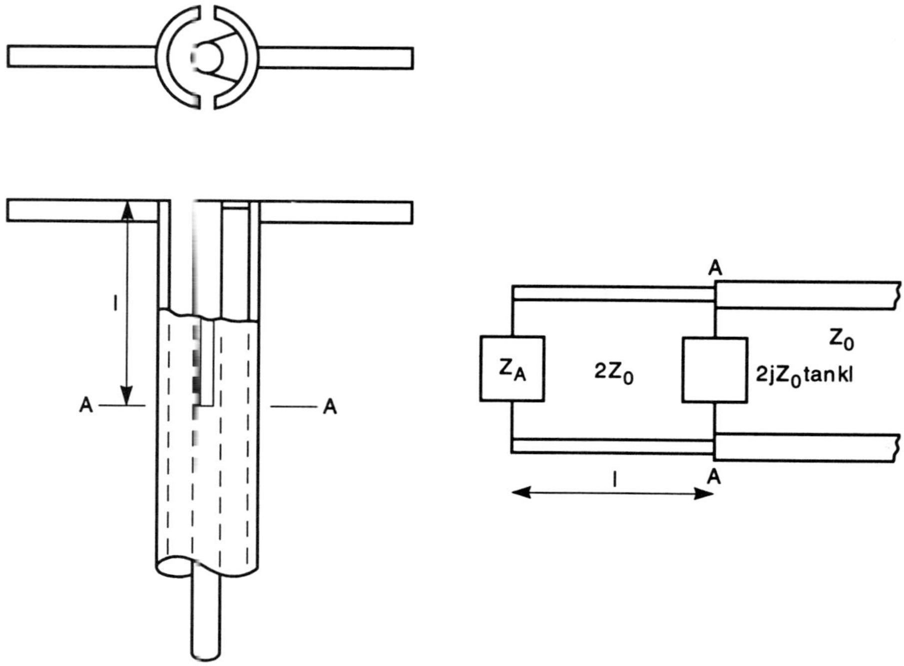

Figure 2.9 *Split balun*

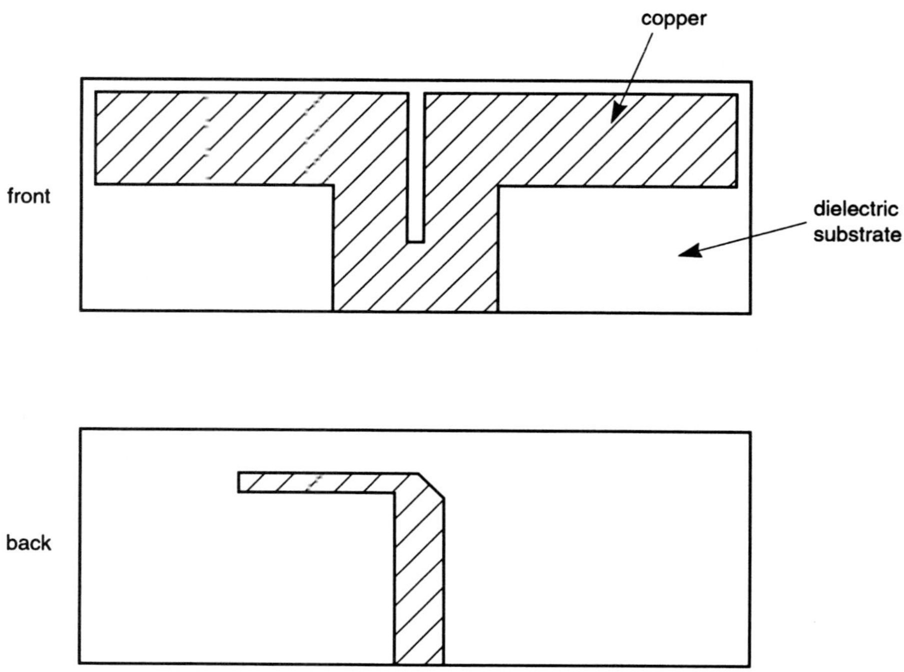

Figure 2.10 *Printed circuit dipole and balun*

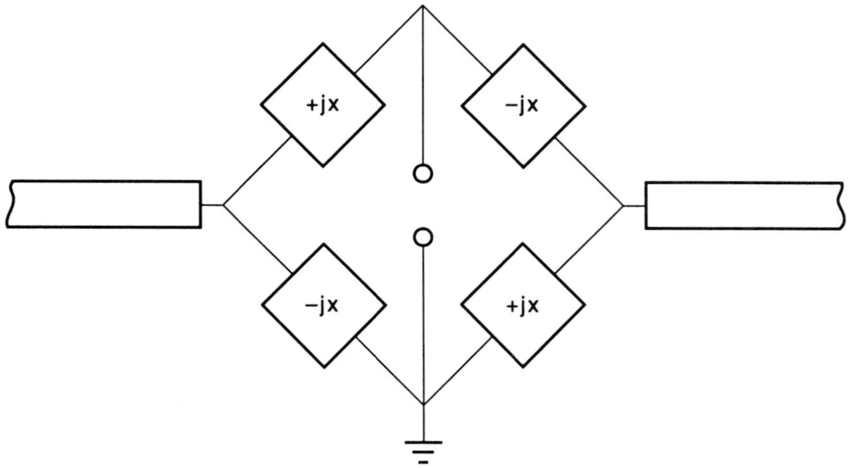

Figure 2.11 *Lumped circuit balun*

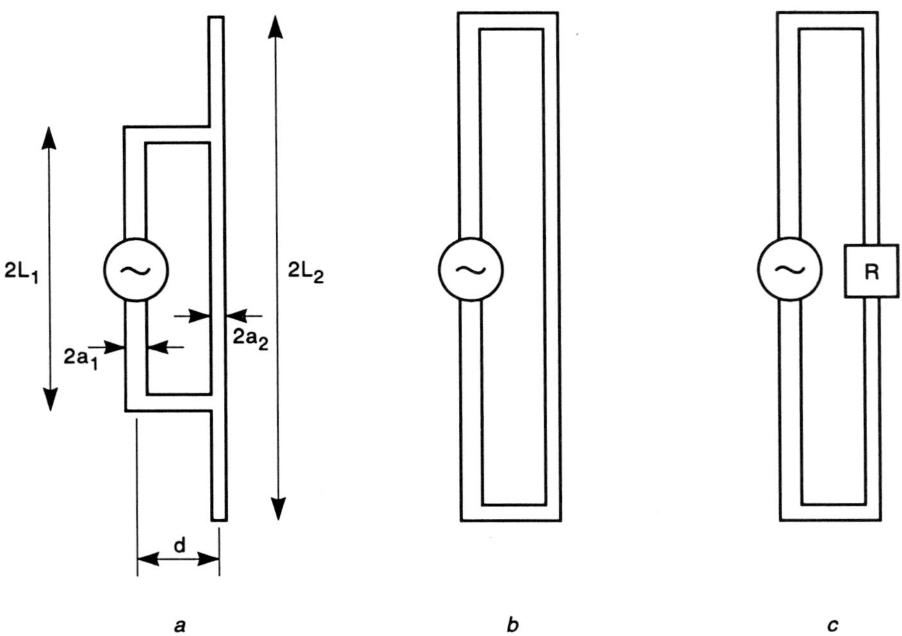

Figure 2.12 *Tee-match dipole*

a Tee-match antenna
b Folded dipole
c Terminated folded dipole

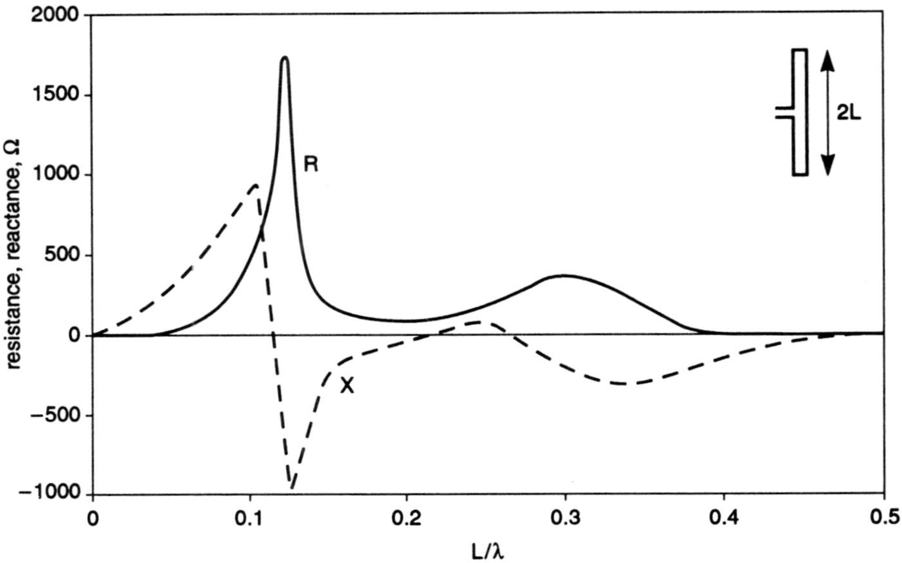

Figure 2.13 *Impedance of a typical folded dipole as a function of element length*

2.1.3 *The folded dipole*

The folded dipole is a particular form of the tee-match dipole (Fig. 2.12) in which the tee extends to the ends of the antenna. In this antenna two components, one radiating and one non-radiating, have to be considered. The current is divided between the two parallel conductors the 'current division

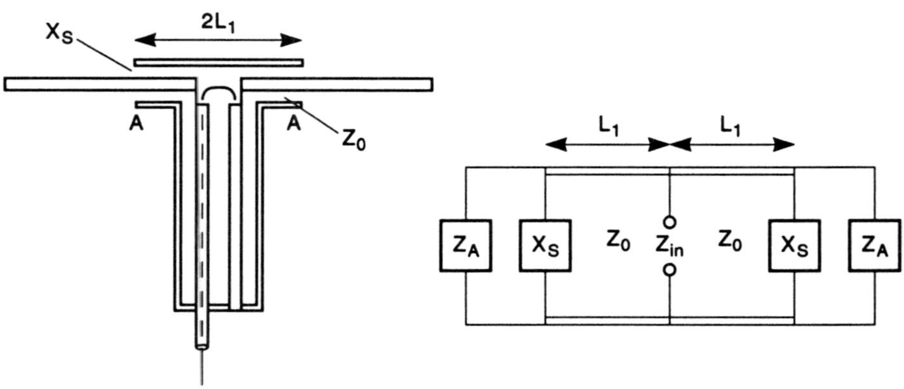

Figure 2.14 *Sleeve dipole*

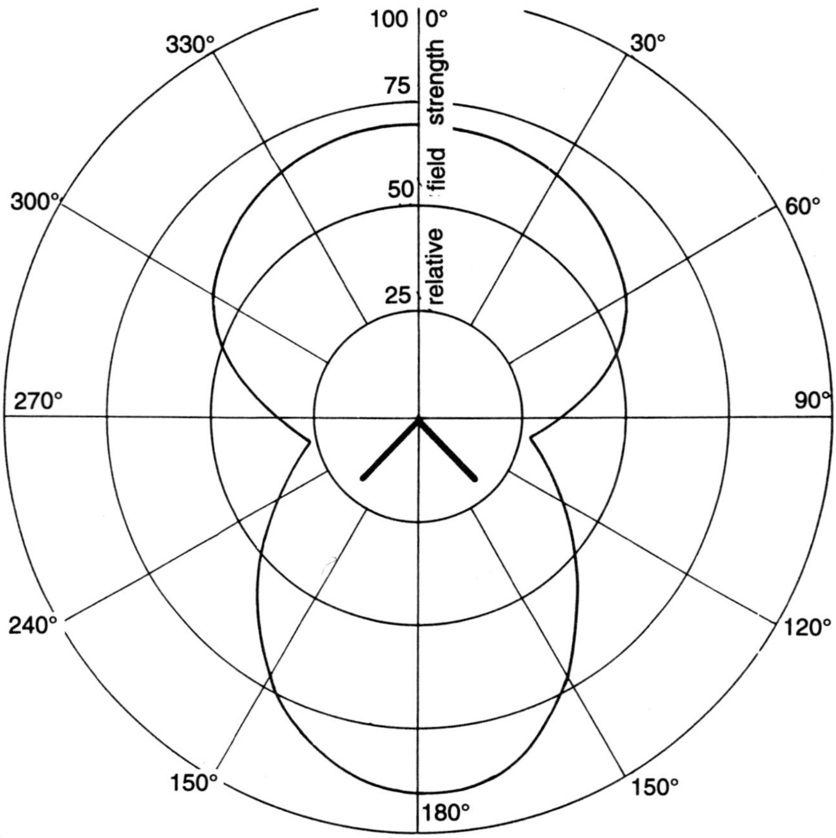

Figure 2.15 *Azimuth pattern of horizontal vee dipole*
Courtesy RAeS

factor' a being given by

$$a = \frac{\cosh^{-1}\left(\dfrac{v^2 - u^2 + 1}{2v}\right)}{\cosh^{-1}\left(\dfrac{v^2 + u^2 - 1}{2uv}\right)} \approx \frac{\ln v}{\ln v - \ln u}$$

where $u = a_2/a_1$ and $v = d/a_1$.
The two conductors are equivalent to one conductor thicker than either with radius a_e where

$$\ln a_e \approx \ln a_1 + \frac{1}{(1+u)^2} \cdot (u^2 \ln u + 2u \ln v)$$

Let Z_r be the input impedance of this equivalent dipole.

The two parallel conductors form a two-wire transmission line short-circuited at its outer end. Each of the two sections then has an impedance Z_F at the dipole

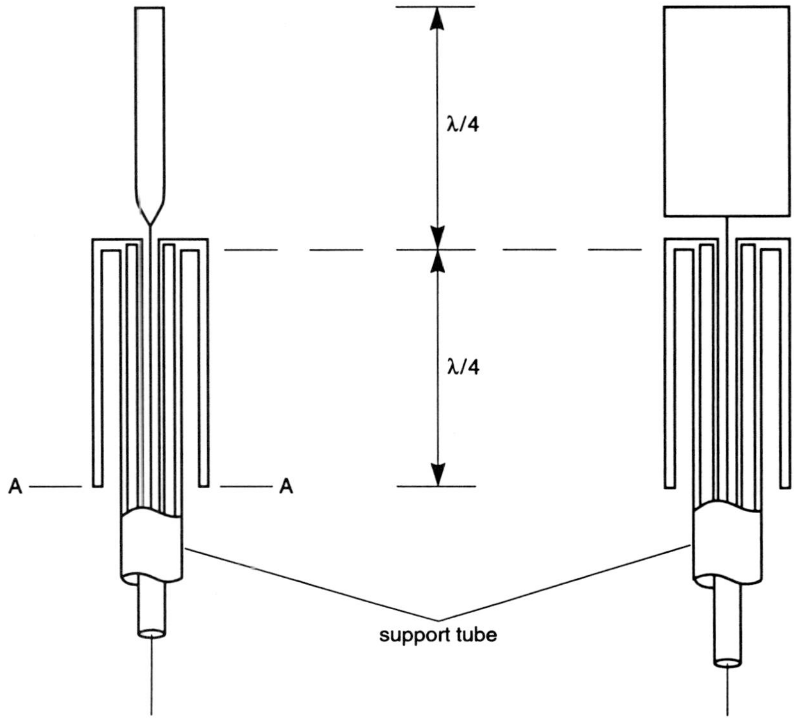

Figure 2.16 *Cylindrical coaxial dipoles*

centre where

$$Z_F = jZ_0 \tan kL$$

Z_0 being the characteristic impedance of the parallel line.

It can be shown (for example in Wolff [11]) that the input impedance of the antenna is

$$Z_{in} = \frac{2(1+a)^2 Z_r Z_f}{2Z_f + (1+a)^2 Z_r}$$

Effectively the impedance Z_r is stepped up by a ratio $(1+a)$, which is a function of the conductor diameters and spacing, and shunted by the non-radiating impedance.

In the special case where $L_1 = \lambda/4$, Z_f becomes very large and

$$Z_{in} = (1+a)^2 Z_r$$

It might appear that folding would offer a method of increasing the resistance of a dipole shorter than $\lambda/2$. This is not so because of the effect of the shunting reactance. Fig. 2.13 shows the impedance of a typical folded dipole as a function of element length. It will be seen that the first resonance occurs at about $L = 0{\cdot}125\lambda$ and the resistance is extremely high. These were measured figures.

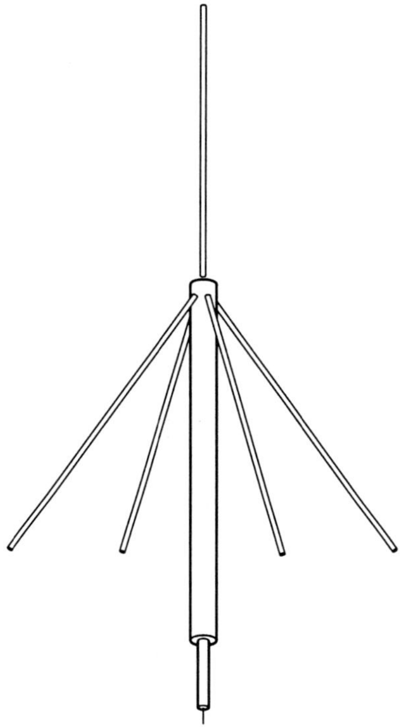

Figure 2.17 *Conical coaxial dipole with radials*

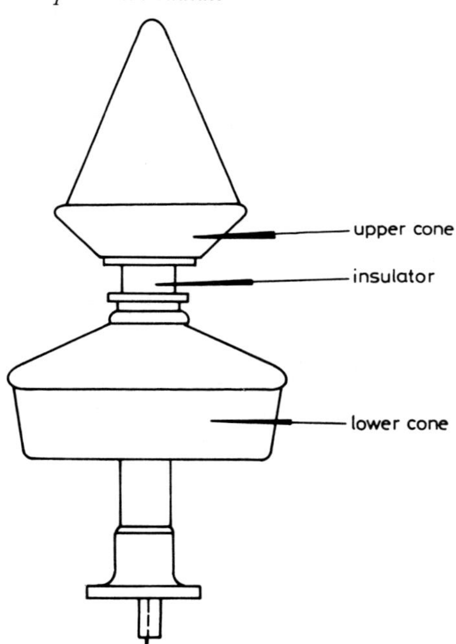

Figure 2.18 *UHF Naval bicone*

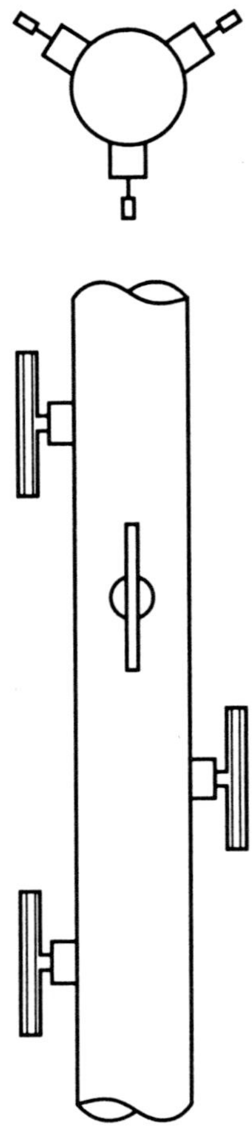

Figure 2.19 *Stack of vertical dipoles round a mast*

The terminated folded dipole shown in Fig. 2.12*c* is somewhat controversial. It was originally developed by MacPherson as a variant of the rhombic antenna with short arms and was claimed to have good broadband properties in the HF band. Details are given in Williams [10] and the writer confirms that it did have good broadband impedance characteristics although at what cost in efficiency is not known. Josephson [5] produced a theory for the antenna and measured impedance on scale model monopoles at 1:30 and 1:60 scales. His results do not agree with Williams. Josephson's results oscillate so wildly that it seems probable that his terminating loads were incorrect: this type of antenna is known to be sensitive to load impedance and there would have been problems in 1957 in obtaining good high value pure resistive loads at frequencies up to 600 MHz.

The value of this antenna for broadband operation must be considered non-proven but is worth further investigation.

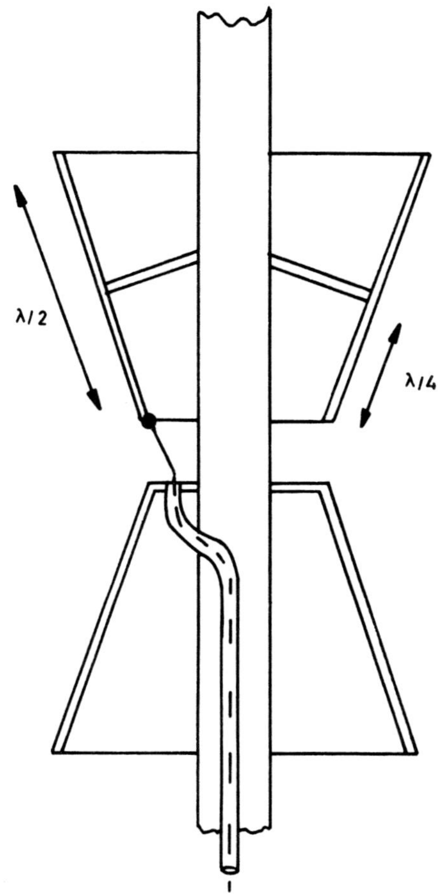

Figure 2.20 *Element of stacked full-wave dipoles*

2.1.4 *The sleeve dipole*

To provide mechanical support, the can surrounding a balun as in Fig. 2.8*b* is often extended to cover the base of the dipole arms as in Fig. 2.14. This sleeve is effectively an extension of coaxial line which modifies the dipole impedance but not, for dipoles up to about a half-wavelength, the radiation patterns. In this diagram, Z_A is the impedance of each dipole arm at the top of the sleeve, X_S is the reactance, generally capacitive, at the same point and Z_0 is the characteristic impedance of the coaxial line formed by the sleeve and the dipole arm. In practice there will be dielectric between the two conductors; this must be considered in determining Z_0 and the electrical length of the line. When each dipole arm is approximately $\lambda/4$ the resonance impedance at A is $R_A = R_0/\cos^2 kL$ where R_0 is the base impedance of the monopole without the sleeve. Since the sleeve generally increases the radiation resistance and the capacitive reactance there are few electrical advantages in sleeved dipoles unless they are used with a

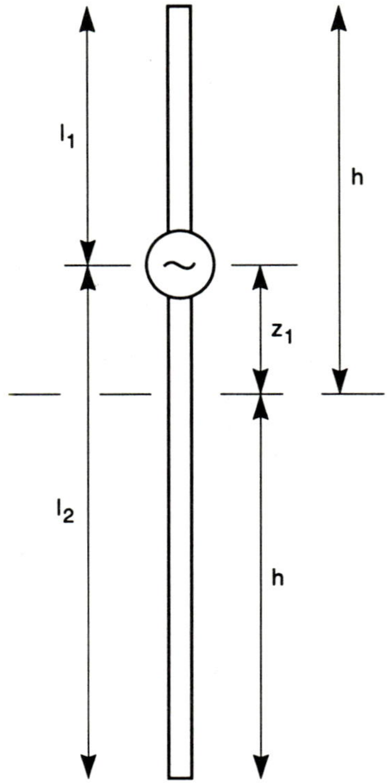

Figure 2.21 *Asymmetrical dipole*

parasite or reflector which would reduce their input impedances. On the other hand, the sleeve is of considerable practical significance with monopoles as is demonstrated in Chapter 3.

2.1.5 *Vee dipole*

The vertical dipole has an omniazimuth radiation pattern but the horizontal pattern of a horizontal dipole is a figure-of-eight (Fig. 2.1). An omniazimuth pattern is sometimes required with horizontal polarisation and a bent dipole will give a reasonable approximation. The arms may be curved to give a U-dipole or simply bent back to form a vee. An azimuth pattern of a half-wave dipole with 90° included angle is shown in Fig. 2.15.

2.1.6 *The coaxial dipole*

2.1.6.1 *Cylindrical form*

In a vertical dipole with a horizontal central feed the down lead should be at least a half-wavelength away to avoid significant disturbance of the radiation pattern by re-radiation. This is not always convenient and it may be better to use a coaxial dipole which is designed to decouple the down lead. Fig. 2.16 shows typical cylindrical coaxial dipoles. The fatter dipole will have a better bandwidth but as shown the top section puts a high wind-load on the inner conductor. If this type is to be used it may be better to enclose the whole dipole in a thin dielectric tube which will add materially to its strength. As the coaxial dipole is often clamped to the top of a stouter tube it will usually have a metal tube coaxial with the lower sleeve, the coaxial cable being inside this support tube.

The lower sleeve functions as the outer of a short-circuited transmission line, chosen to be a quarter-wavelength long to provide a high impedance between the outer sleeve and the outer of the coaxial cable. If it is thought necessary to close the bottom of the sleeve with a dielectric bung to prevent insects sheltering inside, the bung must be of low-loss material. It will in any case act as a capacitance across the end of the line and will increase its electrical length.

For frequencies much below 100 MHz this arrangement is not very good mechanically unless the transmission line can be made of sufficiently large diameter to give adequate bending strength. Whilst filling it with dielectric reduces the length required it also reduces the line Z_0 so the bandwidth over which the choke is effective will decrease. Even with a quarter-wave line the impedance will only remain high over about a 10% bandwidth as it is not practical to obtain a high Z_0 for the line. It is possible to use capacitive loading at the open end but this increases the complexity of the antenna.

2.1.6.2 *Conical form*

For wider bandwidths it is necessary to increase the choke Z_0 by making the outer conical in form. For frequencies above about 200 MHz a solid cone is possible but for lower frequencies it is more usual to employ a series of radials.

(*a*) *Cone formed of radials*

Fig. 2.17 shows a typical arrangement with four radials. The number and the angle of the radials has some influence on impedance. If the radials are horizontal they are effectively a ground plane and the antenna becomes a monopole with a base resistance of between 30 and 35 ohms depending on its length/diameter. If, on the other hand, the radials are vertical forming a vestigial cylinder then the resistance will lie between 60 and 70 ohms, assuming the top element and the radials to be a quarter-wavelength. Thus by adjusting the depression angle the impedance can be optimised for either 50 ohm or 75 ohm coaxial cable.

There is some evidence that, over a wide frequency band, resonances in the radials can result in a very variable VSWR. To overcome this, the radials should be joined together at intervals by circular conducting rings. These should be closer together at the upper end of the radials to improve the performance at the higher frequencies.

An obvious extension of the antenna in Fig. 2.17 is to make the upper element also as a number of radials. The resulting biconical dipole has a very flat impedance characteristic if the flare angle between opposing radials is of the order of 90°. This topic will be considered in more detail in Chapter 3 since most of the measured results relate to monopoles.

(*b*) *Solid cone*

Usually the solid cone has a smoother impedance curve than the cone constructed by radials as explained above. One disadvantage is increased capacitance at the base of the cones so it may be necessary to increase the spacing. For broadband operation it is preferable to work around the full-wave rather than the half-wave condition. In this case the cones do not extend linearly outward but are brought back towards a point. Fig. 2.18 depicts a bicone used for the UHF communications band, 225–400 MHz, on British warships.

The development from the bicone to the discone is described in Chapter 8.

2.2 Stacked dipoles

To increase the gain of an antenna system and yet retain an omniazimuth pattern vertical stacking is often used. With centre-fed vertical dipoles the usual problem arises of disposing of the downleads without perturbing pattern for impedance. One method is to mount dipoles around the support structure as shown in Fig. 2.19. Three or four dipoles around the circumference should be adequate to produce a near omnidirectional pattern. The phasing can be adjusted to produce a beam tilt if desired.

An alternative arrangement in Fig. 2.20 uses full-wave dipoles through which the support tube passes. Current on the support tube is minimised by the high impedance chokes. An octave bandwidth should be achievable for each element.

2.3 The asymmetrical dipole

Few people would set out to use an asymmetrical dipole, Fig. 2.21, deliberately because its radiation patterns are not generally very useful. It does, however, occur naturally in a number of circumstances. Fig. 15.3 shows two examples: the man pack and the aircraft fin-cap (or wing tip) antenna. It is probable also that VHF telemetry and command antennas on drum-shaped satellites should also be regarded as asymmetric dipoles if the drum diameter is not very large in terms of wavelengths.

From Fig. 2.21 it can be seen that the current on the dipole can be expressed as

$$I(z) = \begin{cases} I_1 \sin[k(h-z)] & z > z_1 \\ I_2 \sin[k(h+z)] & z < z_1 \end{cases}$$

It can be shown that the current is sinusoidal if the dipole is thin, regardless of the position of the feed point, and is always zero at the ends. This will be modified slightly if the dipole is fat. It would be possible to calculate the radiation patterns in the z-plane. A number of measured patterns are presented in Jasik [4] and demonstrate that long asymmetrical dipoles are very frequency conscious. For overall lengths up to about one wavelength the patterns will not differ significantly from those of symmetrical antennas. King [6] proposed a mean value theorem for the impedance of an asymmetrical dipole. According to this theorem the impedance Z is approximately $\frac{1}{2}(z_1 + z_2)$ where z_1 and z_2 are the impedances of two centre-fed dipoles of length $2l_1$ and $2l_2$ (Fig. 2.21). One practical point arising from this theorem concerns fin-cap antennas on aircraft. One arm of the asymmetrical dipole is the remainder of the fin plus the fuselage.

Reference to the impedance curves of fat dipoles shows that if the equivalent diameter is greater than 0·05λ then the impedance remains substantially constant for arm lengths greater than 0·5λ. So if the fin-cap antenna itself is broadband the impedance seen at the feed point of the asymmetrical dipole will also have broadband characteristics. This would certainly not be the case for a thin asymmetrical dipole of considerable electrical length. It also means that in the design phase no mock-up, for frequencies above about 100 MHz, need extend below the cap for more than a half-wavelength.

2.4 References

1 BLACKBAND, W.T.: 'Coaxial transmission lines and components' *in* RUDGE, A.W. *et al.* (Eds.): 'Handbook of antenna design, Vol. 2' (Peter Peregrinus Ltd, 1983)
2 BROWN, A.H., and STANIER, H.M.: 'Recent developments on VHF ground-communication aerials for short distances', *JIEE*, 1947, **94** Pt IIIA, pp. 637–643
3 BUSCHBECK, : 'Aus Theorie und Technik der Antennen', *ZWB*, 1943, **11**, p. 72
4 JASIK, H. (Ed.): 'Antenna engineering handbook' (McGraw-Hill Book Co., New York, 1961) chap. 3
5 JOSEPHSON, B.: 'The quarter-wave dipole'. IRE Wescon Conv. Rec., 1957 pp. 77–90
6 KING, R.W.P.: 'Asymmetrically-driven antennas and the sleeve dipole', *Proc. IRE*, 1950, **38**, p. 1154
7 SCHELKUNOFF, S.A.: 'Electromagnetic waves' (D. van Nostrand Inc., New York, 1945)
8 SHNITKIN, H., and LEVY, S.: 'Getting maximum bandwidth with dipole antennas', *Electronics*, 31 Aug 1962, pp. 40–42
9 SMITH, R.A.: 'Aerials for metre and decimetre wavelengths' (Cambridge University Press, 1949)
10 WILLIAMS, H.P.: 'Antenna theory and design, vol II' (Isaac Pitman, London, 1950)
11 WOLFF, E.A.: 'Antenna analysis' (John Wiley, 1966)

Chapter 3

Monopole antennas

In practice the monopole is not simply half a dipole; such a situation is only true when the ground plane is infinite and, as will be demonstrated below, even very large ground planes give radiation patterns significantly different from that on an infinite plane. The ground plane affects the performance of the monopole not only because it is finite in size but because the capacitance between the base of the monopole and the ground plane differs from that between two halves of a dipole.

3.1 Effects of flat ground plane

3.1.1 *Impedance*

Storer [21] calculated the change of input impedance of a thin base-fed monopole on a large finite circular ground plane from the impedance of the same antenna on an infinite ground plane:

$$\Delta Z = Z - Z_0 = j\frac{60}{kd}\exp(-jkd)\left| k\int_0^h \frac{I(z)dz}{I(0)}\right|^2$$

where Z_0 = impedance of monopole on an infinite ground plane
d = diameter of circular ground plane
h = height of monopole
$I(z)$ = current distribution on monopole
$I(0)$ = base current

For a quarter-wave monopole this reduces to

$$\Delta Z = j\frac{60}{kd}\exp(-jkd)$$

This is for large diameter screens ('large' not defined); Wolff [24] proposes an additional factor for the general case:

$$\Delta Z = j\frac{60}{kd}\exp(-jkd)\left[1 + \exp\frac{[j(kd + 3\pi/4)]}{(2\pi kd)^{1/2}}\right]^{-1}$$

Awadalla and Maclean [1] calculated this factor for small ground planes and the impedance for a monopole is shown in Fig. 15.1 for ground plane diameters down to 0·5 λ. Meier and Summers [15] plotted results for various monopole and ground plane sizes and also showed that the variation with a square ground plane of side S was about half that of a circular ground plane of diameter S.

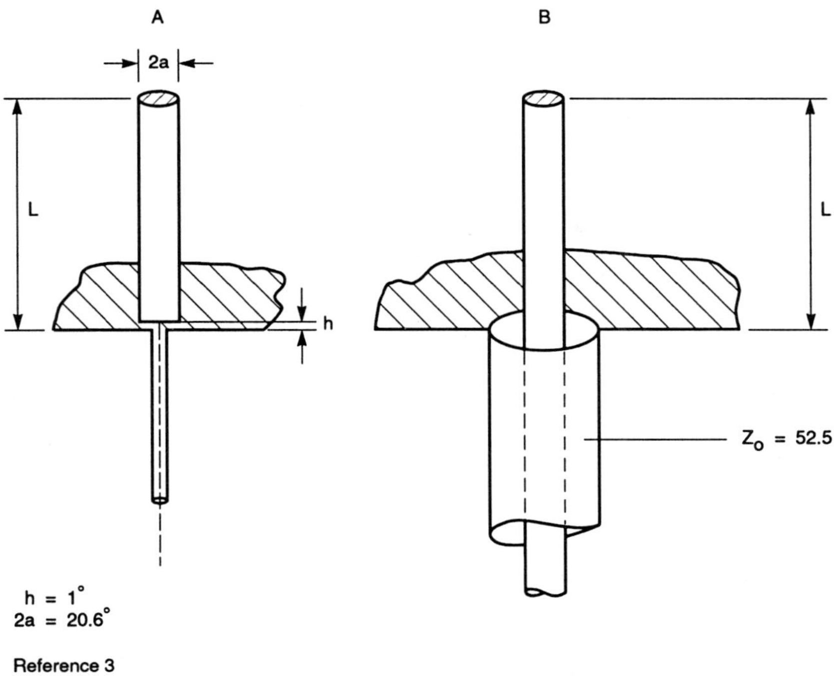

Figure 3.1 *Fat monopole with different feed arrangements*

The effect of base capacitance is shown neatly by experiments by Brown and Woodward [3]. They measured the impedance of the same fat monopole with a large discontinuity between antenna and feed cable and with a feed line whose inner was an extension of the monopole, Fig. 3.1. Resistance and reactance are shown in Fig. 3.2. The base capacitance is $0{\cdot}00885\ \varepsilon\pi a^2/h$ picofarads when a and h are measured in mm. Reducing the base capacitance by tapering the base of the antenna may be an advantage in raising the antenna resistance. There is still the near-base capacitance which is $C_{nb} = 0{\cdot}0354\ \varepsilon a \ln a/h$ picofarads. This increases as the outer diameter of the coaxial line decreases and the maximum resistance decreases.

3.1.2 *Radiation pattern*

On an infinite ground plane the peak radiation occurs at 0° elevation but this is never true on a finite ground plane. The half-dipole pattern is modified by diffraction from the edge of the ground plane. Various methods have been proposed for computing patterns, some of them restricted to circular ground planes of small dimensions. Of the more general methods, the use of the Cornu spiral was favoured for many years. Although this correctly gives the elevation angle it also gives a smooth curve which is at variance with measured patterns. Best results are achieved using the method of moments for the smaller ground planes and the Geometrical Theory of Diffraction (GTD) for the larger.

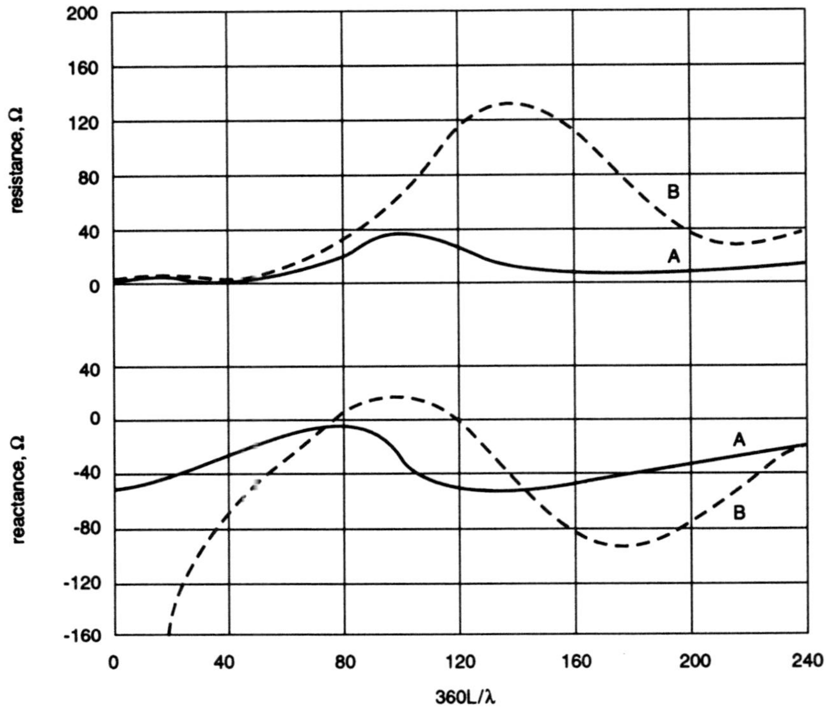

Reference 3

Figure 3.2 *Resistance and reactance of the antennas of Fig. 3.1*

Consider the radiation components in the plane of the paper for the monopole on the finite ground plane of Fig. 3.3. This ground plane is assumed to be rectangular and it is assumed that there is negligible contribution from the edges orthogonal to those shown. In fact it can be seen from experiments on

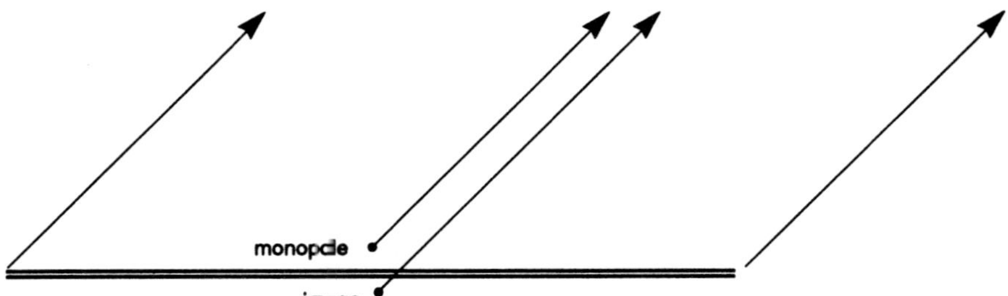

Figure 3.3 *Radiation components from monopole on a finite ground plane*

sheets of side 5λ or more that only the components shown are significant. Clearly the two edges will have mirror-image contributions if the monopole is central as shown, so the two contributions create an interference pattern determined by their spacing. The greater the spacing the more numerous the lobes will be and their impact on the direct radiation will be reduced. This is demonstrated in Figs. 3.4, 3.5 and 3.6, the first and second for circular ground planes and the third for square ones. It can be seen from Fig. 3.6 that even for a monopole-to-edge distance of 100λ the peak radiation occurs at a few degrees above the ground plane. This diagram also shows clearly the decreasing depth of nulls as the ground plane size increases. In fact the null depths for sheets of side 2 or 3λ can be significant where wide angle elevation coverage is required.

It will be noted that at high angles there is more radiation from circular ground planes than from square ones. This is because the contributions from the rim are all in phase from a circular sheet but not from a square one. Fig. 3.7 gives the angle of peak signal as a function of sheet size.

It had been long assumed that the signal in the plane of the ground plane was always −0·85 dBi and the peak gain +5·15 dBi. This work on radiation patterns by Foster and Miller [9] showed that the peak gain was not

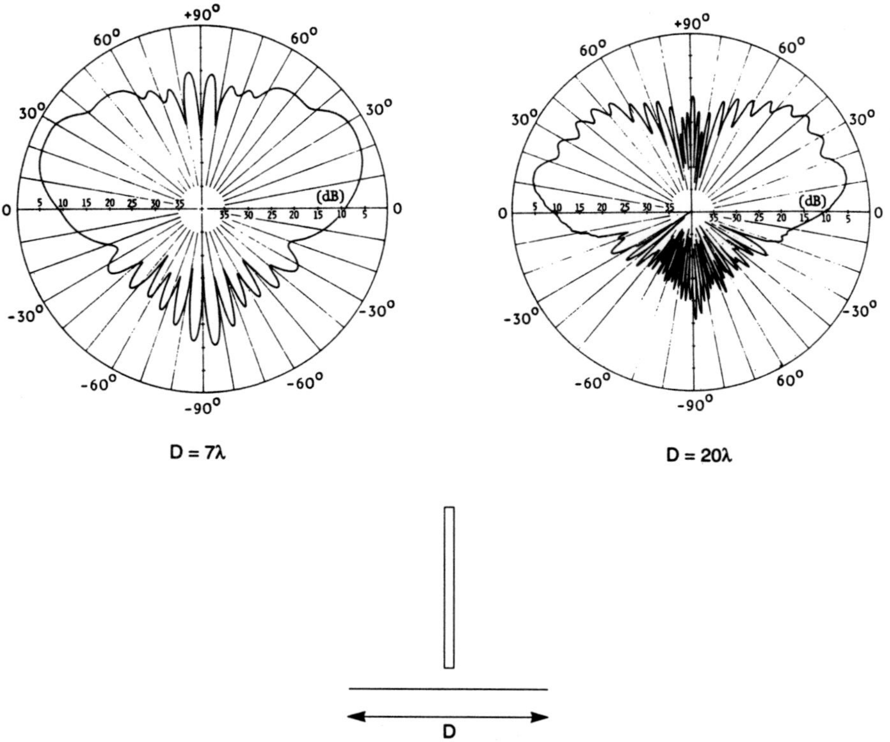

Figure 3.4 *Measured radiation patterns of monopoles on circular ground planes, D = 7λ and 20λ*

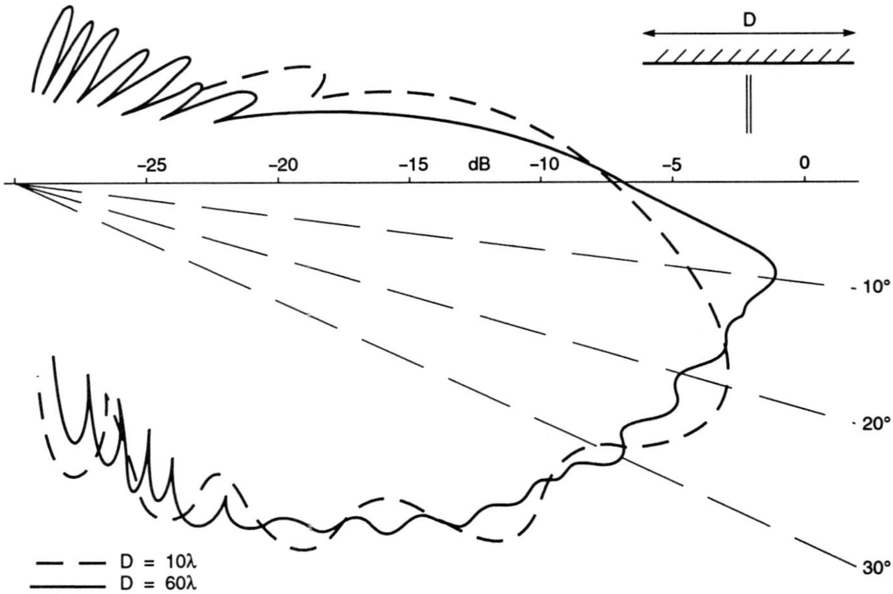

Figure 3.5 *Measured radiation patterns for monopoles on circular ground planes* $D = 10\lambda$ and 60λ

independent of sheet size and the 0° level was not 6 dB below the peak. Fig. 3.8 plots gain of the 0° signal compared with the peak.

Several important points come from this work:

(i) For large ground planes the rate of change of gain close to the plane of the sheet is high so that any gain measurements involving antennas on ground planes require very careful alignment of the ground plane.

(ii) As the elevation plane pattern is a function of distance to the edge it is possible to construct a complete spherical pattern from a series of elevation planes. Only the distance to the edge is required at each angle.

(iii) The azimuth pattern of a vertical monopole on a non-circular sheet will have maxima where the distance to the edge is least and minima where it is greatest.

(iv) Very large ground planes are needed to reduce the radiation on the shadow side to negligible proportions. This means that an elevated monopole on a small ground plane will still illuminate the ground strongly.

Computed and measured peak gains were determined by Foster and Miller for square ground planes. These are given in Table 3.1. The computed figures for ten sources used ten elementary sources equally spaced along the monopole. The measured results were obtained by sampling at 2° intervals the co-polar and cross-polar patterns over the complete sphere and integrating.

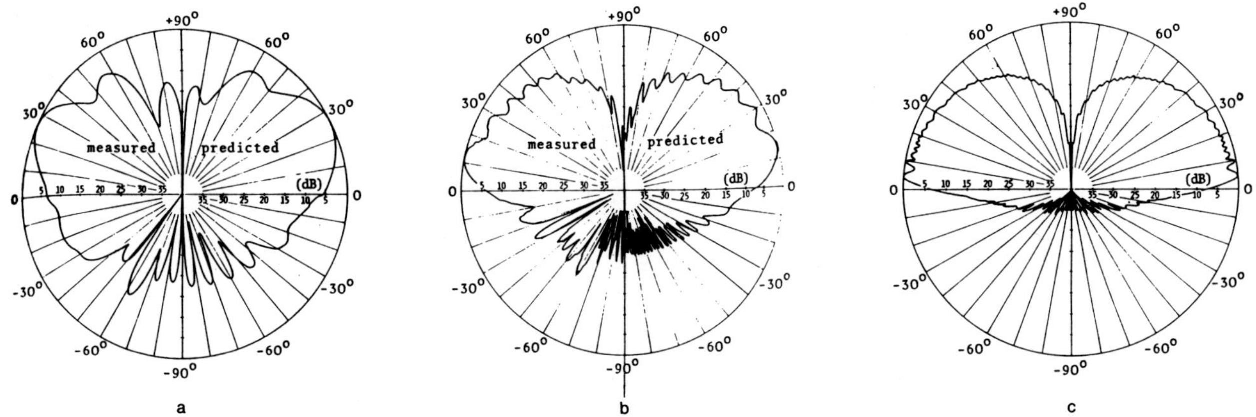

Figure 3.6 *Radiation patterns for monopoles on square ground planes of side 5λ, 20λ and 200λ*

a Side = 5λ
b Side = 20λ
c Side = 200λ (predicted)

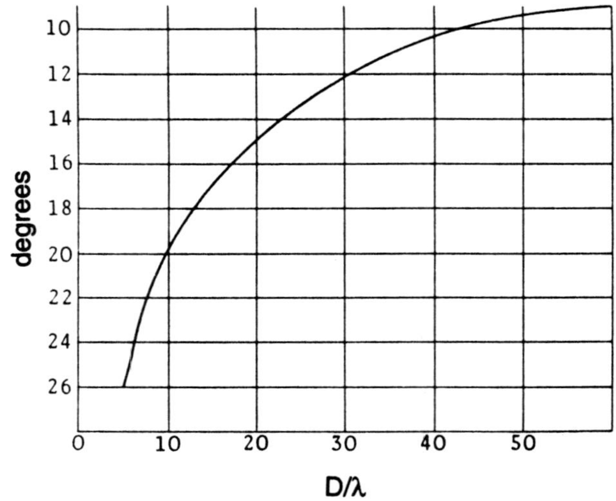

Figure 3.7 *Angle of peak signal as a function of sheet size for monopole on a ground plane*

3.2 Top-loaded monopole

The simplest modification to the cylindrical monopole is to reduce its height by top loading. A number of different forms of top loading are shown in Fig. 3.9: of these (*a*), (*b*) and (*c*) have symmetrical non-radiating tops whilst that in (*d*) produces radiation polarised orthogonally to the main radiation. This last form has its uses in producing near-omniazimuth coverage on aircraft: this will be discussed in Chapter 12.

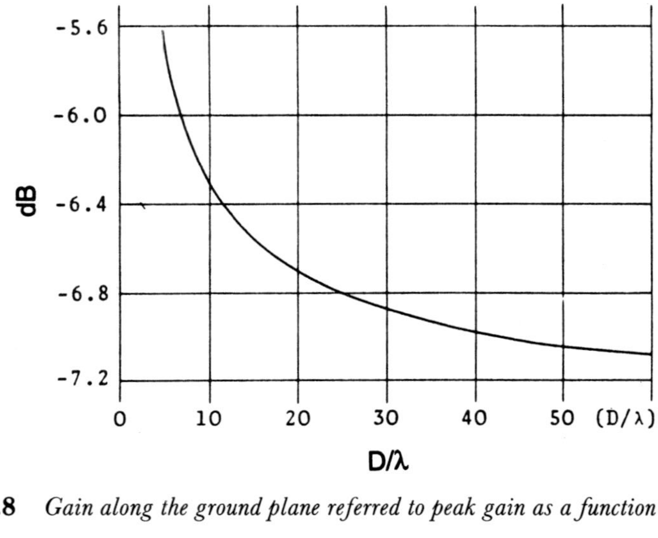

Figure 3.8 *Gain along the ground plane referred to peak gain as a function of sheet size*

Table 3.1 *Computed and measured directivities*

Dimensions of ground plane wavelengths	Directivity (dBi)		
	Computed (1 source)	Computed (10 sources)	Measured
5 × 5	5·3	5·5	6·36 ± 0·75
10 × 10	5·9		6·89 ± 0·5
20 × 20	6·2		6·95 ± 0·5
60 × 60	6·45	7·0	
200 × 200	6·5		

3.2.1 *Impedance*

Of the various formulae evolved to compute the impedance of a top-loaded monopole only that due to Laport [14] holds for antennas of electrical length approaching $\lambda/4$. According to Laport the radiation resistance is given in terms of the area A of the plot of current distribution on the radiating surface. A is in degree–amperes; assuming 1 A base current, then

$$R = 0{\cdot}01215\ A^2 \text{ ohms}$$

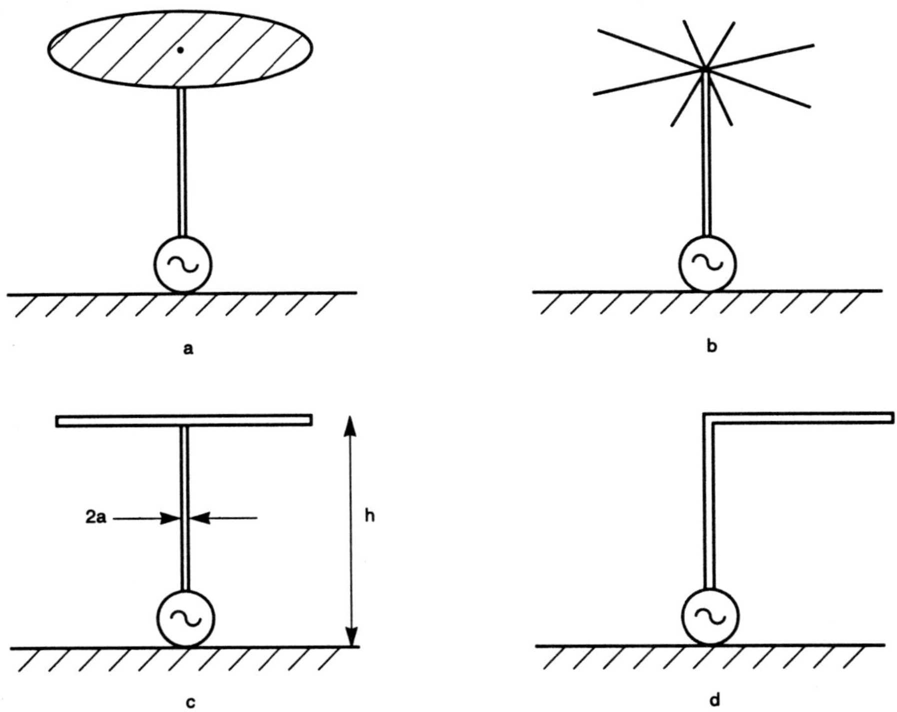

Figure 3.9 *Forms of top-loading for monopole*

This holds for any shape of in-phase current distribution.
Three cases can be considered:
Case 1: No loading

$$A = \frac{180}{\pi} \frac{1 - \cos G_A}{\sin G_A}$$

where $G_A = (360\ h/\lambda \times \mathrm{M})$ = electrical length of radiator

h = physical height of the radiator

M is a multiplying factor depending on the ratio h/d of the antenna where d is the equivalent diameter of cross-section. M is shown in Table 3.2.

Table 3.2 Effect of thickness on electrical length

h/d	G°	M
10	81·5	1·10
20	81·5	1·10
40	82·5	1·09
100	84·2	1·07
200	85·0	1·06
400	85·7	1·05
1000	86·4	1·04
2000	87·1	1·03
4000	87·4	1·03

G = physical length at resonance
M = 90/*G*

Case 2: Top-loaded

$$A = \frac{180}{\pi} \frac{\cos(G_A - G_V) - \cos G_A}{\sin G_A}$$

where G_A = electrical length of whole antenna
G_V = electrical height of vertical portion
Case 3: Top-loaded at resonance

$$A = \frac{180}{\pi} \sin G_V$$

3.2.1.1 *Reactance*
The reactance X_b of the vertical portion is given by

$$X_b = -jZ_0^v \cot G_A$$

For radiators of constant circular cross-section of radius a,

$$Z_0^v = 60(\ln h/a - 1)$$

If X is the reactance due to top loading, then the feed point reactance is

$$X_b = Z_0^v \frac{X \cos G_V + jZ_0^v \sin G_V}{Z_0 \cos G_V + jX \sin G_V}$$

Alternatively for small electrical heights the reactance of the vertical portion can be considered as an inductance in series with the reactance due to the top loading. Both methods give similar results certainly up to heights of 20°.

If Z_0^T is the characteristic impedance of the horizontal portion as in Fig. 3.9*c* or *d* against the ground plane, then

$$X = -jZ_0^T \cot G_T$$

where G_T is the electrical length of the top.

Hence $G_T = \cot^{-1} \dfrac{Z_0^v}{Z_0^T} \tan G_V$ at resonance

since $X \cos G_V = -jZ_0^v \sin G_V$ from the equation above for X_b.

It should be noted that, at resonance,

$$G_V + G_T \neq G_A \text{ unless } Z_0^T = Z_0^v$$

G_A is obtained by equating X_b with the reactance due to a straight vertical monopole of characteristic impedance Z_0^v:

$$Z_v \times \frac{X \cos G_V + jZ_0^v \sin G_V}{Z_0^v \cos G_V + jX \sin G_V} = -jZ_0^v \cot G_A$$

from which

$$G_A = \frac{\cot^{-1} \dfrac{Z_0^T}{Z_0^v} \cos G_T - \tan G_V}{1 + \dfrac{Z_0^T}{Z_0^v} \tan G_V}$$

$$X_b = \frac{-jZ_0^V \times \dfrac{Z_0^T}{Z_0^V} \cot G_T - \tan G_V}{1 + \dfrac{Z_0^T}{Z_0^V} \tan G_V}$$

From these, R and X can be determined for a range of frequencies off-resonance.

At resonance an antenna $\lambda/8$ high and top-loaded has a resistance of approximately 20 ohms according to Laport's theory. It is instructive to look at other theories for the top-loaded monopole. For antenna $\lambda/8$ high and with a top-loading rod $\lambda/8$ long the following figures are obtained:

Wanselow and Milligan [23] $R = 1575(h/\lambda)^2 = 24{\cdot}6\ \Omega$

Gouillou [10] $R = 60(kh)^2 = 37\ \Omega$

Wolff [24] $$R = \frac{30k^2h^2}{4 \sin ks}\left(1 - \frac{\sin 2ks}{2ks}\right) \simeq 38\ \Omega$$

(s = length of top-loading)

Burton [6] $$R = \frac{30k^2h^2}{4} = 74\ \Omega$$

It appears therefore that these equations cannot hold for such large vertical dimensions.

Because the resistance of a top-loaded monopole will always be well below that of any usual transmission line Z_0, some method of increasing R is necessary. Some of these methods are examined in subsequent sections of this chapter. They include the folded monopole, the shunt-fed monopole and the sleeve monopole.

3.3 Shunt-fed monopole

The shunt-fed monopole is one-half of the tee-match dipole described in Section 2.1.3. Its impedance may be derived from the latter and is

$$Z_{in} = \frac{(1+a)^2 Z_v Z_f}{Zf + (1+a)^2 Z_v}$$

where Z_v is the base impedance of the plain monopole and Z_f is the reactance of the two-wire short-circuited transmission line.

3.3.1 *Shunt-fed top-loaded monopole*

If the top loading is fixed there will only be one frequency for optimum VSWR. It is, however, possible to tune the antenna over a range of frequencies by incorporating a variable capacitance. This would normally be placed at the end of the top loading as in Fig. 3.10 as this requires the smallest capacitance for a given set of dimensions. It is, however, perfectly possible to place the capacitance at some other point along the line; larger values will be required but the voltage across the capacitance will be reduced. It must be remembered that the vertical member will have a significant inductance so the capacitance values required will be smaller than if there were no inductance. Using this arrangement the antenna could be tuned over a range of frequencies. For an antenna 0·45 m high it was possible to tune over the band 30–80 MHz with a VSWR better than 3:1. The limitation is set by the variation of the basic antenna resistance and by the instantaneous bandwidth, which becomes very small when the antenna height is small in terms of wavelength. This topic is covered in more detail in Chapter 9.

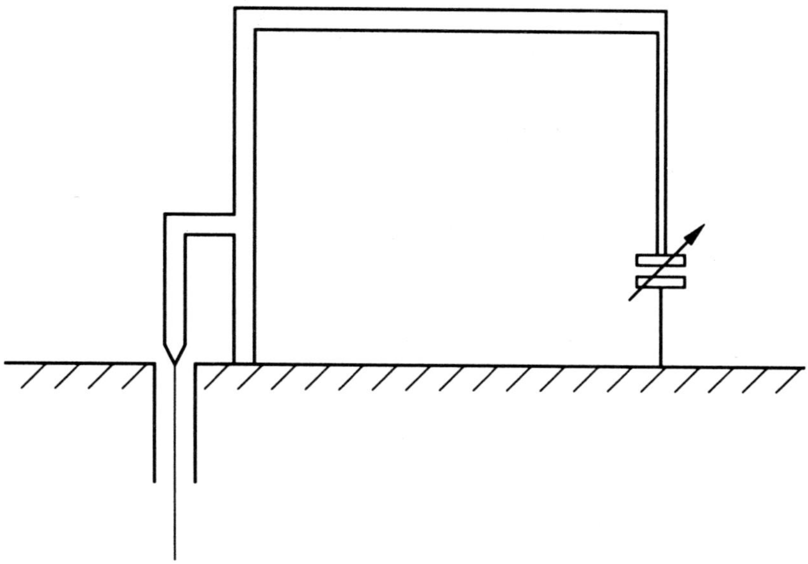

Figure 3.10 *Capacitance-tuned top-loaded shunt-fed monopole*
Courtesy RAeS

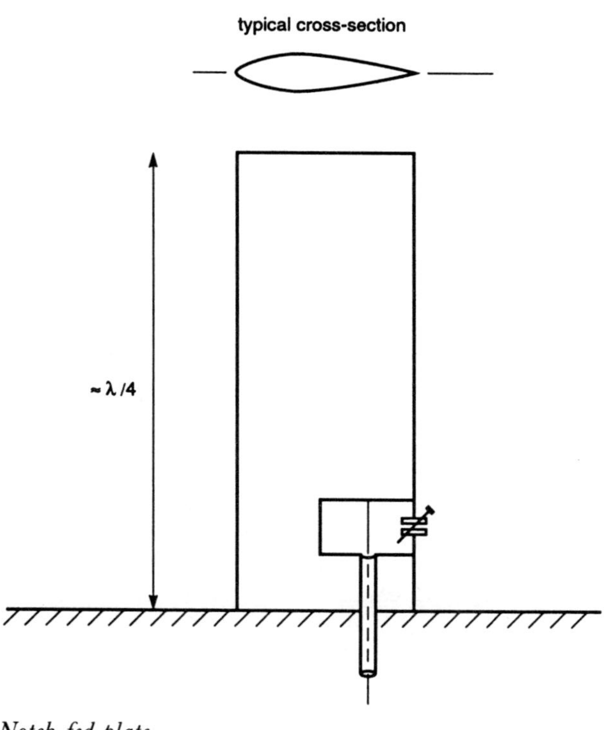

Figure 3.11 *Notch-fed plate*

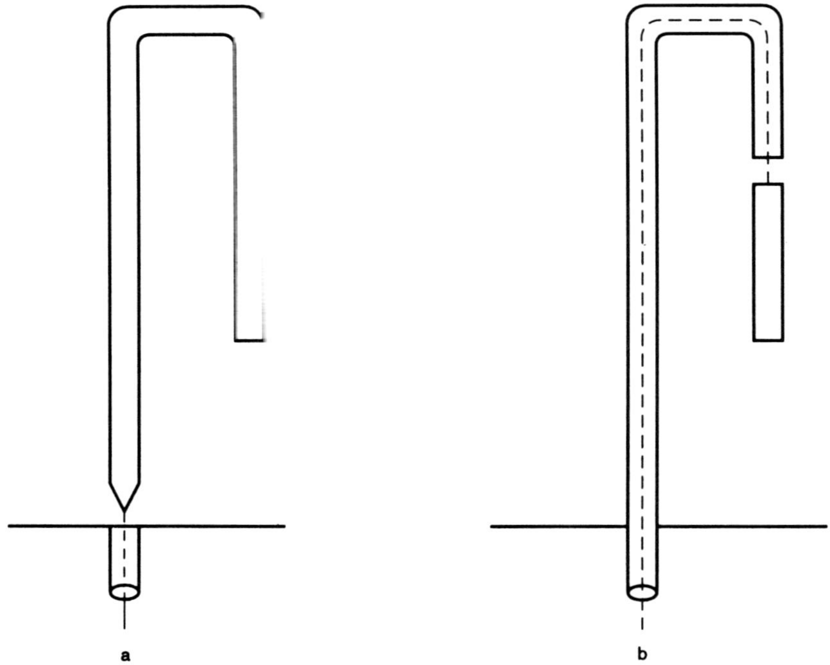

Figure 3.12 *Open folded monopole*

a Base fed
b Displaced feed

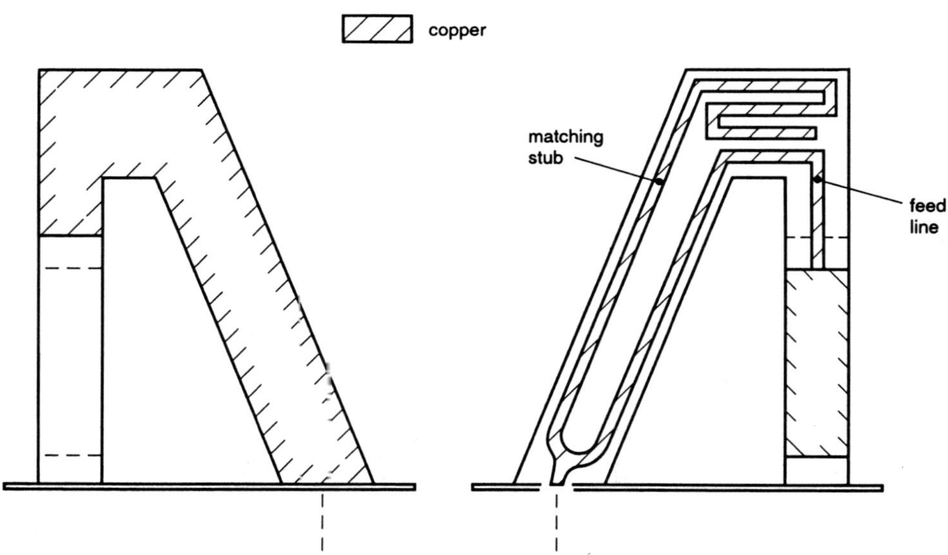

Reference 13

Figure 3.13 *Printed circuit form of open folded monopole*

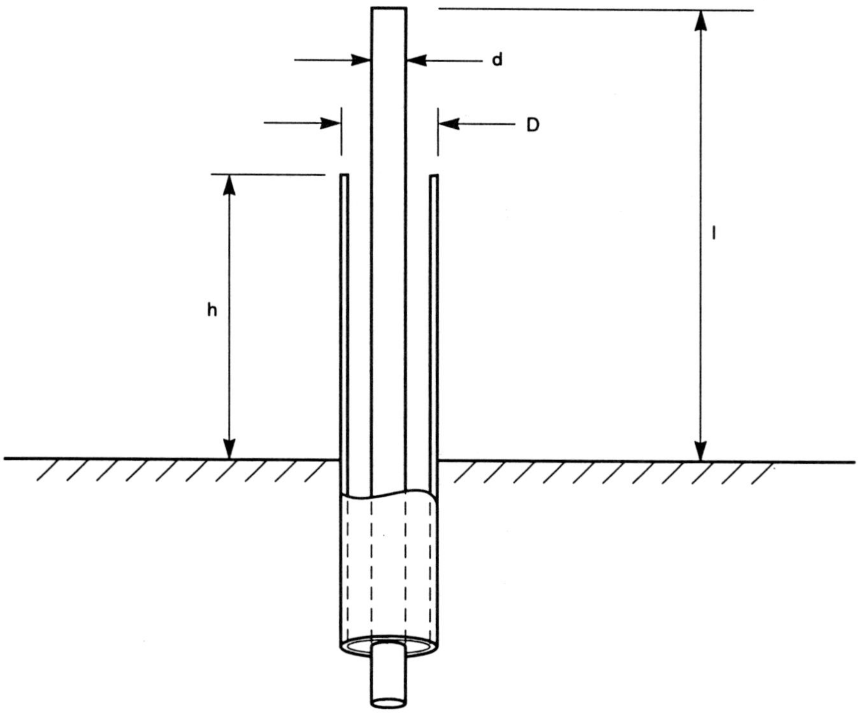

Figure 3.14 *Basic sleeve monopole*

3.3.2 *Notch-fed plate*

Fig. 3.11 shows a flat plate $\lambda/4$ high and earthed at its base. It is fed by a short, capacitance-loaded notch. The antenna element could equally well be of aerofoil section. It is clear that this is a form of shunt-fed antenna. A theory for the impedance of this antenna has not been developed but it is simple to arrive at its dimensions by experiment. Provided the plate width is not more than about $\lambda/8$ the azimuth radiation pattern for a vertical antenna will be essentially circular.

3.4 Folded monopole

The folded monopole is a special case of the tee-match antenna and has been dealt with, in the dipole form, in Chapter 2. The main use of the folded monopole is likely to be in a directional antenna using a director and reflector where the resistance of a plain monopole could be as low as 10–15 ohms. Folding is also of value in stepping up the impedance of a top-loaded monopole; this topic is addressed in Chapter 9. One special form of the folded monopole is properly known as the open folded monopole.

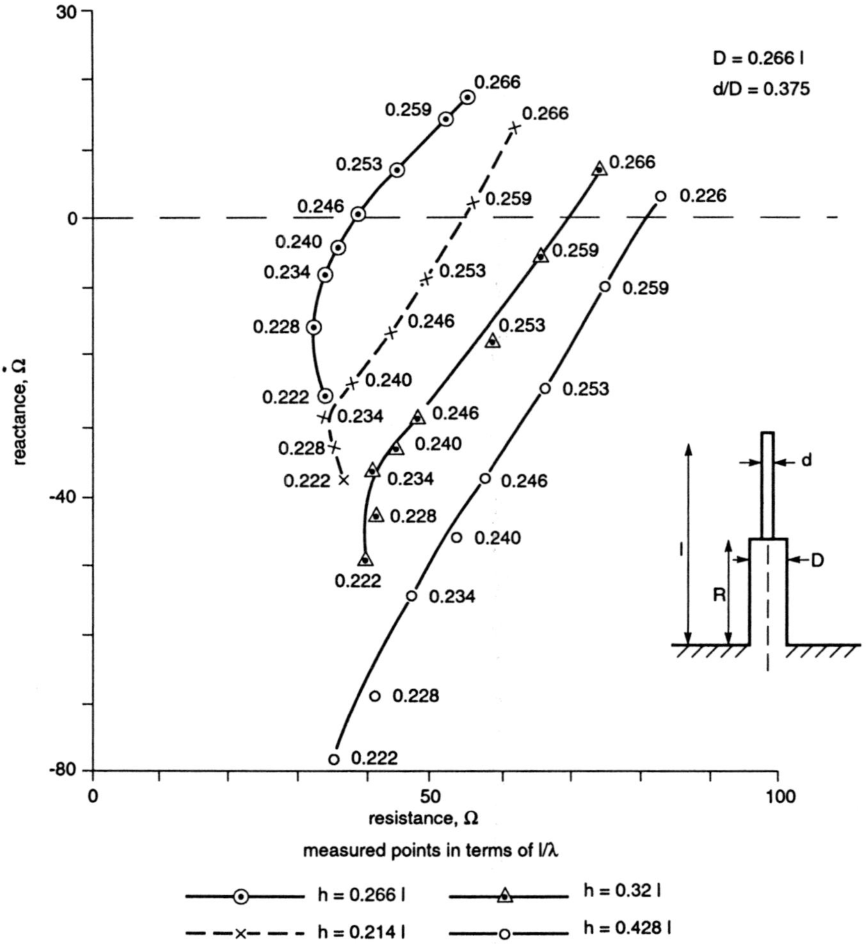

Figure 3.15 *Impedance of sleeve monopole as a function of sleeve height*

3.4.1 *Open folded monopole*

This antenna was first described by Josephson [13]. Essentially it is a quarter-wave monopole bent over so that the normally horizontal top loading lies parallel to the vertical element whose height is approximately $\lambda/8$. The base resistance of the simple antenna of Fig. 3.12*a* is of the order of 10 ohms. To increase this the feed point may be displaced as in Fig. 3.12*b*.

There are obvious advantages in making the elements fat and in a printed circuit version, Fig. 3.13, a frequency range of 195–250 MHz was achieved for VSWR less than 2:1 referred to 50 ohms.

One of the advantages of using a folded antenna or indeed a shunt one is that there is a DC path to ground from the top of the antenna. Not only is this

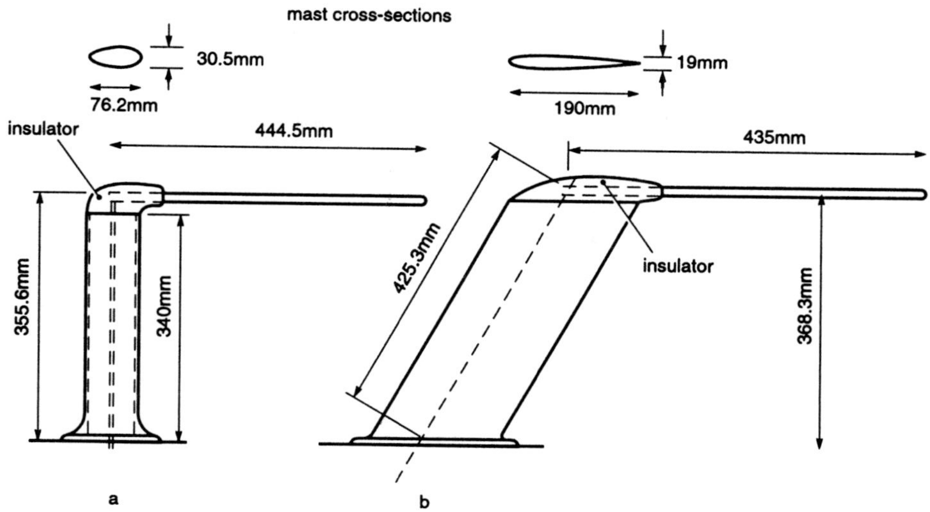

Figure 3.16 *Bent sleeve VHF monopoles*
a For helicopters and light transport aircraft
b For high performance aircraft

important for protecting the radio equipment from lightning strikes but it can also be used to carry cables to a warning light mounted at the top of the antenna, the cables being shielded from pick-up inside a hollow tubular conductor.

3.5 Sleeve monopoles

The basic sleeve monopole is shown in Fig. 3.14 where the coaxial feed is extended above the ground plane. At resonance the current distribution and the radiated power are not appreciably altered by the position of the feed-point. If h is the height of the feed-point above the ground plane in an antenna of total height $l=\lambda/4$, then

$$R_h = \frac{R_0}{\cos^2 kh}$$

Then if in a practical monopole $R_0 = 30$ ohms a sleeve of just under 0·1 λ will give a feed-point impedance of 50 ohms. Fig. 3.15 shows impedance for different sleeve heights. If the sleeve is 0·125 λ the impedance becomes $2R_0$. This will normally be too high for a 50 ohm system but is of considerable value if the physical height of the antenna has to be reduced.

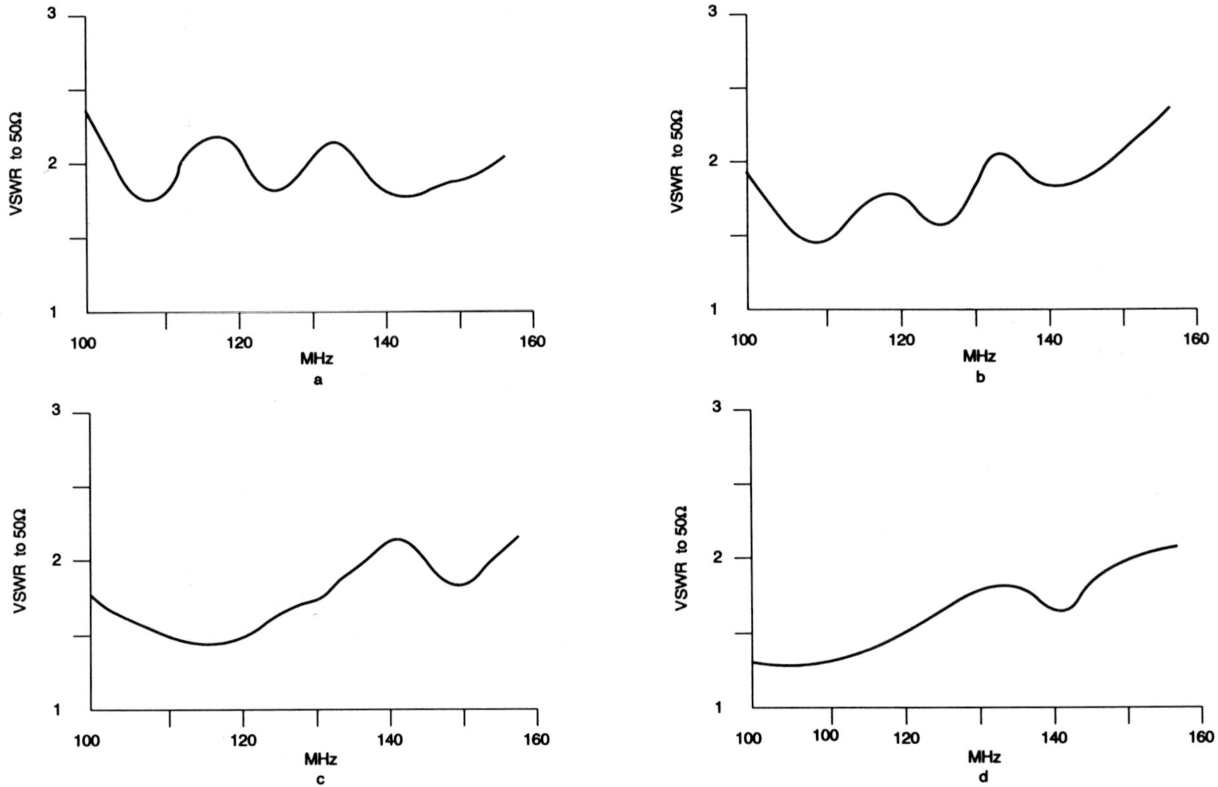

Figure 3.17 *VSWR curves of Fig. 3.16b on various ground planes*

a On 2·5m diameter ground plane *b* On 2·5m diameter fuselage *c* On 0·25m diameter cylinder *d* On 0·17m high fairing on 2·5m diameter fuselage

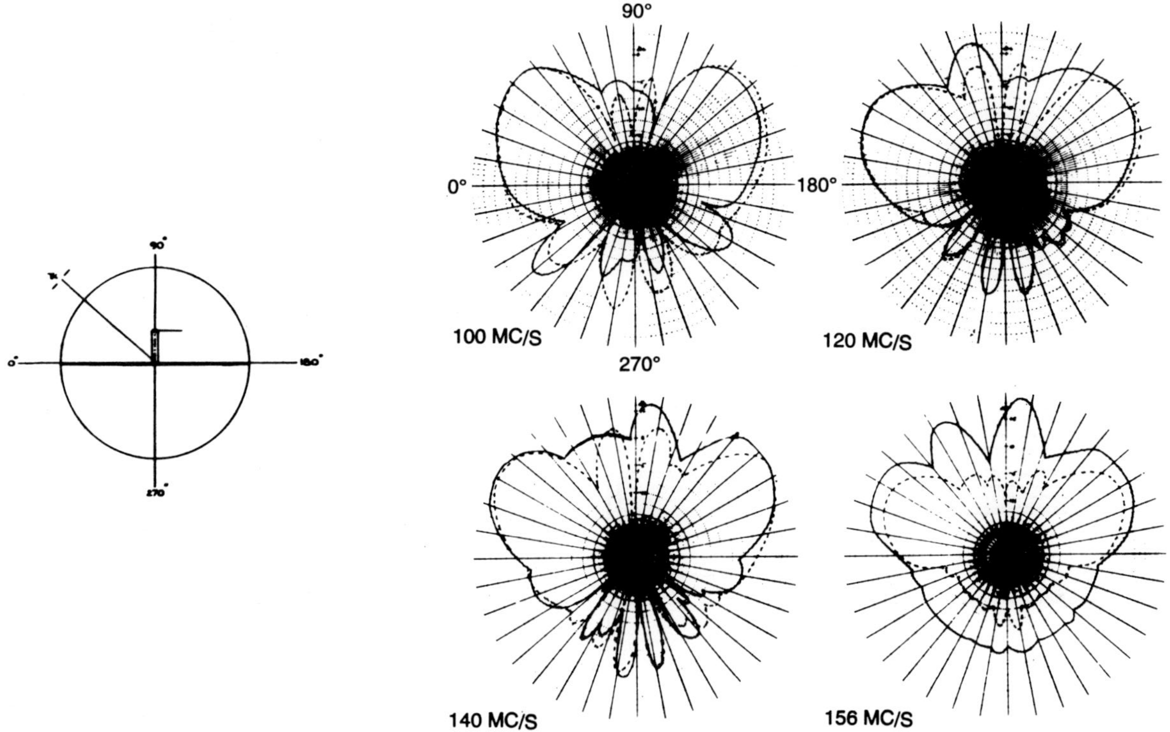

Figure 3.18 *Elevation patterns of bent sleeve monopole*

Bent sleeve monopole compared with $\lambda/4$ vertical monopole, both mounted at centre of 7·5m diameter ground plane, measured at $\frac{1}{10}$ scale
—— bent sleeve
– – – vertical monopole

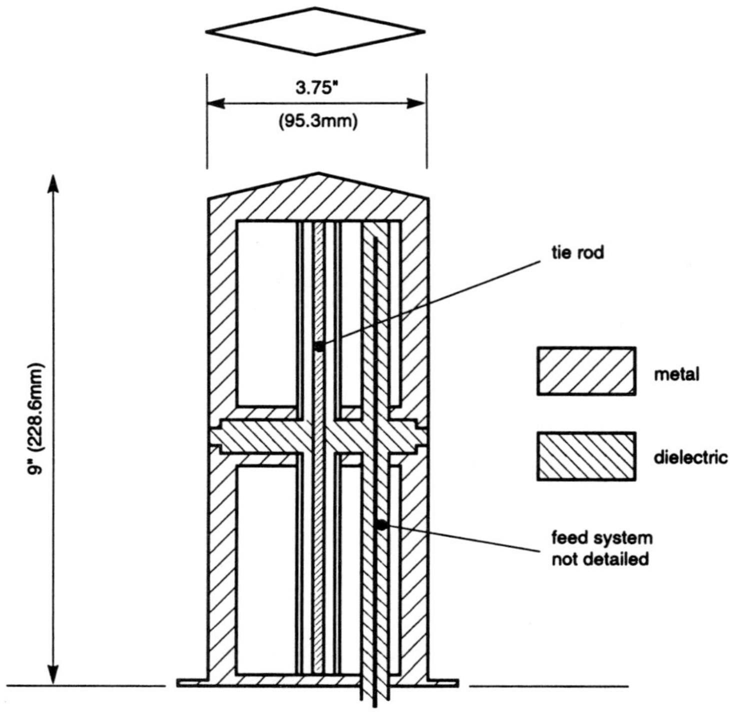

Figure 3.19 *Broadband sleeve, US type AT256A*

3.5.1 *Bent sleeve*

This was the basis of a design developed by the author in 1954 to overcome the problem of low ground clearance on helicopters. The antenna shown in Fig. 3.16*a* was still in service in 1990 as was the streamlined form of Fig. 3.16*b* designed for low-level transonic fighter aircraft.

These antennas are characterised by high mechanical strength and very good flutter characteristics unlike some of the straight blades and whip antennas they were designed to replace. Both cover the VHF communication band 100–156 MHz with a VSWR better than 2·5:1 referred to 50 ohms. In each the sleeve is the outer of the coaxial line, a tubular inner being used to provide the appropriate Z_0 for optimum bandwidth. Fig. 3.17 shows VSWR curves of Fig. 3.16*b* for a variety of mounting positions and demonstrates the versatility of the antenna.

Subsequent designs in which a 50 ohm coaxial cable was fitted inside the sleeve covered the UHF communication band [225–400 MHz, 2:1 VSWR] and other bands up to 1200 MHz. A microstrip version of the 960–1220 MHz antenna was built but found to suffer from radiation from the inner of the stripline as the width of the ground plane was inadequate. A triplate line was, however, successful; the whole assembly including the horizontal rod was moulded in high density polythene to give an 85 °C top working temperature.

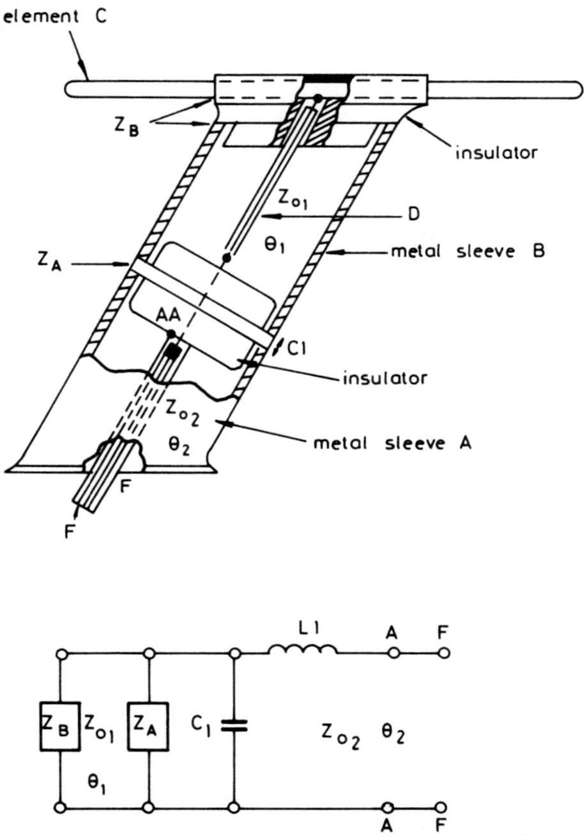

Figure 3.20 *Double-band sleeve antenna*

This type of antenna has some radiation from the horizontal rod as Fig. 3.18 shows. The levels for the bent sleeve and for the monopole have been set equal at 0°: in fact the bent sleeve antenna must have a slight loss compared with the monopole at 0° to account for the upward radiation.

3.5.2 *Broadband sleeve*

Using a large diameter sleeve has considerable effect on the bandwidth as Fig. 3.17 illustrates. Equally good results may be obtained by making the upper half of the antenna with the same cross-section as the sleeve, Fig. 3.19. There is a greater mechanical problem, compared with the basic antenna of Fig. 3.14 or the bent sleeves of Fig. 3.16, in transferring the load of the top section across the insulation separating it from the lower section. Various means have been used to overcome this problem. In the US type AT 256A antenna the two halves are held together by a rod which puts the insulator in compression. Because the rod is insulated from the antenna except at its ends it forms two shunt stubs whose impedance appears across the feed-point. This rod also provides a DC path to ground for the top element. With the addition of a series capacitance and some

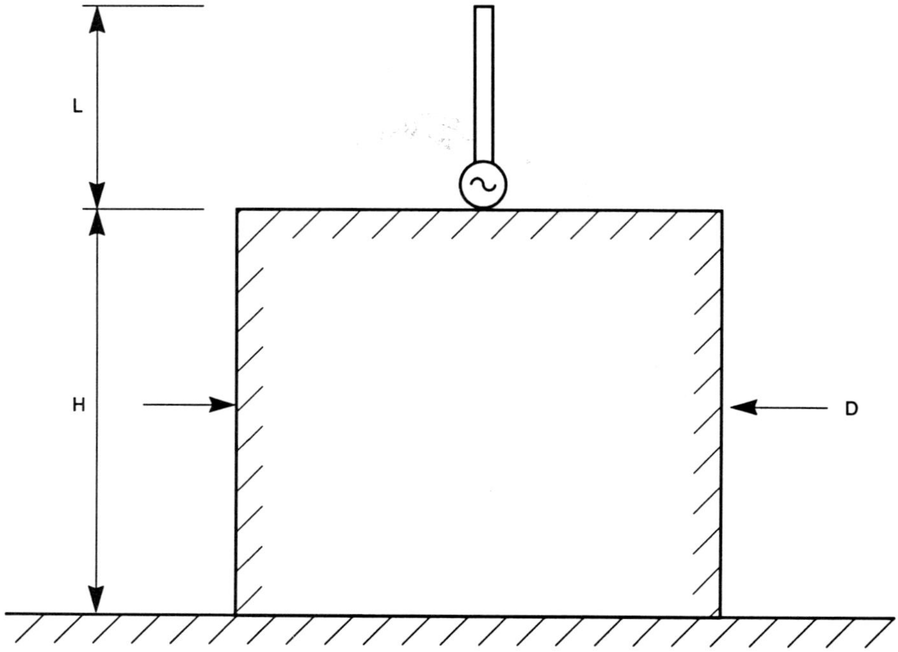

Figure 3.21 *Monopole on large sleeve*

transformer sections the antenna covers the band 225–400 MHz with a VSWR better than 2:1 to 50 ohms. Further details of the design can be found in Jasik [11].

For this frequency band it is practical to use a printed circuit design with microstrip feed. The upper and lower sections are printed on opposite sides of the board and overlapped to give the appropriate capacitance across the feed-point. The whole board is then encapsulated inside a dielectric shell using a closed-cell foam to stiffen the shell and support the board.

3.5.3 *Double-band sleeve antenna*

The sleeve antenna with symmetrical top loading is occasionally used as an electrically short antenna but a variant with a double sleeve, Fig. 3.20, can operate over two distinct frequency bands. The conventional sleeve antenna consisting of sections A and B operates in the UHF range 225–400 MHz. The horizontal top element C is electrically connected to the base of section B but is elsewhere insulated from it. Rod D and the inside surface of sleeve B form a transmission line which presents a high impedance between the top of the sleeve and the top rod C over the frequency range 225–400 MHz. This line section acts as a series inductance over the VHF band 115–140 MHz and the length of the top element is adjusted to obtain a good impedance match over this band. The VSWR figures are

115–140 MHz 3:1

225–400 MHz 2:1

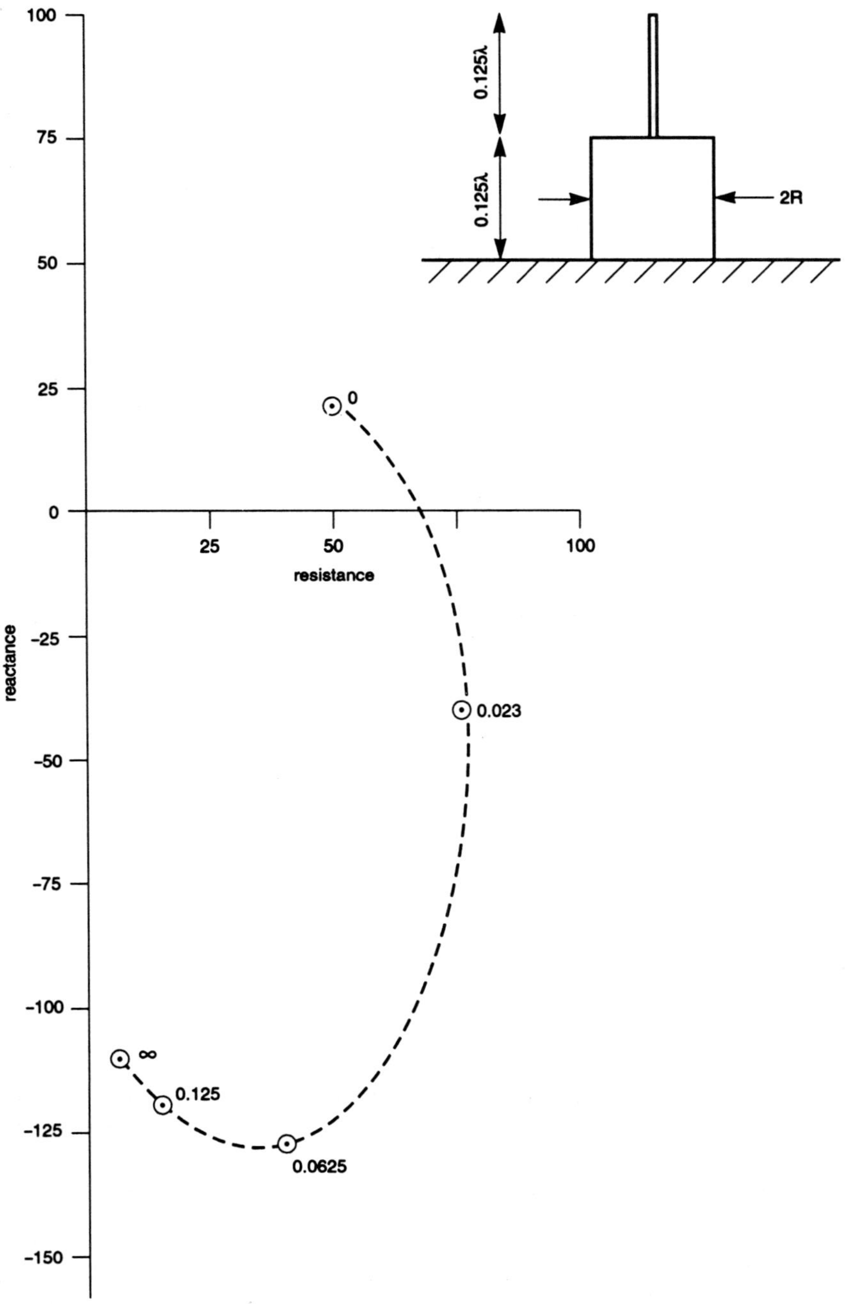

Reference 8

Figure 3.22 *Impedance of λ/8 monopole on a thick cylinder on a ground plane*
Values of R/λ shown on curve

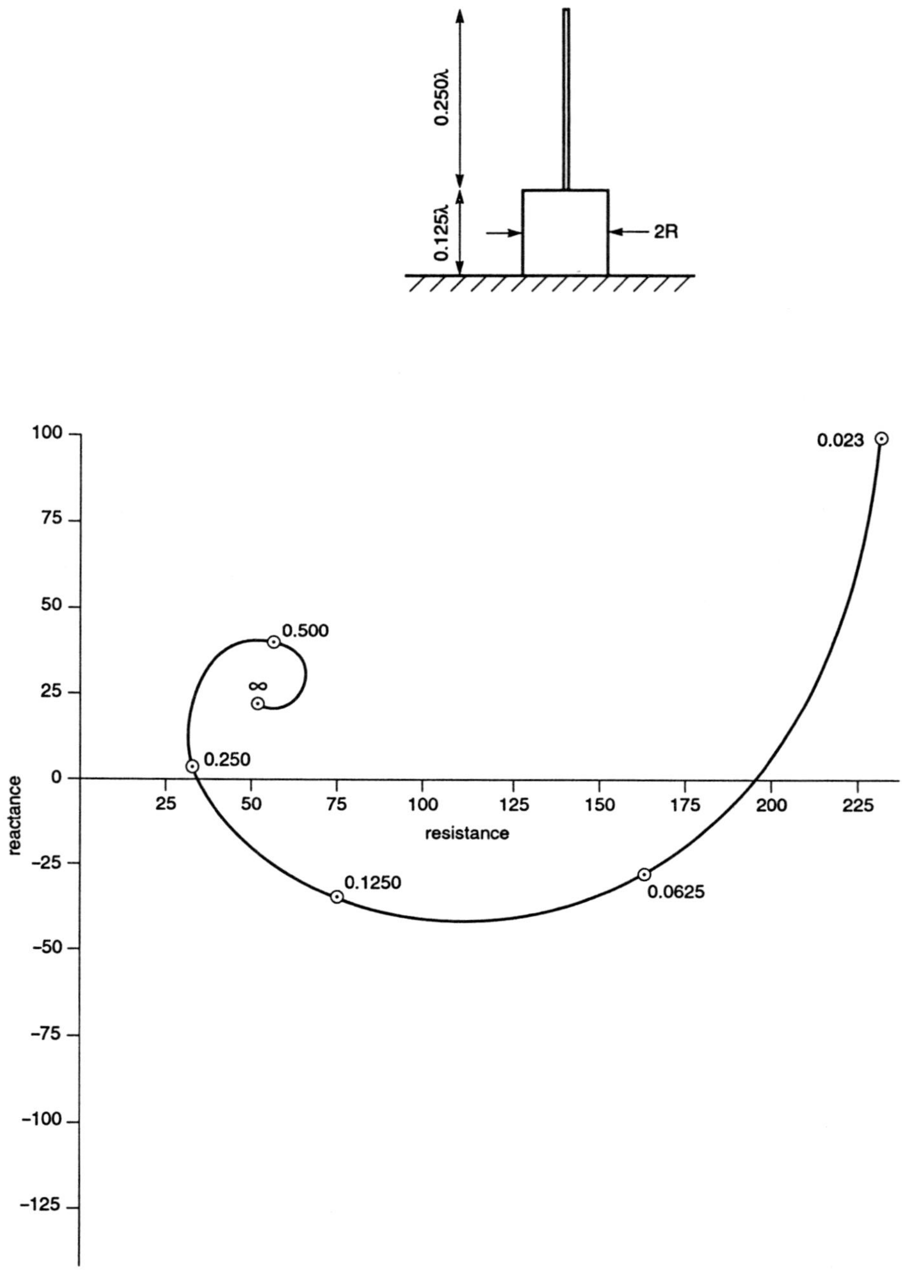

Reference 8

Figure 3.23 *Impedance of λ/4 monopole on a thick cylinder on a ground plane*
Values of R/λ shown on curve

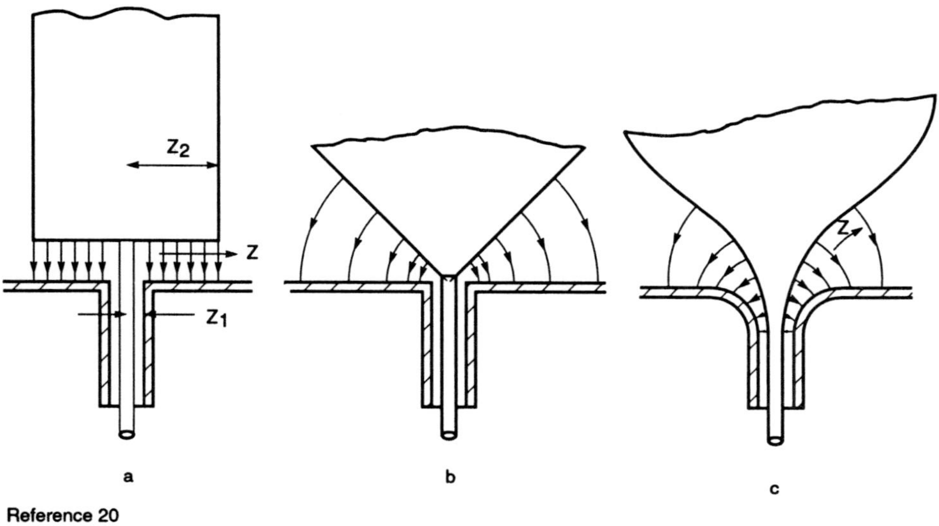

Figure 3.24 *Field distribution around the feed-point of antennas of different shapes*
a Radial feed
b Conical feed
c Arcuate feed

This particular version was constrained to have a base identical with the antenna of Fig. 3.19 thus allowing the easy addition of a VHF facility. Another design, developed but not put into production, had a mast chord of 7·5 in (190 mm) instead of 5 in (127 mm). This achieved a VSWR of 2·6 from 118 to 140 MHz and 1·8 from 220 to 400 MHz.

3.5.4 *Monopole on large sleeve*

As the previous sections have shown, increasing the equivalent diameter of the sleeve can improve the impedance bandwidth. There is clearly a point at which the sleeve has become so large in diameter that its top has to be regarded as a ground plane. Fig. 3.21 shows a typical arrangement which is not uncommon in practice. Cooper [8] in 1975 made a number of measurements of impedance and radiation patterns of monopoles of different lengths on cylinders with various ratios of height to diameter sitting on a large ground plane. The impedance measurements were given in terms of conductance and susceptance and are somewhat difficult to interpret. The figures have been translated to series resistance and reactance at the base of the monopole and are plotted in Figs. 3.22 and 3.23 for two different heights of monopole. In both figures the cylinder height is maintained at $H/\lambda = 0{\cdot}125$. In Fig. 3.22 the monopole extends $0{\cdot}125\lambda$ and the cylinder radius R varies from 0 (a monopole of height $0{\cdot}25\lambda$ above the ground plane) to ∞ (a monopole of height $0{\cdot}125\lambda$ above the ground plane). In Fig. 3.23 the monopole height is $0{\cdot}25\ \lambda$ with the same range of R. More points were available in the latter diagram so the curve joining the points is likely to be more accurate.

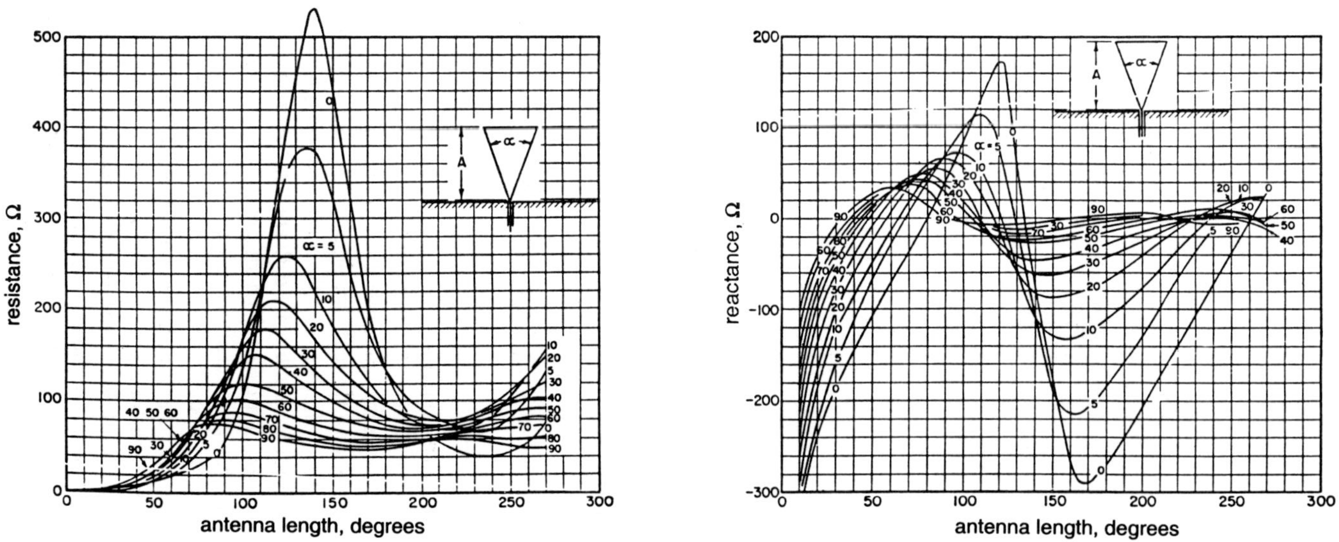

Reference 3

Figure 3.25 *Resistance and reactance of conical monopoles*

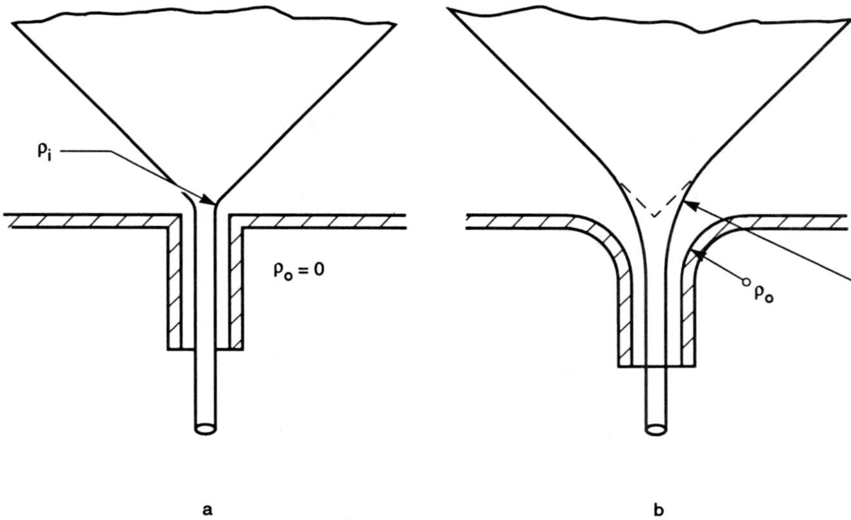

Figure 3.26 *Relationship between conical and arcuate feed systems*
a Conical feed
b Arcuate feed

These results tend to confirm measurements by the author in which a metal cube of side about 1 m standing on the ground was found to give a much better representation of the impedance of a 30–90 MHz monopole on a vehicle than did a large ground plane. Both figures show that a comparatively small diameter cylinder is required to give an adequate simulation of a very large ground plane.

Theoretical work on this topic was reported by Yung and Butler [25].

3.6 Non-circular monopoles

It was recognised long ago that in order to achieve a broad bandwidth the transition from coaxial feed to antenna should cause the minimum field perturbation. The problem is illustrated in Fig. 3.24 where the fields are shown for a fat cylindrical monopole, a conical feed and an arcuate feed. In the first case the feed is a radial transmission line between the ground plane and the base of the cylinder. The characteristic impedance of this line is $Z_0 = 60H/Z$ for $Z_1 < Z < Z_2$. If H is chosen to make Z_0 equal to the coaxial line Z_0 at $Z = Z_1$ then clearly at the edge of the cylinder the Z_0 will be significantly smaller and will be mismatched to the antenna impedance. There are therefore limitations to the use of the radial line as a method of feeding the monopole.

The Z_0 of a conical line is only a function of the cone angle. The field lines are arcs of circles centred on the apex of the cone and the characteristic impedance

$$Z_0 = 60 \ln \frac{\cot(V_i/2)}{\cot(V_0/2)}$$

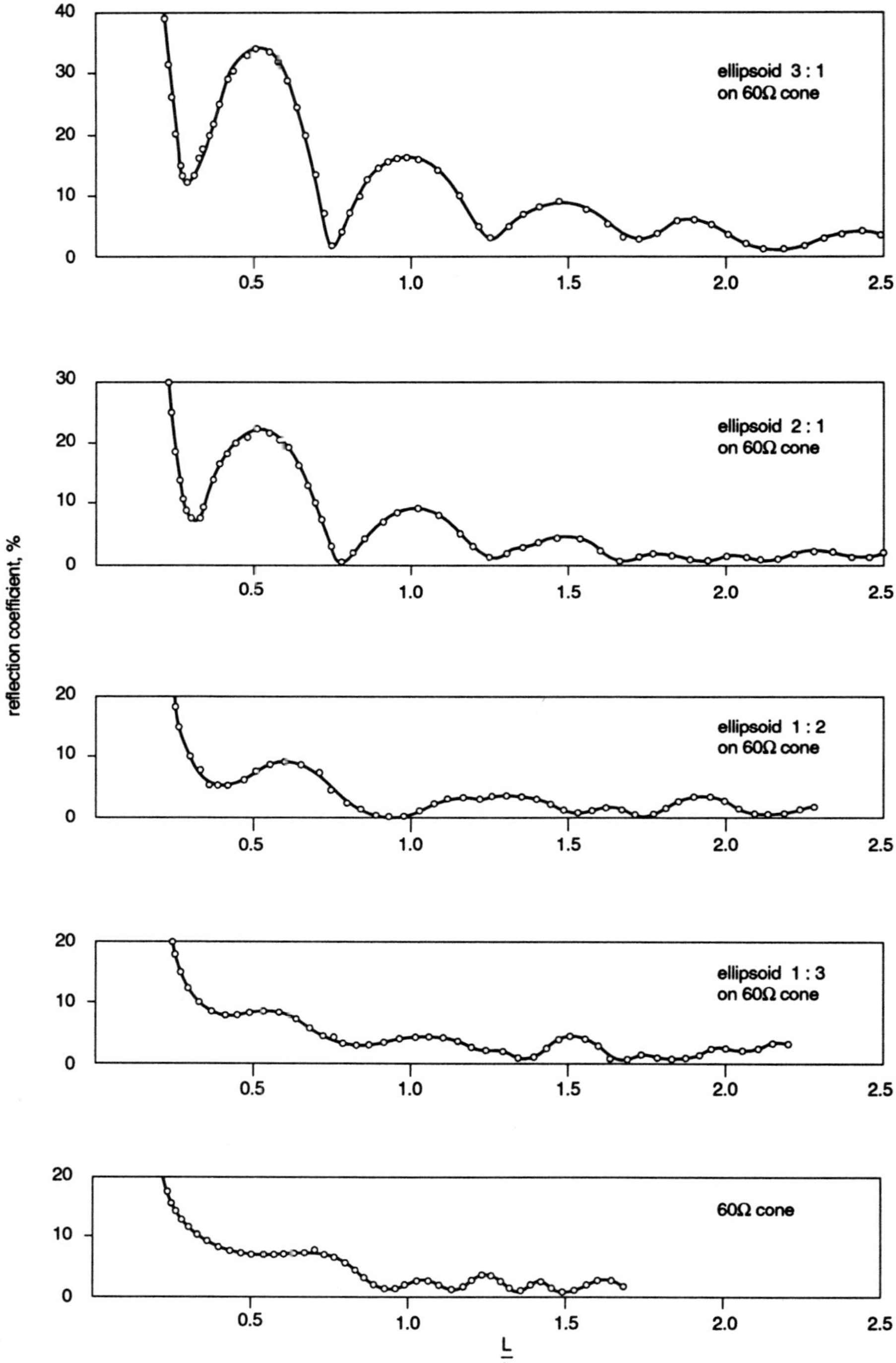

Reference 20

Figure 3.27 *Reflection coefficient of various antennas with arcuate feed*

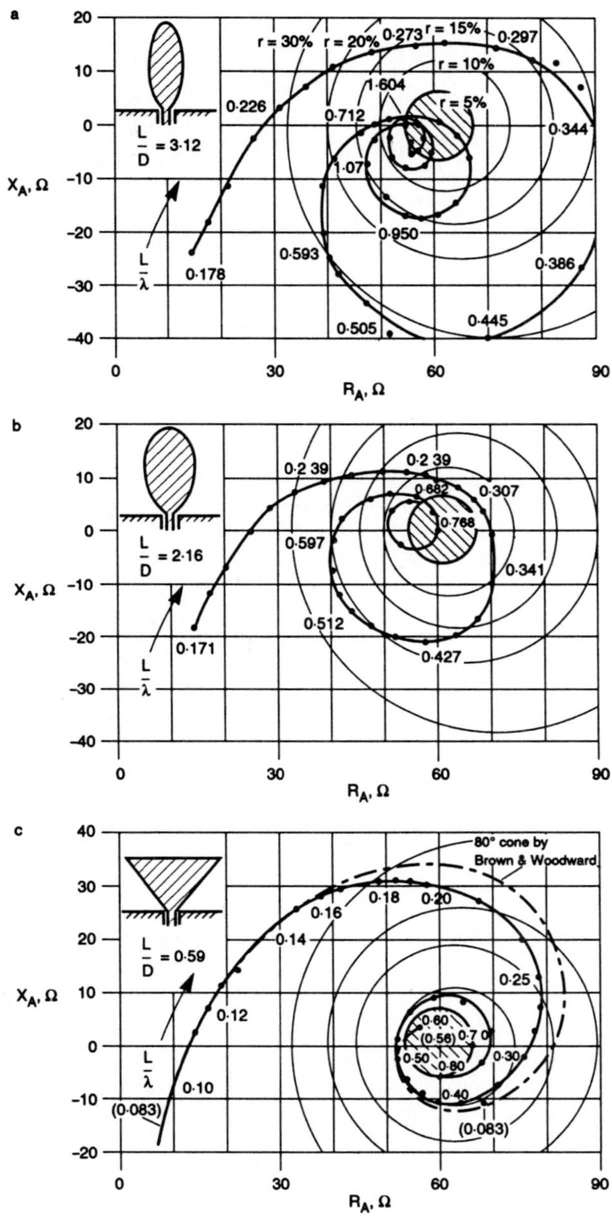

Figure 3.28 *Impedance curves of various antennas with arcuate feed*

a 3:1 ellipsoid
b 2:1 ellipsoid
c 60Ω cone
d 1:2 ellipsoid on 60Ω cone
e 1:3 ellipsoid on 60Ωcone

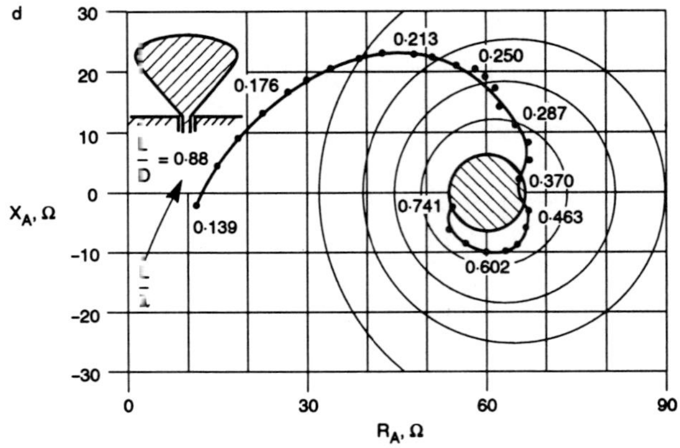

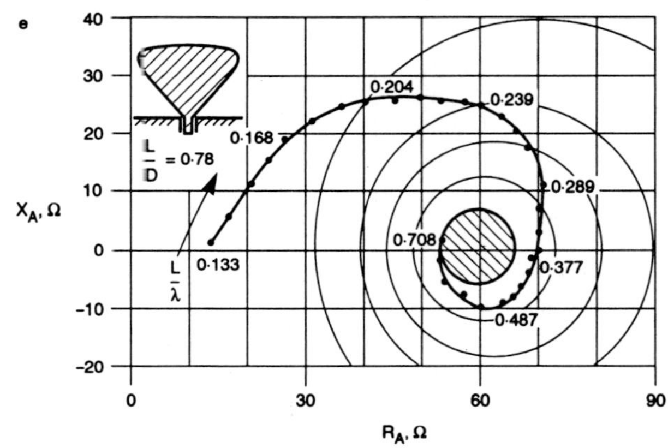

Figure 3.28 *Continued*

where V_i is the half angle of the inner cone and V_0 is the half angle of the outer cone. When the outer cone becomes a flat surface ($V_0 = 90°$) as in Fig. 3.24*b* then $Z_0 = 60 \ln \cot(V_i/2)$. For a conical Z_0 of 60 ohms, $V_i = 40{\cdot}4°$ and for $Z_0 = 50$, $V_i = 47°$.

Impedances for conical monopoles have been calculated by Papas and King [16] and measured by Brown and Woodward [4]. Fig. 3.25 gives reactance and resistance as a function of cone angle and electrical length.

Another feed arrangement uses conductors whose shapes follow arcs of circles, Fig. 3.24*c*. This is described in some detail by Stöhr and Zinke [20] who use the term Kreisbogenleitung which may be translated as 'Arcuate conductors'. As Fig. 3.26 shows, the conical feed is in essence a degenerate form of the arcuate feed. The reflection coefficient (to 60 ohms) of various shapes of antenna are depicted in Fig. 3.27. In each case the antenna has an arcuate feed

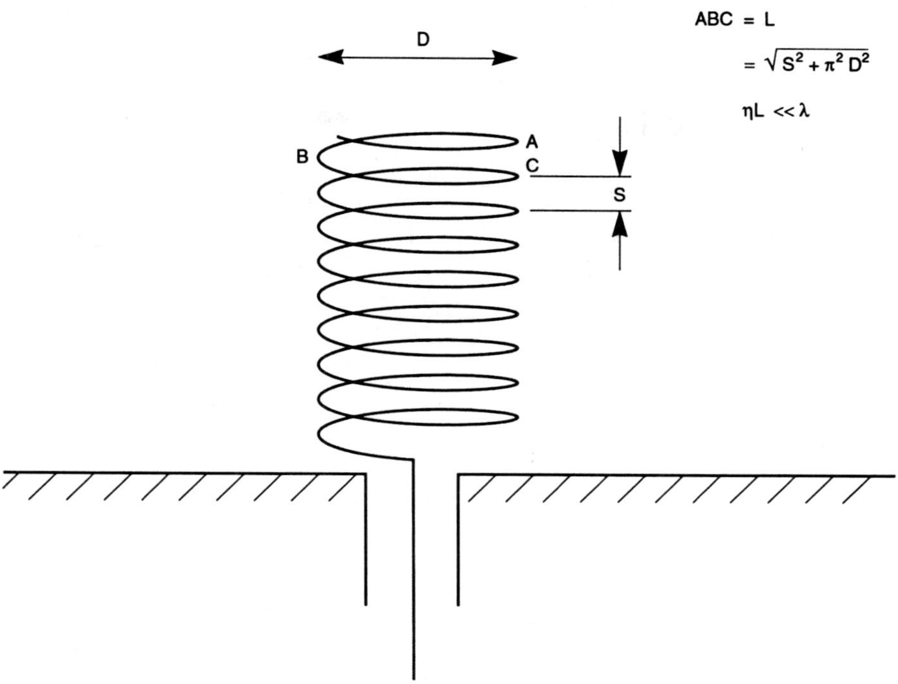

Figure 3.29 *Sidefire helix*

followed by a conical feed both of $Z_0 = 60$ ohms. Fig. 3.28 shows the shapes referred to in Fig. 3.27 and also the impedance curves as a function of L/λ.

Whilst the arcuate feed provides a better transition into a coaxial line than does the conical feed it is clearly more difficult to manufacture. A compromise using a coned section of coaxial line might be acceptable if the extreme bandwidth requirements are essential. What is evident from these results is that the region around the base of the antenna and its connection to the feed line is very important in giving a broadband impedance characteristic.

3.7 Sidefire helix

Consider the antenna shown in Fig. 3.29. If the length round one turn of the helix is L then the total length of the helix is nL where n is the number of turns. Provided that $nL \ll \lambda$ the antenna will radiate in the 'normal' mode, that is, with maximum radiation normal to the axis and zero field on axis. This is of the same form as a dipole.

If the axial distance between successive turns is S then the pitch angle α is $\sin^{-1} S/L$. When $\alpha = 0°$ the helix degenerates to a loop and when $\alpha = 90°$ the helix becomes a straight dipole. It is reasonable therefore to expect that in general the radiation will have components due to a horizontal loop and a

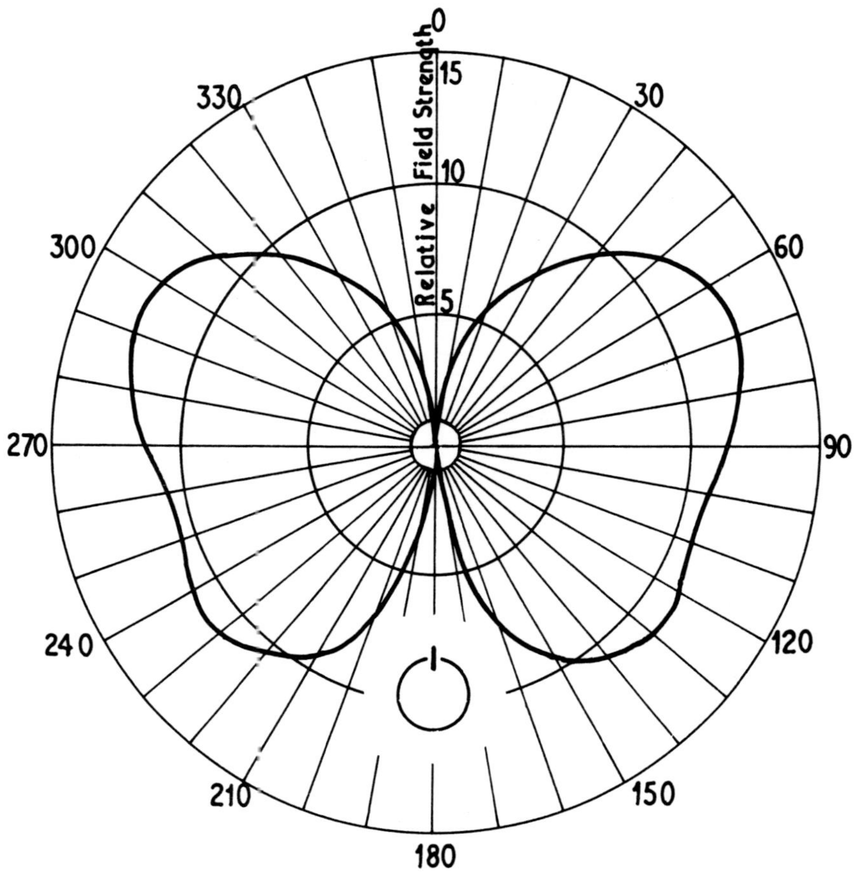

Figure 3.30 *Roll plane pattern of monopole on cylinder of radius λ/4*

vertical dipole, the proportions varying according to the diameter of the helix and its spacing. There does not appear to have been much theoretical work done on the normal mode antenna possibly because the simplifying assumption of uniform, in-phase current results in impractical antennas. A good approximation of the theoretical antenna can be achieved by fitting a capacitance plate to the top of the helix (see Bach Anderson and Hansen [2] which gives impedance data and efficiency for thin and thick helices; the writers concluded that the short, fat helix gave the better performance).

Gouillou [10] describes a conical spiral antenna for the nose cone of a space vehicle: it is essentially a modification of the helix. It appears that the antenna consisted of a conductor of length $\lambda/4$ wound in the form of a conical spiral having a height of $0{\cdot}04\lambda$ and base diameter of $0{\cdot}025\lambda$. It had five turns, not of constant pitch. This antenna was shunt fed giving a resistance of 1·8 ohms, presumably on the vehicle mock-up. An additional shunt capacitance was used for matching.

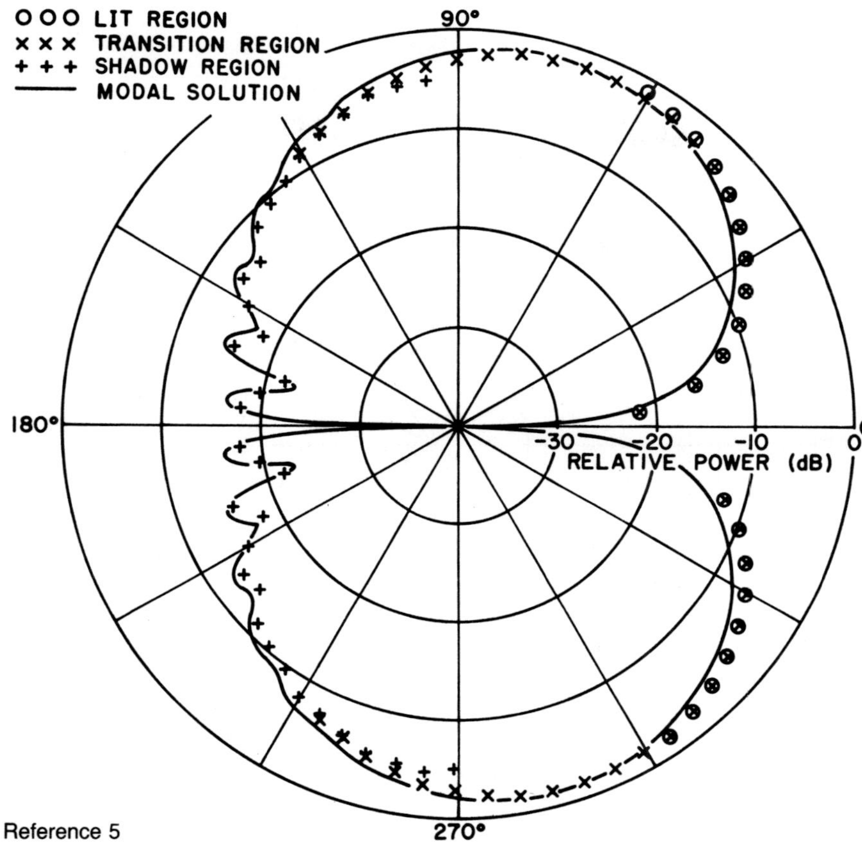

Figure 3.31 *Roll plane pattern of monopole on cylinder of radius 2λ*

3.8 Monopole on cylinder

3.8.1 *Radiation patterns*

Among the earliest papers published on this topic, Carter [7] on circular cylinders and Sinclair [19] on elliptic cylinders have remained classics. Sinclair *et al.* [18] also included patterns of monopoles on circular cylinders. Only the roll-plane patterns are of interest as patterns along the axis of the cylinder are similar to those on flat ground planes. Fig. 3.30 shows a pattern for a monopole on a cylinder of radius $\lambda/4$ — not very different from a dipole pattern. Further details on this topic are given in Wait [22].

Early work was based on modal analysis. More recently the Geometrical Theory of Diffraction (GTD) has provided a powerful tool for modelling antenna patterns on complex structures. Figs. 3.31 and 3.32 are for a monopole on a circular and an elliptic cylinder respectively. The first of these shows the good agreement with the modal solution. In using GTD three zones are considered:

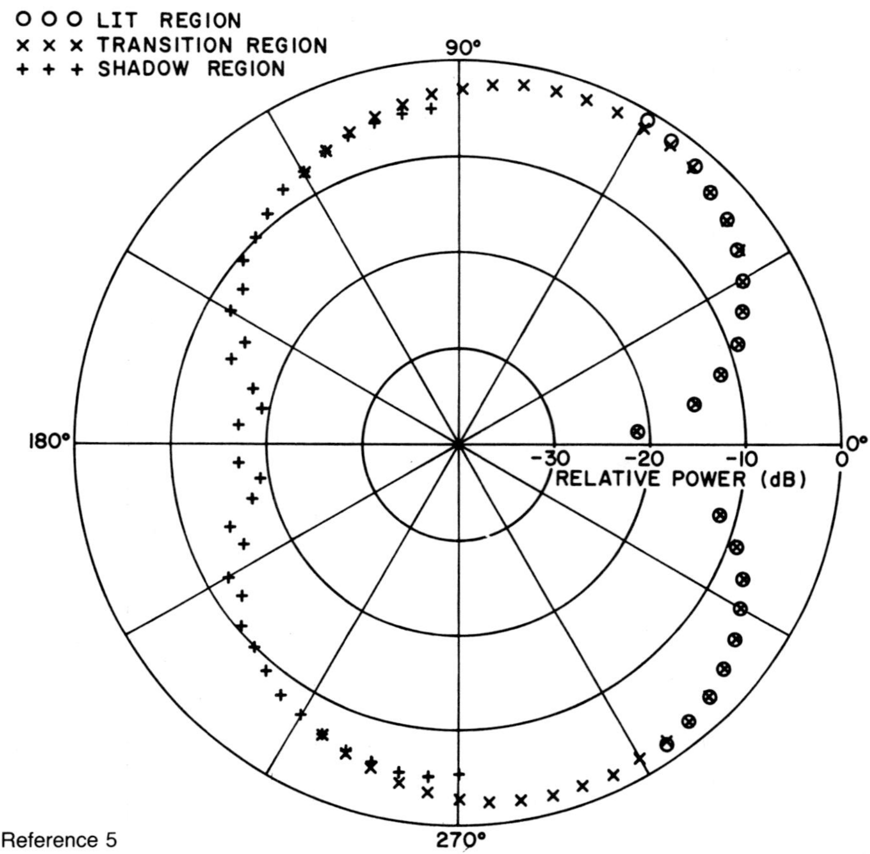

Reference 5

Figure 3.32 *Roll plane pattern of monopole on elliptic cylinder, $a = 4\lambda$, $b = 2\lambda$*

- Lit region
- Transition region
- Shadow region

It can be seen from Fig. 3.31 that the pattern in the shadow region essentially assumes two virtual sources at the ends of a diameter or major axis, the resulting pattern being one of interference with a periodicity dependent on the separation. These figures were taken from Burnside *et al.* [5].

3.8.2 *Impedance*

Pavey [17] has shown that a monopole of height h mounted on a cylinder of radius a is equivalent to a dipole of total height $l = h[1 + a/a + h]$. Clearly as $a \rightarrow \infty$, $l \rightarrow 2h$ which is as expected.

When the cylinder has elliptical cross section, the impedance will depend on the position of the monopole relative to the major and minor axes. When the cylinder is highly elliptical ($a \ll b$) then an antenna at the minor axis, Fig. 3.33, will have an impedance approaching that of a monopole on a flat sheet of the

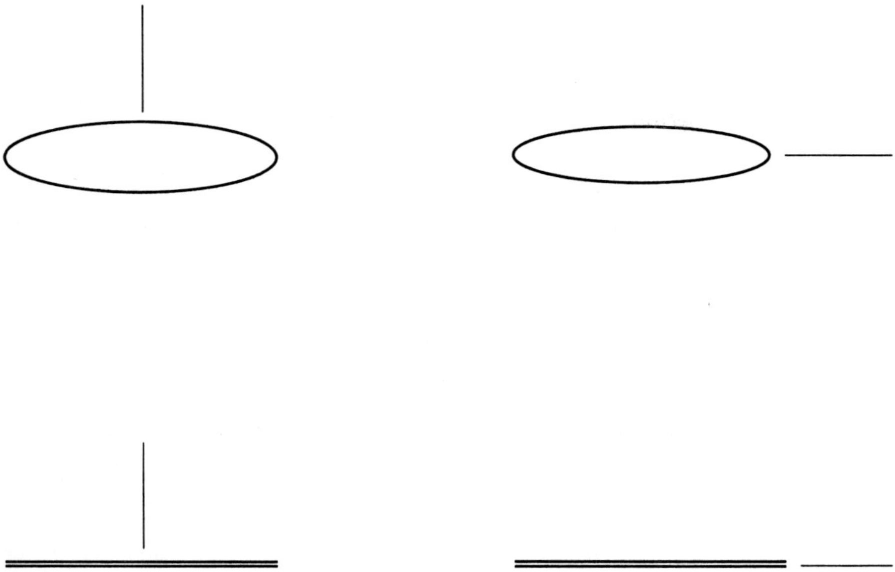

Figure 3.33 *Monopole on highly elliptical cylinder — impedance approximations*

same width. When the antenna is at the major axis, the impedance will approach that of a monopole in the plane of a flat sheet and on its edge. Johnson [12] gives the radiation resistance as

$$R_F = \tfrac{3}{8}\,120\pi[C(kl) - \cot(kl) \times S(kl)]^2$$

where

$$C(kl) = \int_0^{kl} \frac{\cos t}{\sqrt{(2\pi t)}}\,dt \text{ and } S(kl) = \int_0^{kl} \frac{\sin t}{\sqrt{2\pi t}}\,dt$$

(There is an inconsistency in Johnson's paper for the value of R_F; this is believed to be the correct formula). When l is short so that the current may be considered as linear,

$$R_F = \tfrac{2}{3}\,120\pi\frac{l}{\lambda}$$

It is interesting to note that, for short antennas, R_F is directly proportional to l/λ whereas for normal short monopoles R would be proportional to $(l/\lambda)^2$. The reactance can be calculated from the average characteristic impedance $Z_{AV} = 60 \ln (8l/d - 1)$ for a monopole of length l and diameter d. The input reactance will therefore be $-jZ_{AV} \cot kl$.

Jasik [11] gives a value of 86·3 ohms for the radiation resistance of a monopole of length 0·25 λ on the edge of a sheet. This agrees well enough with the value derived from Johnson.

3.9 References

1 AWADALLA, K.H., and MACLEAN, T.S.M.: 'Input impedance of a monopole antenna at the centre of a finite ground plane' *IEEE Trans.*, 1978, **AP–26**, pp. 244–248

2 BACH ANDERSEN, J., and HANSEN, F.: 'Antennas for VHF/UHF personal radio: a theoretical and experimental study of characteristics and performance', *IEEE Trans.*, 1977, **VT–26**, pp. 349–357.

3 BROWN, G.H., and WOODWARD, O.M.: 'Experimentally determined impedance characteristics of cylindrical antennas', *Proc. IRE*, Apr. 1945, pp. 257–262

4 BROWN, G.H., and WOODWARD, O.M.: 'Experimentally determined radiation characteristics of conical and triangular antennas', *RCA Review*, 1952, **13**, p. 425

5 BURNSIDE, W.D., MARHEFKA, R.J., and YU, C.L.: 'Roll plane analysis of on-aircraft antennas', AGARD Conference Publication, CP139, November 1973, paper 41

6 BURTON, R.W., and KING, R.W.P.: 'Theoretical considerations and experimental results for the hula-hoop antenna', *Microwave J.*, 1963, **6**, pp. 89–90

7 CARTER, P.S.: 'Antenna arrays around cylinders', *Proc. IRE*, 1943, **31**, pp. 671–693

8 COOPER, L.J.: 'Monopole antenna on electrically-thick conducting cylinders'. Tech Report 660, Harvard University, 1975

9 FOSTER, P.R., and MILLER, T.: 'Radiation patterns of a quarter-wave monopole on a finite ground plane'. IEE Conf. Pub. 195, April 1981 pp. 451–455

10 GOUILLOU, R.: 'Antenna ultra courte à spirale conique' *in* 'Radio antennas for aircraft and aerospace vehicles'. Technivision Services, Maidenhead, England, Nov. 1967, pp. 251–264

11 JASIK, J. (Ed.): 'Antenna engineering handbook' (McGraw Hill Book Co., 1961)

12 JOHNSON, W.A.: 'The notch aerial and some applications to aircraft radio installations', *Proc. IEE*, 1955, **102** Pt. B, pp. 211–218

13 JOSEPHSON, B.: 'The quarter-wave dipole'. IRE Wescon Conv. Record, 1957, pp. 77–90

14 LAPORT, E.A.: 'Radio antenna engineering', (McGraw-Hill Book Co., 1952)

15 MEIER, A.S., and SUMMERS, W.P.: 'Measured impedance of vertical antennas over finite ground planes', *Proc. IRE*, June 1949, pp. 609–616

16 PAPAS, C.H., and KING, R.: 'Input impedance of wide-angle conical antennas fed by a coaxial line', *Proc. IRE*, 1949, **37**, p. 1269

17 PAVEY, N.A.D.: 'Guide to aircraft high frequency communication antenna systems'. RAE Technical Report 80156, Dec. 1980

18 SINCLAIR, G., JORDAN, E.C., and VAUGHAN, E.W.: 'Measurement of aircraft antenna patterns using models', *Proc. IRE*, 1947, **35**, pp. 1451–1462

19 SINCLAIR, G.: 'The patterns of antennas located near cylinders of elliptical cross-section', *Proc. IRE*, 1951, **39**, pp. 660–668

20 STÖHR, W., and ZINKE, O.: 'Wege zum optimalen Breitband-Rundstrahler', *Frequenz*, 1960, **14**, pp. 26–35

21 STORER, J.E.: 'The impedance of an antenna over a large circular screen', *J. Appl. Phys.*, 1951, **12**, p. 1058

22 WAIT, J.R.: 'Electromagnetic radiation from cylindrical structures' (Pergamon Press 1959)

23 WANSELOW, R.D., and MILLIGAN, D.W.: 'A compact, low profile transmission line antenna – tunable over greater than octave bandwidth', *IEEE Trans.*, 1966, **AP–14**, pp. 701–707

24 WOLFF, E.A.: 'Antenna analysis' (John Wiley, New York, 1966)

25 YUNG, E.K., and BUTLER, C.M.: 'Coaxial-line driven monopole on an electrically-thick conducting cylinder over a ground plane' *IEE Proc.*, 1984, **131**, Pt. H, pp. 54–60

Chapter 4
The loop antenna

Much has been written about the theory of loop antennas but most of it refers to loops in which the current is constant round the loop and in phase. In practice this applies only to small loops with a single feed or to large loops with a multiplicity of in-phase feeds.

Storer [5] computed the current distribution on loops of radius b and wire radius a from $Kb=0{\cdot}1$ to $Kb=2{\cdot}5$. These results show that the current is near constant in amplitude up to $Kb=0{\cdot}1$ and substantially constant in phase up to $Kb=0{\cdot}2$. The patterns in the plane of the loop would not depart much from a circle up to $Kb=0{\cdot}3$.

The radiation resistance of a small thin single-turn loop is $320\pi^4\ A^2/\lambda^4$, where A is the area of the loop. For circular loops this becomes $20\pi^2(Kb)^4$. Storer

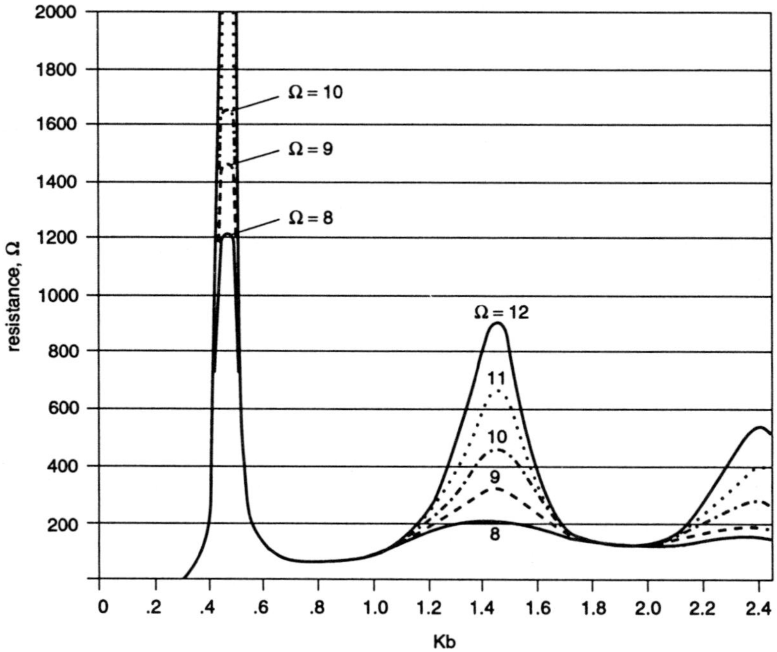

Reference 5

Figure 4.1 *Radiation resistance of loop antennas as a function of diameter and thickness*

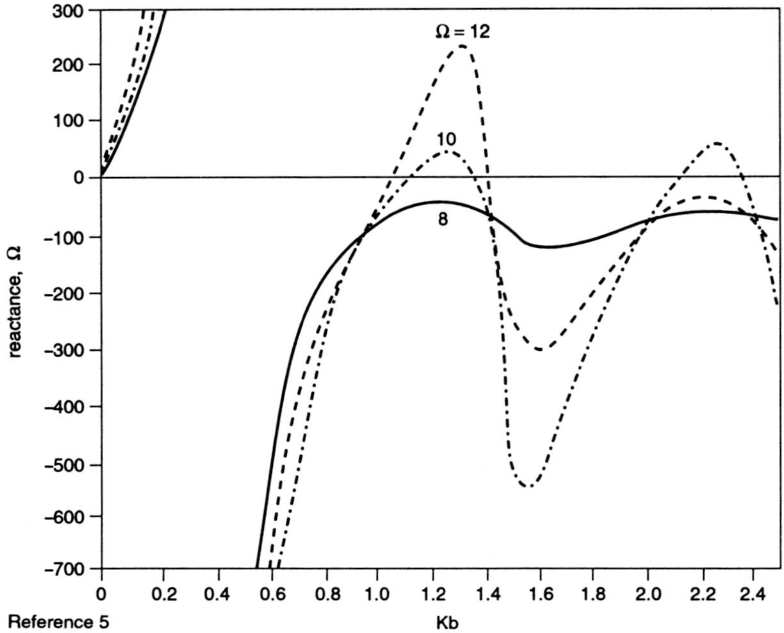

Figure 4.2 *Reactance of loop antennas as a function of diameter and thickness*

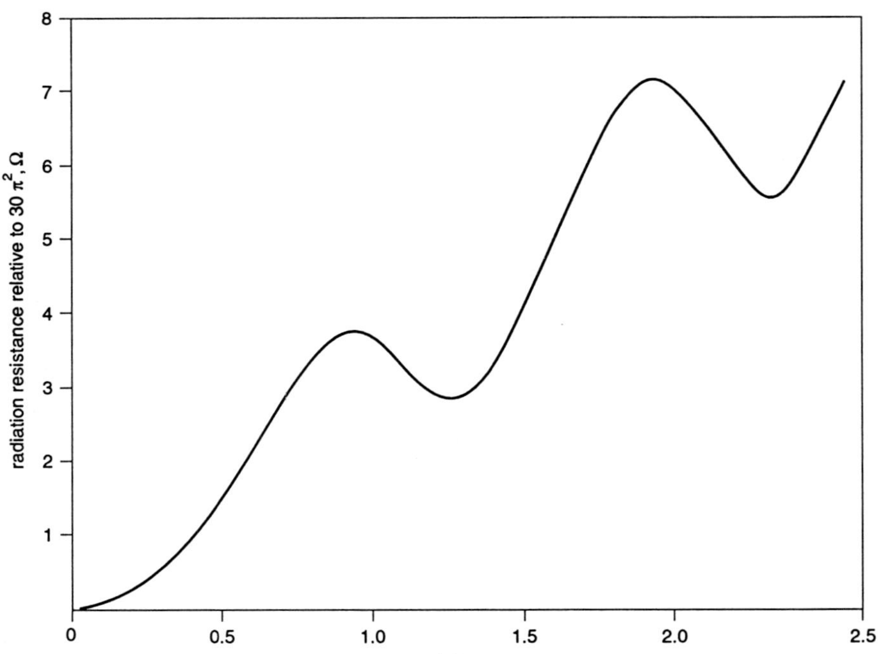

Figure 4.3 *Radiation resistance of loop antennas with constant current*

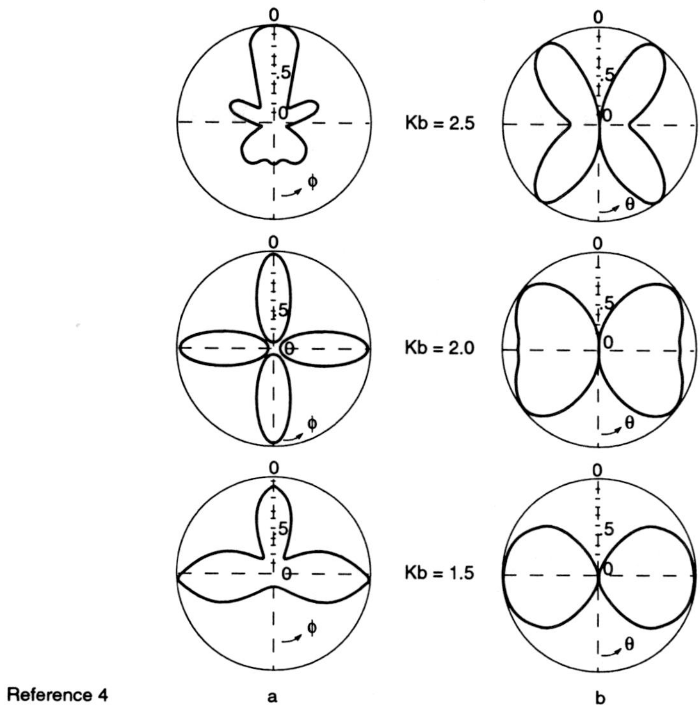

Figure 4.4 *Azimuth and elevation patterns of large circular loops*
a Horizontal field pattern
$\theta = \pi/2$ plane
$\Omega = 2 \ln b/a = 10$
b Vertical field pattern
$\phi = 0$ plane
$\Omega = 2 \ln b/a = 10$

computed radiation resistance for loops having ratios of b/a between 8·69 and 64·21 for Kb between 0·05 and 2·50. Unfortunately his figures for small loops do not agree with the simple theory and, in addition, do not show any trend as the ratio b/a is altered. They must, therefore, be considered with some suspicion. As an example, at $Kb = 0{\cdot}05$, simple theory would give $R = 0{\cdot}0012$ while Storer's figures are around 0·0048. Similarly his reactance values appear high, at least for small loops.

If Storer's impedance values for small loops appear dubious they appear to agree with his measured results for larger loops. These results show that the resistance rises to a very high figure when Kb is of the order of 0·5, which is to be expected. It can also be seen that the resistance peaks are highest for the largest b/a values. For broad bandwidth one would choose Kb in the range 0·8 to 1·2 and one would make the loop as fat as possible i.e. low b/a. Storer's curves for resistance and reactance are shown in Figs. 4.1 and 4.2. The parameter $\Omega = 2 \ln 2\pi b/a$. His figures should be compared with those of Moullin [3] for large loops with constant current shown in Fig. 4.3.

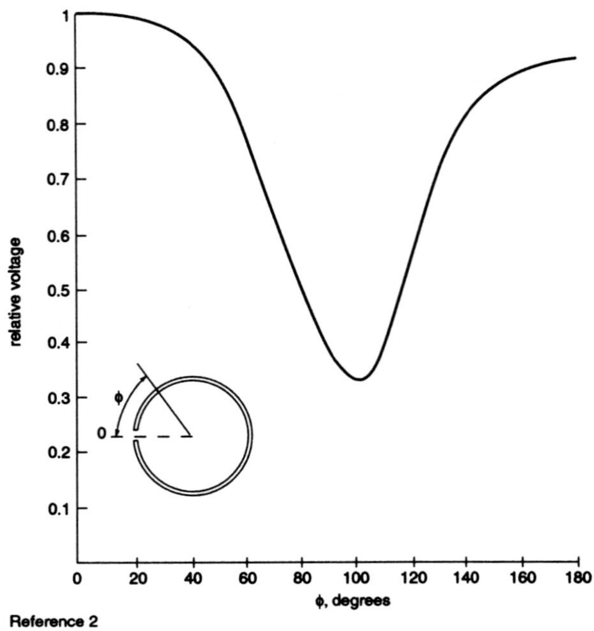

Figure 4.5 *Azimuth pattern for Kb = 1*

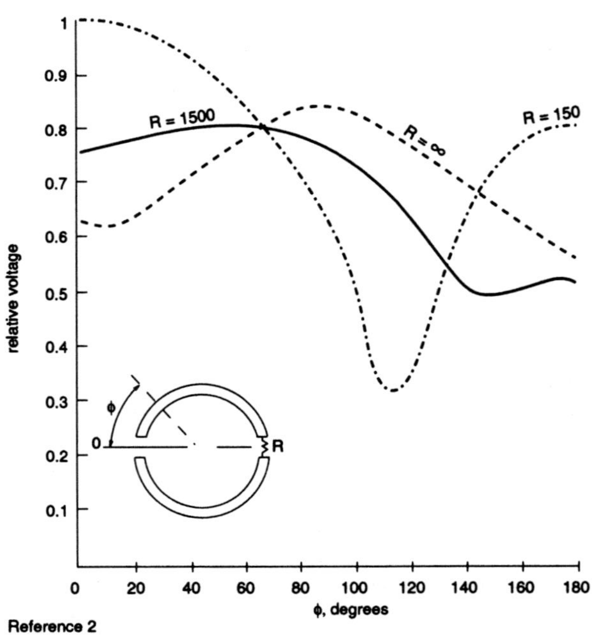

Figure 4.6 *Azimuth patterns of loaded loops for Kb = 1*

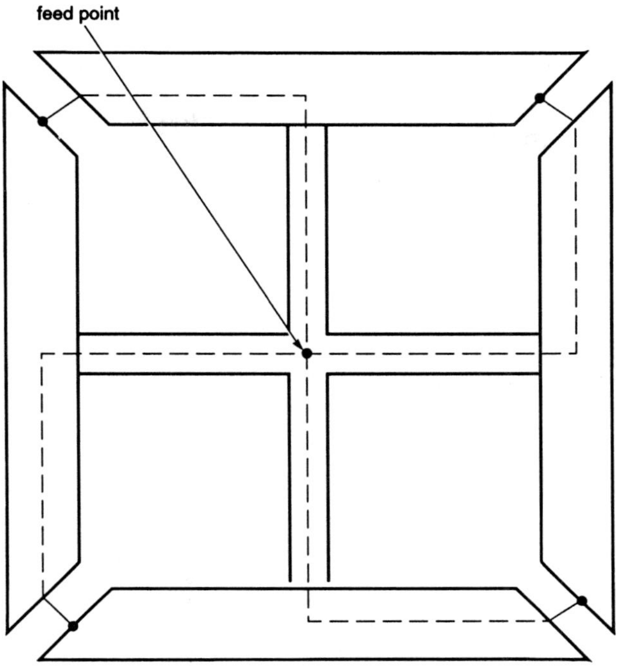

Figure 4.7 *Alford loop*

Confirmation of Storer's theory has been shown by Rao [4] who calculated and measured azimuth and elevation patterns of horizontal circular loops for $Kb = 1{\cdot}5$, $2{\cdot}0$ and $2{\cdot}5$, with $\Omega = 10$ ($b/a = 23{\cdot}6$). Good agreement was obtained. Some of these patterns are reproduced in Fig. 4.4. For a small loop the patterns are those of a vertical dipole with the E and H vectors reversed — circular in the azimuth plane and figure-of-eight in elevation. A measured pattern for $Kb = 1{\cdot}0$ is given in Fig. 4.5. The value of Ω is not known.

Iizuka [2] has explored the possibility of altering the characteristics of a loop by resistive loading at various points around the perimeter. He used both positive and negative resistances and found that the biggest effects occurred when the loads were placed at the position of maximum current, a half-wavelength from the feed-point. What should come as no surprise is that the conductance and susceptance curves are both flattened by using positive resistances and have increased variation with Kb when the loading is negative. Placing positive resistances at points of maximum current is clearly to make the current more uniform and in consequence the radiation pattern should alter. This is demonstrated in Fig. 4.6 which shows azimuth patterns measured on a loop of $Kb = 1$ with various loads opposite the feed-point. It will be seen that the position of the maxima and minima shift as the load resistance is increased. The ratio b/a for this experimental loop is not stated. The author's own experiments suggest that the ratio has some influence on the value of load required for the best radiation pattern. They also showed that the patterns for a loaded loop remained more or less constant up to $Kb = 1$.

The admittance is not well suited for matching to a transmission line, the conductance and susceptance varying very considerably. There is, however, a region around $Kb = 0{\cdot}8$ where the susceptance is near zero and the conductance is small, i.e. almost a pure high resistance. This suggests that an antenna of this size might be more useful for reception than transmission and in fact such an antenna has been used as the basis of a broadband direction-finding system for which, unusually, horizontal polarisation was required. It was found possible to put one loop inside another without significant interaction and hence cover a wide frequency band.

Another method of obtaining a near-omnidirectional pattern and at the same time achieving a reasonable impedance match derives from work by Alford and Kandoian [1] and is shown in Fig. 4.7. This particular arrangement is designed for coaxial feed whereas the original arrangements were fed from twin lines. Each side of the loop is about a half-wavelength.

4.1 References

1 ALFORD, A., and KANDOIAN, A.: 'Loop antennas', *Trans. AIEE*, Supplement, 1940, **59**, p. 843

2 IIZUKA, K.: 'The circular loop antenna multi-loaded with positive and negative resistors', *IEEE Trans.*, 1965, **AP–13**, pp. 7–20

3 MOULLIN, E.B.: 'Radiation from large circular loops', *J. IEE*, 1946, **93** Pt. III, pp. 345–351

4 RAO, B.R.: 'Far field patterns of large circular loop antennas: Theoretical and experimental results', *IEEE Trans.*, 1968, **AP–16**, pp. 269–270

5 STORER, J.E.: 'Impedance of thin-wire loop antennas', *Trans. AIEE*, Communications and Electronics, 1956, **75** Pt. 1, pp. 606–619

Chapter 5

Slot antennas

5.1 Introduction

Impetus to the development of the slot antenna was given by the need for VHF and UHF antennas with low aerodynamic drag for military aircraft in World War Two and subsequently for the first generation of civil airliners post war. The cavity-backed half-wave slot antenna was among the first 'suppressed' antennas, today generally referred to as 'flush mounted'. As aircraft engines have increased in power the desperate need to keep aerodynamic drag to a minimum has largely disappeared and the undoubted structural problems posed by slot antennas, particularly in pressurised fuselages, has led to a diminution in the use of slot antennas in this particular field. Nevertheless, slot antennas still have their uses, sometimes in surprising areas. If this chapter appears to have a large proportion of historical applications this is because much of the development work occurred between 1940 and 1960. The results are still valid and it is important that this body of knowledge should not be lost to today's engineers.

Three classes of practical slot antenna can be distinguished

- Cavity-backed slot in a flat ground plane
- Slotted cylinder
- Annular slot

5.2 The basic slot antenna

Consider a thin, straight, half-wave slot cut in an infinite conducting sheet. Whether the slot is directly fed or excited by an incident electric field the voltage is clearly zero at the ends and maximum in the middle. It is also obvious that the electric vector which produces this voltage distribution must be at right angles to the length of the slot, i.e. across its narrow dimension. Compare this with a half-wave strip dipole of the same size as the slot and similarly oriented in space, Fig. 5.1. Here the voltage distribution has a minimum at the centre and maxima at the ends — the reverse of the slot distribution. Furthermore the electric vector is parallel to the length of the dipole instead of transverse. For these reasons the slot is sometimes described as a 'magnetic dipole'.

Booker [2] first showed the complementary nature of slot and dipole and demonstrated the important relationship

$$Z_1 Z_2 = \tfrac{1}{4}(120\,\pi)^2$$

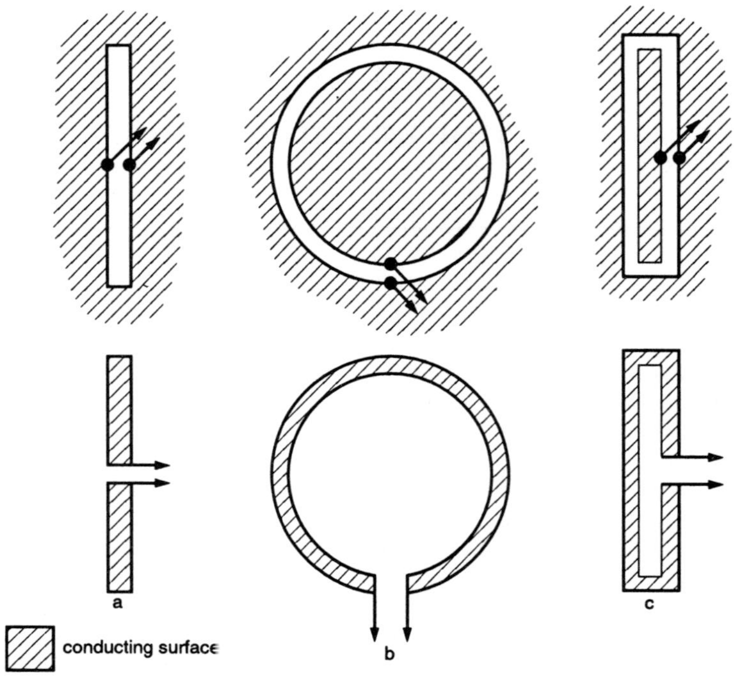

Figure 5.1 *Basic slot and equivalent dipole*

a Half-wave dipole
b Loop
c Folded dipole

Courtesy RAeS

where Z_1 is the impedance of a dipole and Z_2 of its complementary slot. The direct equivalent of a slot is a strip dipole but this can be easily related to a cylindrical dipole whose diameter d is half the strip width W. If the centre point impedance of a thin half-wave dipole is $73{\cdot}2+j42{\cdot}5$ ohms then that of the equivalent slot is $365-j212$ ohms. For more practical antennas Kraus [11] gives two examples:

(i) A dipole of length $0{\cdot}475\lambda$ and diameter $0{\cdot}005\lambda$ has a resonant resistance of 67 ohms. The equivalent slot of the same length but width $0{\cdot}01\lambda$ has a centre point impedance of 530 ohms.

(ii) A dipole of length $0{\cdot}925\lambda$ and diameter $0{\cdot}033\lambda$ has an input impedance of $710+j0$. The equivalent slot of width $0{\cdot}066\lambda$ has an input impedance of $50+j0$ ohms.

It will be noted that the sign of the slot reactance is opposite to that of the dipole. This can be appreciated if one half of the slot is considered as a short-circuited strip transmission line. Then clearly for lengths less than $\lambda/4$ the input reactance of each half will be inductive whilst that of the equivalent dipole will be capacitive.

As expected from similar effects with monopoles on finite ground planes, the impedance of the slot is affected by the size of the ground plane. Frood and Wait [8] show values of slot conductance and susceptance for a waveguide-fed slot as a function of sheet width. These results are normalised to the guide Z_0 so it is difficult to obtain actual values but the oscillatory nature of the curves can be seen. Provided the sheet extends a reasonable distance beyond the ends of the slot, this dimension has negligible effect, only the slot width being important. Kraus [11] suggests that the slot should be at least a wavelength from the edge of the sheet in the width direction to make the effect negligible.

The radiation patterns of a horizontal slot in an infinite sheet are identical with those of the complementary electrical dipole but with the E and H components reversed. Thus in the horizontal plane the pattern falls to zero in the plane of the sheet whilst the pattern is circular in the vertical plane through the slot centre and normal to the sheet.

5.3 Cavity-backed slots

If one side of the slot is covered by a conducting box forming a cavity behind the slot, Fig. 5.2, then the radiation is restricted to the open side. The radiation resistance then becomes double that of the slot radiating on both sides. The reactance will depend on the susceptance introduced by the cavity; this susceptance will be negligible if the cavity depth is of the order of a quarter-wavelength. Some attempts have been made to compute the susceptance by assuming the cavity to be a section of short-circuited waveguide. However, in

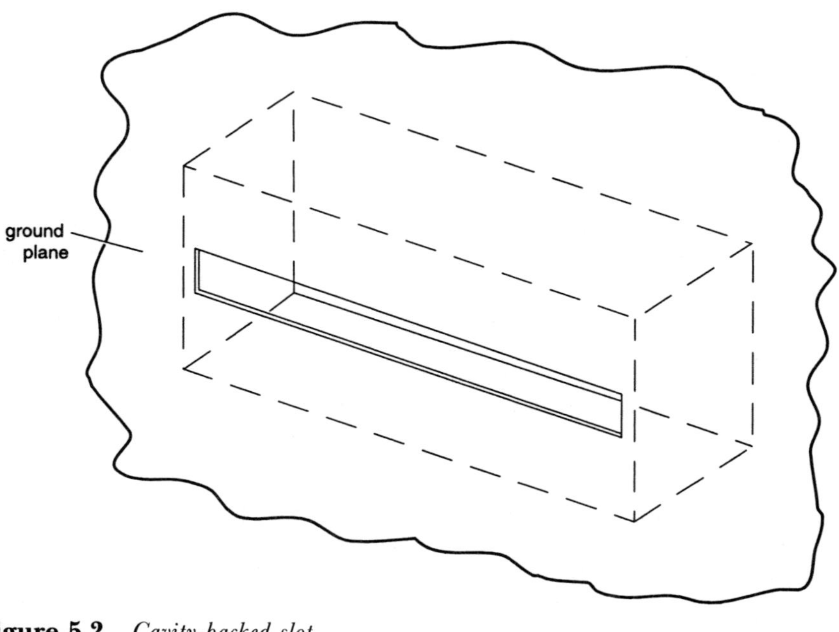

Figure 5.2 *Cavity-backed slot*

many instances the transverse dimension of the 'guide' is too small and the waveguide has to be considered as evanescent. Calculation of the slot impedance in a cavity-backed slot is likely to be time-consuming and of dubious accuracy because of the effects of the feed itself.

If the sheet containing the slot is infinite then the radiation patterns are simply one half those of the slot radiating on both sides. If, however, the sheet width is finite then diffraction effects from the edges will create lobed patterns in the E plane normal to the slot. The pattern in the H plane, through the slot length, will not be affected. Frood and Wait [8] show that on the shadow side of the sheet the field is proportional to

$$K|F[-(2kd)^{1/2}\sin\phi_0/2]+F[-(2kd)^{1/2}\cos\phi_0/2]|$$

where

$$F(x)=\exp j\pi/4\times\pi^{-1/2}\int_{-\infty}^{0}\exp -jx^2\,dx$$

and the slot is centrally situated in a sheet of width $2d$. Φ_0 is the angle measured from the sheet.

On the illuminated side,

$$E=K|F[(2kd)^{1/2}\sin\Phi_0/2]+F[(2kd)^{1/2}\cos\Phi_0/2]-1|$$

K is a constant depending on the slot dimensions.

Experimental and calculated patterns are shown in this reference for values of kd between 4 and 141. A typical voltage pattern in polar form is shown in Fig. 5.3.

5.3.1 *Methods of feeding*

The two methods to be considered here are

(*a*) Direct coaxial feeding
(*b*) Probe feeding

5.3.1.1 *Coaxial feeding*

It has already been shown that the centre-point impedance of a cavity backed slot is very high if the slot is of the order of a half-wavelength. Much longer slots are going to be physically unattractive except at very high frequencies so the half-wave slot is likely to be the most used type. Except, indeed, for the instance where the slot is on the ground and the cavity is a metal lined trench under it, the slot antenna is unlikely to be the most attractive solution for frequencies below 200 MHz. This high impedance is ill-suited to coaxial cable systems so, although it is possible to obtain a match by juggling the slot reactance and the cavity susceptance, the bandwidth will be small.

Just as it is possible to step *up* a dipole impedance by folding, so the slot impedance can be stepped *down* in the same way as Fig. 5.1*c* shows. In this illustration the step down ratio is 1:4 as the two arms are equal but other ratios can be achieved by altering the relative widths of the arms. Supporting the metal strip in the slot may be difficult unless it is put on a dielectric cover. One method is to use a double-sided printed board, Fig. 5.4, which has the

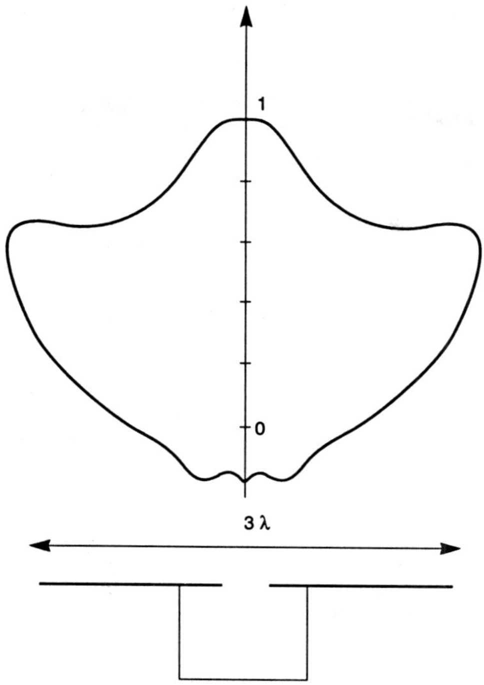

Figure 5.3 *E plane pattern of cavity backed slot in sheet of width* $2d = 3\lambda$

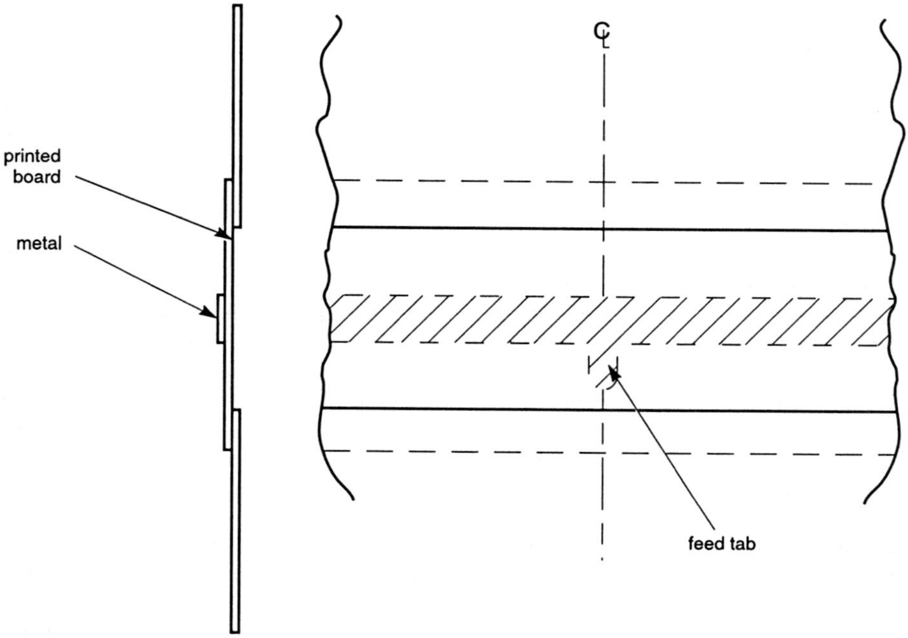

Figure 5.4 *Construction of folded feed system*

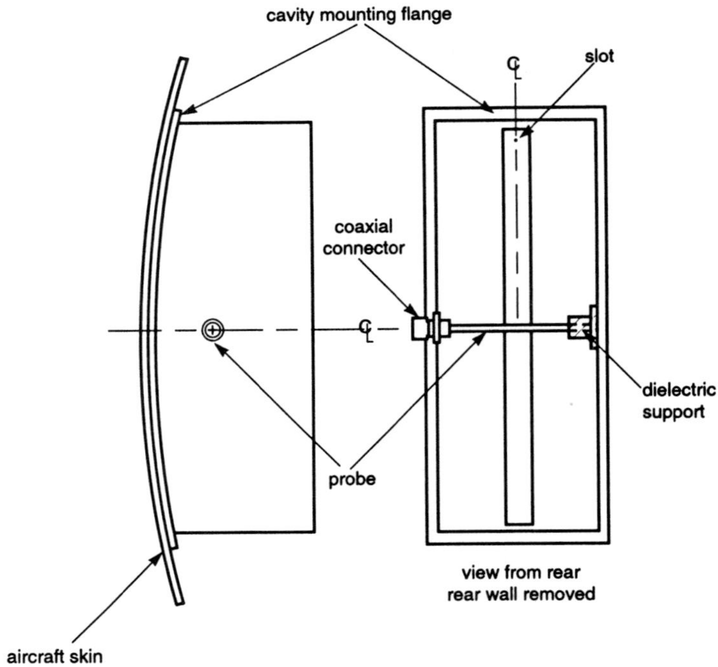

Figure 5.5 *Probe fed slot*

advantage of maintaining a long insulation path between the strip and the sides of the slot.

An alternative method of finding a matched impedance for direct coaxial feed is to tap along the slot towards the closed end. On the analogy of an offset fed dipole the impedance at a distance l from the centre should be $Z_c \cos^2 kl$ where Z_c is the centre point impedance. This suggests that l should be about 0·21λ for a half-wave slot with $\lambda/4$ deep cavity behind.

5.3.1.2 *Probe feeding*

The simplest method is a probe parallel to the electric vector and set close behind the slot, Fig. 5.5. It needs to be as far as possible forward of the back of the cavity to minimise the effect of cavity reactance. In one example of this type the probe was a rod of 0·005λ diameter mounted 0·112λ from the back wall of a rectangular cavity 0·447λ by 0·168λ. The slot was 0·419λ by 0·028λ covered with 1·5 mm GRP. A VSWR of 1·5 to 50 ohms was achieved over a bandwidth of 2·4%. In this example the probe was supported by a polythene block which would have provided a small amount of capacitance to the opposite side of the cavity.

In a variation of this antenna two offset probes were used, Fig. 5.6, to provide two antennas with identical radiation patterns for an automatic landing system. The probes were buffered by a small amount of attenuation so that faults between one probe and its radio receiver had negligible effect on the other system.

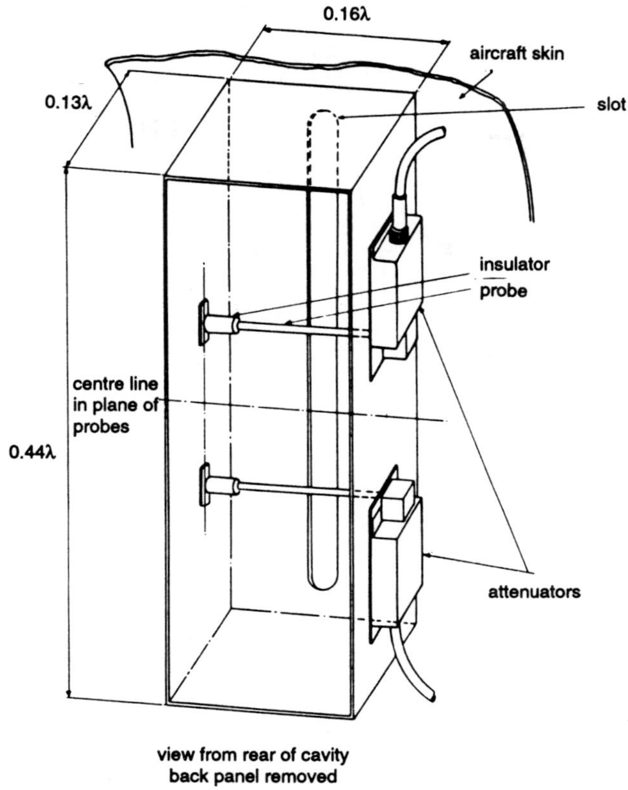

Figure 5.6 *Cavity backed slot with dual feed*

For broadband operation the bar-and-post or Tee-feed has been used very successfully. Much of the pioneer work was done by the staff of the Radio Research Labs. of Harvard University [17] and in particular by Lazarus [12]. Details of one of these antennas are shown in Fig. 5.7. This antenna had a 2:1 VSWR bandwidth of over 120% $[100(F_{max}-F_{min})/F_{min}]$. It should be noted that in these antennas the slot is as wide and as high as the cavity whereas in most antennas the slot is narrow compared with the cavity. With the narrow slot, the ground plane and slot act as a capacitive iris across the cavity, if the latter is considered as a section of waveguide. Further details on the design of T-bar feeds are given by Newman and Thiele [15].

5.3.1.3 *Slot shortening*

Various methods of increasing the electrical length of a slot, or reducing the physical length for a required electrical length have been suggested. One method described by Lopez [13] is to enlarge the ends of the slot thus forming a dumb-bell loaded slot Fig. 5.8. By folding the central portion of the slot the impedance can be reduced to a convenient value for matching to a coaxial cable. Ends of different shapes have been used by other workers in this field.

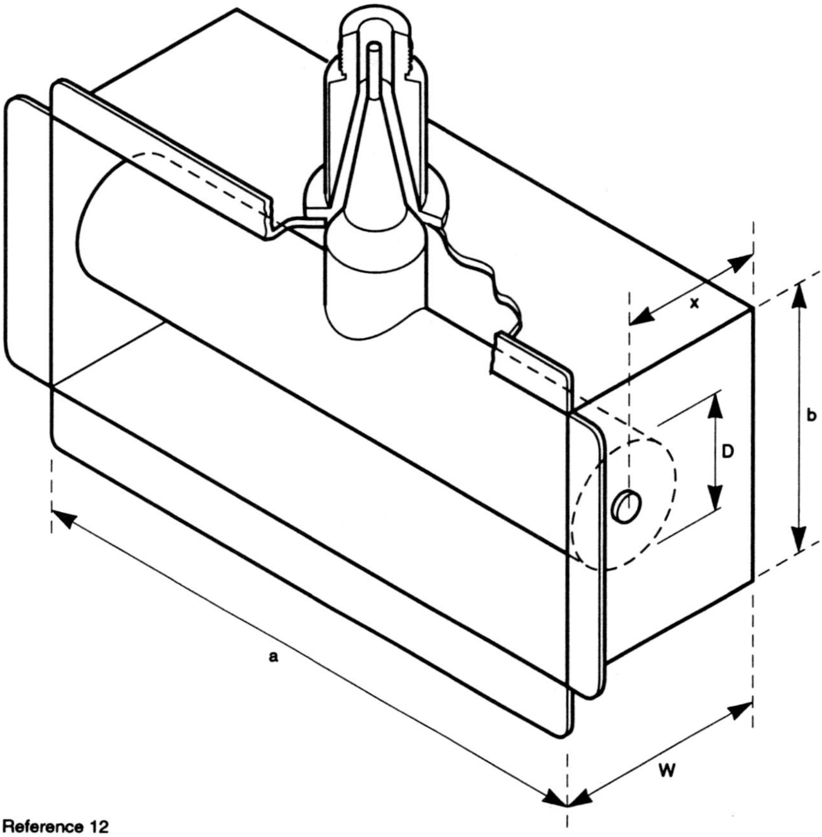

Figure 5.7 *Tee-bar feed cavity backed slot*

At F_{min} $a = 0{\cdot}583\lambda$
$b = 0{\cdot}274\lambda$
$w = 0{\cdot}197\lambda$
$x = 0{\cdot}129\lambda$
$D = 0{\cdot}103\lambda$

As well as end-loading a slot it is also possible to reduce its length by capacitance-loading near the centre. This can take a number of forms:

(i) Narrowing the slot over a short distance
(ii) Adding flanges to the edges of the slot
(iii) Increasing the thickness of dielectric over the central portion.

There is probably no particular advantage of one form over another, the choice being determined by engineering considerations.

It is usually necessary to cover a slot with a dielectric to prevent ingress of rain, snow, dust etc. and also to provide a smooth aerodynamic surface on aircraft. If the cover is in intimate contact with the slot edges it will increase the electrical length. There is, however, the danger that ice or snow on the cover

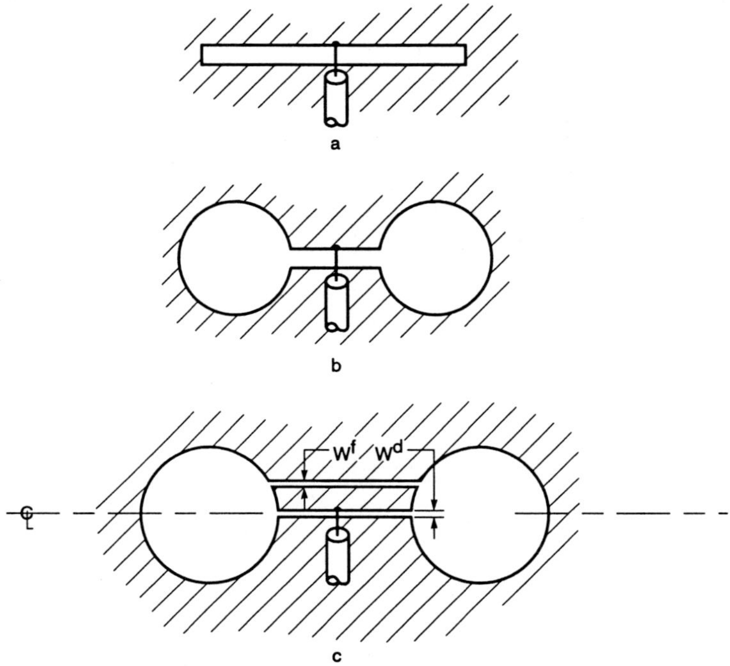

Figure 5.8 *Dumb-bell slot*

a Simple slot
b Dumb-bell loaded slot
c Folded dumb-bell slot

will upset the antenna matching. Two methods of minimising the effect of the cover are shown in Fig. 5.9. In any case the cover should be of low loss material and as thin as physically possible.

5.3.2 *The pocket slot*

It is not necessary for the slot to be central in the cavity cross-section, the main effect of offset being a change in impedance. Equally, the cavity does not have to have its axis normal to the ground plane. The extreme case of this is the pocket slot Fig. 5.10. In this instance one wall of the cavity is the ground plane. One advantage of this arrangement is that it may permit a deeper cavity than would otherwise be possible. The cavity may also be split as in Fig. 5.11 where the reactances of the two cavities are in parallel. At frequencies above 1·5 GHz this form has been constructed in printed circuit form with a rather thick substrate.

In the above examples the cavity is behind the ground plane but with such shallow cavities it will make little difference if the cavity is in front. The resulting antenna can be described as a 'Pannier slot' and Fig. 5.12 shows single and double antennas in a single structure. The double arrangement could be steered in azimuth by adjusting the relative phase to the two antennas or simply used to obtain higher forward gain. One advantage of the Pannier slot is that, as it is external to the mounting structure, it can be added retrospectively and does not affect the integrity of the structure.

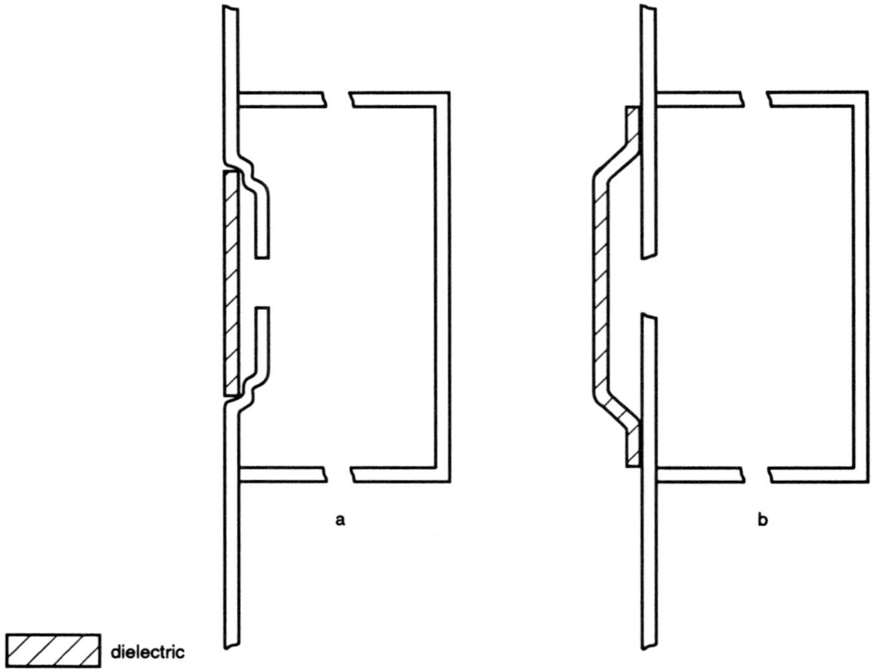

Figure 5.9 *Methods of reducing effect of dielectric cover*
a Aircraft use
b Ground use

5.3.3 *Curved ground planes*

Curving the ground plane in the H plane may be advantageous in widening the coverage in this plane. A number of examples are shown in Lazarus [12], the most extreme being the semicircular front of an aircraft external store housing electronic surveillance equipment. This antenna was fed by a Tee-bar suitably curved to follow the external contour. There is no real advantage in curving in the E plane because the radiation pattern is already wide.

5.3.4 *Miscellaneous applications of cavity-backed slots*

5.3.4.1 *Homing system*

The Beam Approach Beacon System (BABS) used a single feed system to give improved reliability and accuracy in an azimuth approach system for aircraft. Two slots cut round adjacent corners of a rectangular cavity could be alternately short- and open-circuited so that only one slot radiated at a time. The cavity was fed from a central probe and the whole structure mounted in a corner reflector to give two overlapping, mirror-imaged beams. Fig. 5.13 shows the arrangement of the two slots.

5.3.4.2 *Parasitic radiators*

Just as rod elements can be used as parasites in directional antennas such as the Yagi, so parasitic slots can also be used. The slot and cavity dimensions are

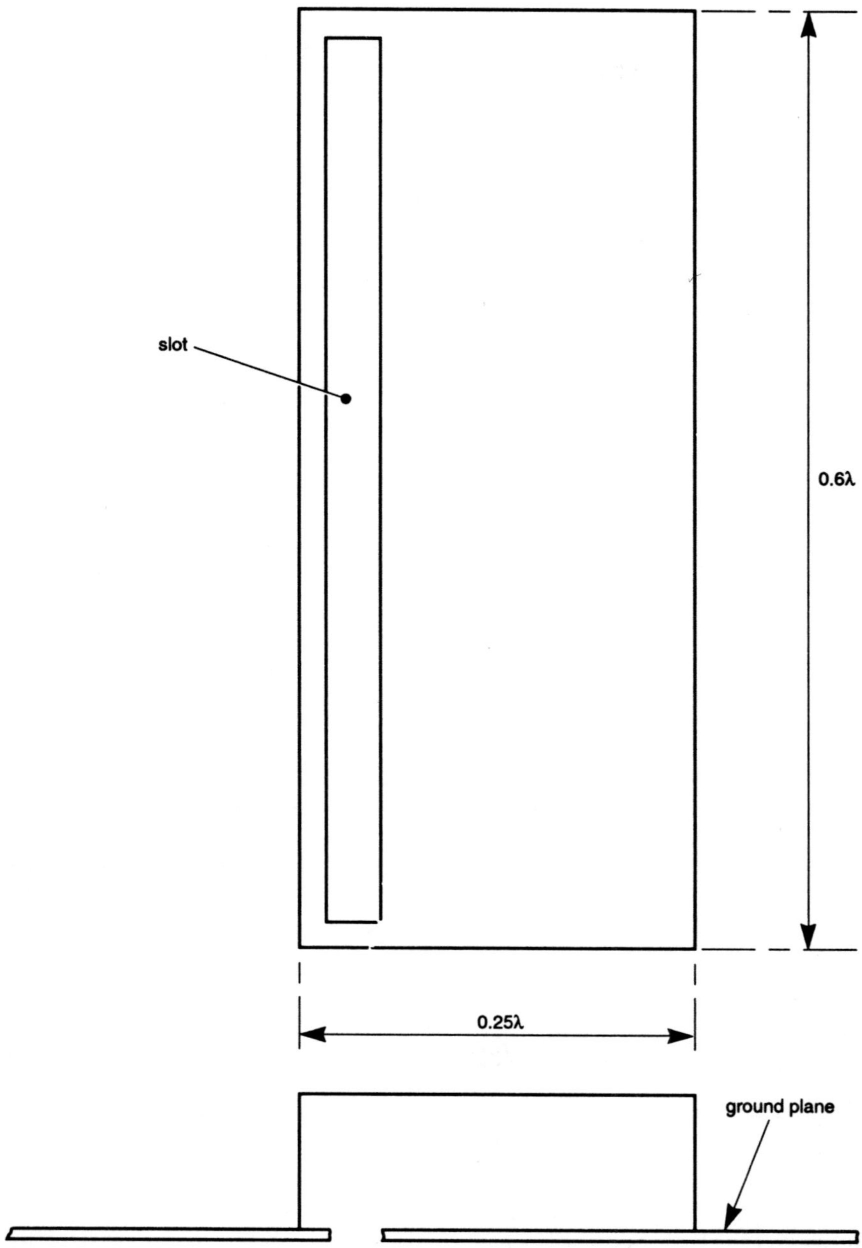

Figure 5.10 *Pocket slot*

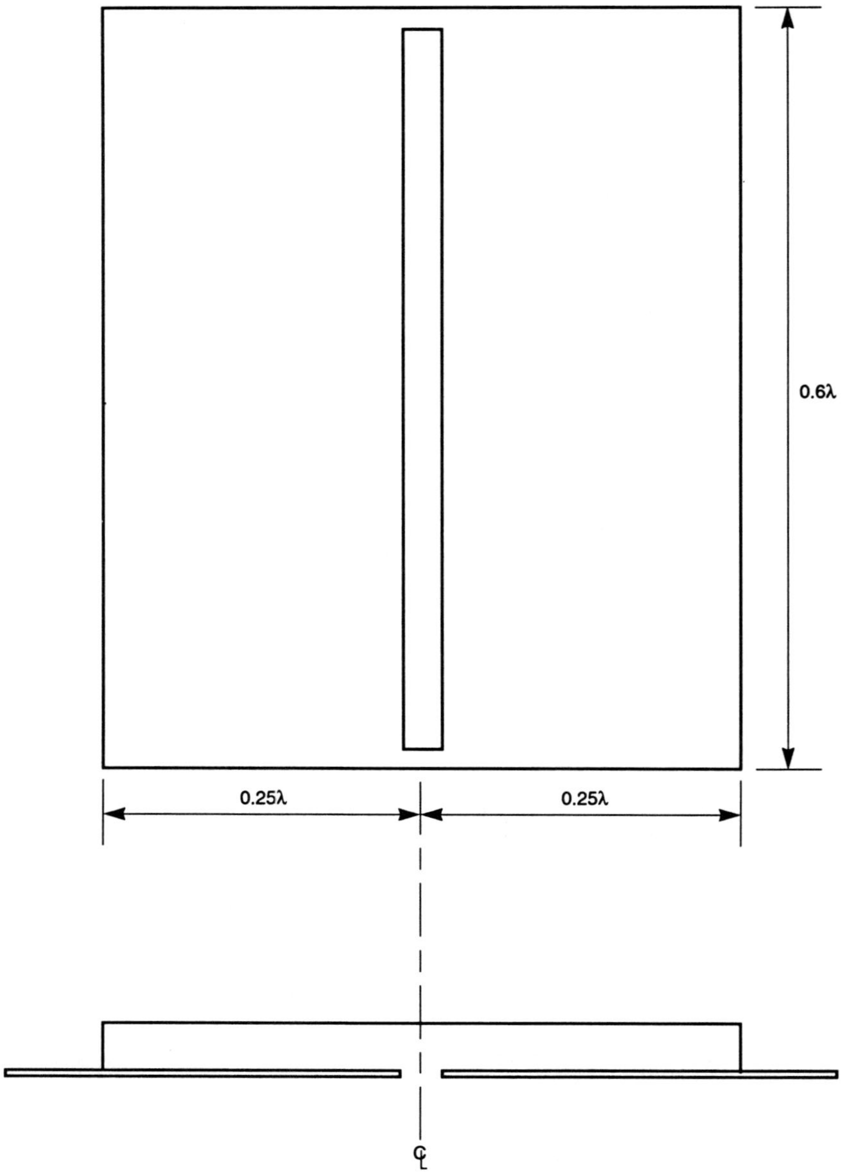

Figure 5.11 *Pocket slot antenna with central slot*

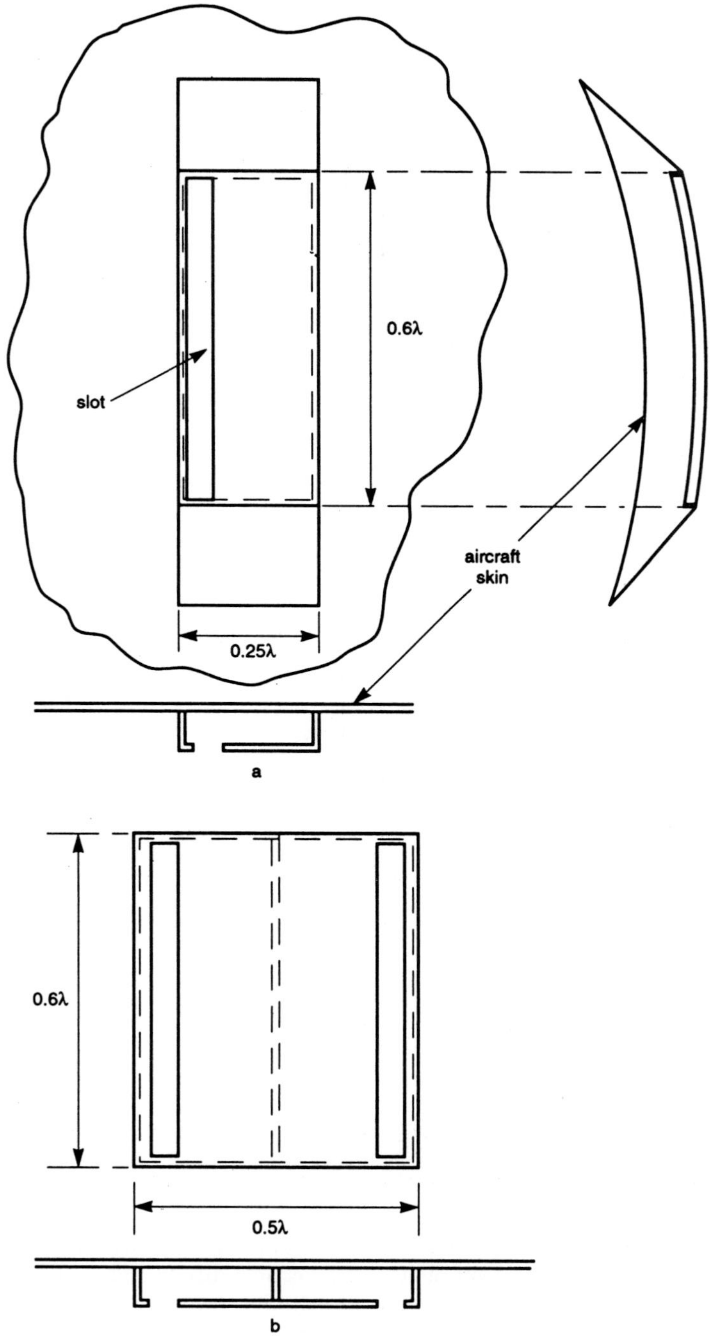

Figure 5.12 *Single and double pannier slots*

a 'Pannier' pocket slot antenna
b Double 'pannier' slot antenna

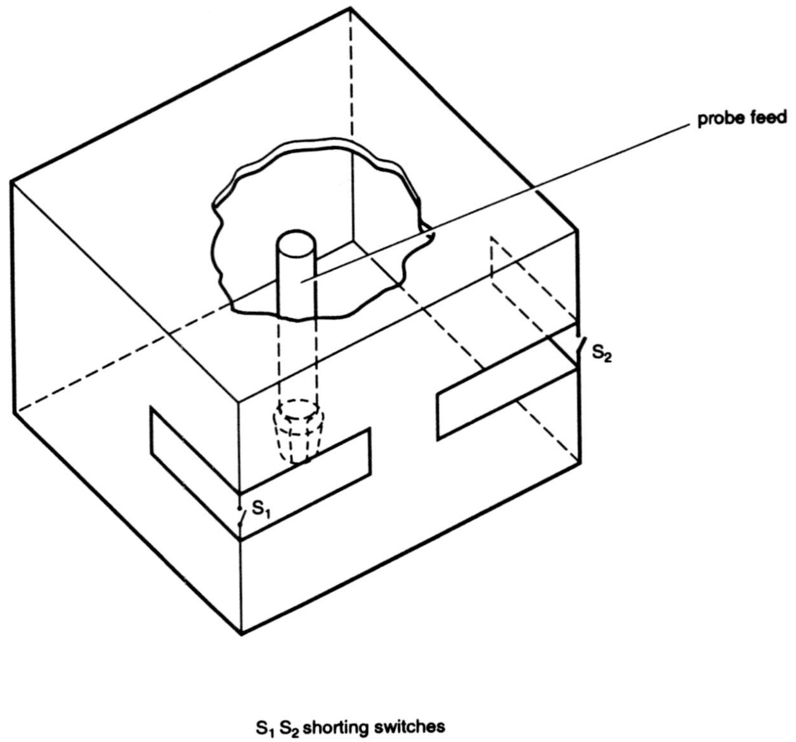

Figure 5.13 *BABS II antenna*

used together to give the correct impedance for the parasite. There is, of course, no feed required. Despite their size parasitic slots have been used as low as 200 MHz (see Cary [5]).

5.4 The slotted cylinder antenna

The ultimate in curved ground planes is the cylinder with axial or circumferential slot. These are shown in Fig. 5.14.

5.4.1 *The circumferential slot*

This form is probably used less than the axial slot because in most possible applications it would be as simple to use a vertical dipole on a support structure than to use a horizontal slot in a vertical cylinder. One exception is the circular array where the more directional pattern of the slot may be an advantage and where the interaction between adjacent elements would be reduced. Against this it must be noted that the length of the slots sets a lower limit to the electrical spacing between them.

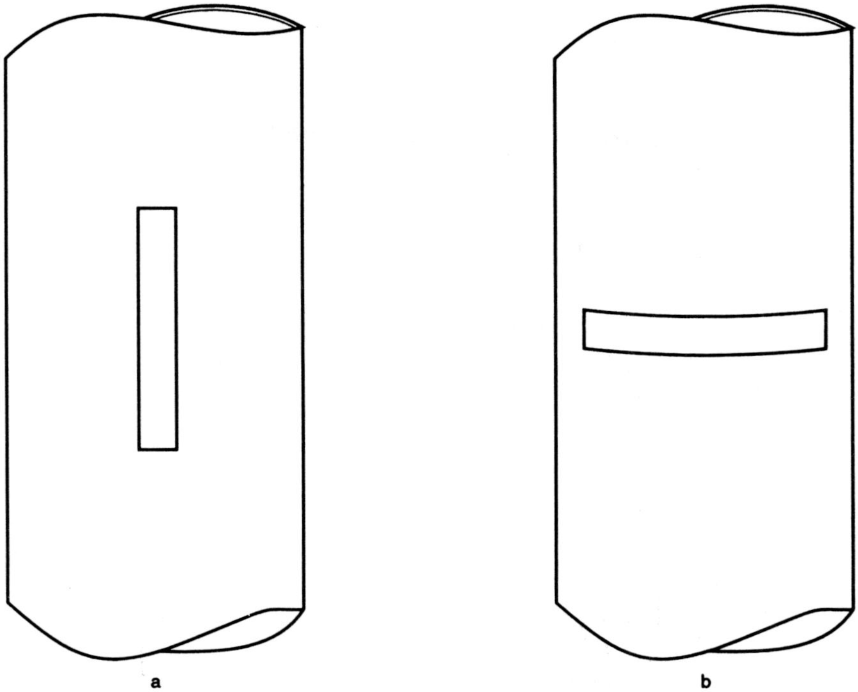

Figure 5.14 *Slotted cylinder antennas axial and circumferential*

a Axial slot
b Circumferential slot

Collin and Zucker [6] have derived formulae for the far field of a circumferential slot. Using the co-ordinate system of Fig. 5.15,

$$E_\theta = \frac{-\exp(-jk_0R)}{R}\,\frac{wE_0}{\pi\phi_0}\,\frac{\sin(k_0w\cos\theta)}{k_0w\cos\theta\sin\theta}\sum_{n=0}^{\infty}\frac{j^{n+1}C_n\cos n\phi\cos n\phi_0}{H_n'(k_0a\sin\theta)\left[\left(\frac{\pi}{2\phi_0}\right)^2-n^2\right]}$$

where w = slot width
$2\phi_0$ = angular length of circumferential slot
k_0 = refers to free space wavelength
a = radius of cylinder

Similarly,

$$E_\phi = -2j\,\frac{\exp(-jk_0R)}{R}\,\frac{\cot\theta}{k_0a\sin\theta}\,\frac{wE_0}{\pi}\,\frac{\sin(k_0w\cos\theta)}{k_0w\cos\theta}$$

$$\times\sum_{n=1}^{\infty}\frac{j^n n\sin n\Phi\cos n\Phi_0}{\Phi_0\left[\left(\frac{\pi}{2\phi_0}\right)^2-n^2\right]H_n^1(k_0a\sin\theta)}$$

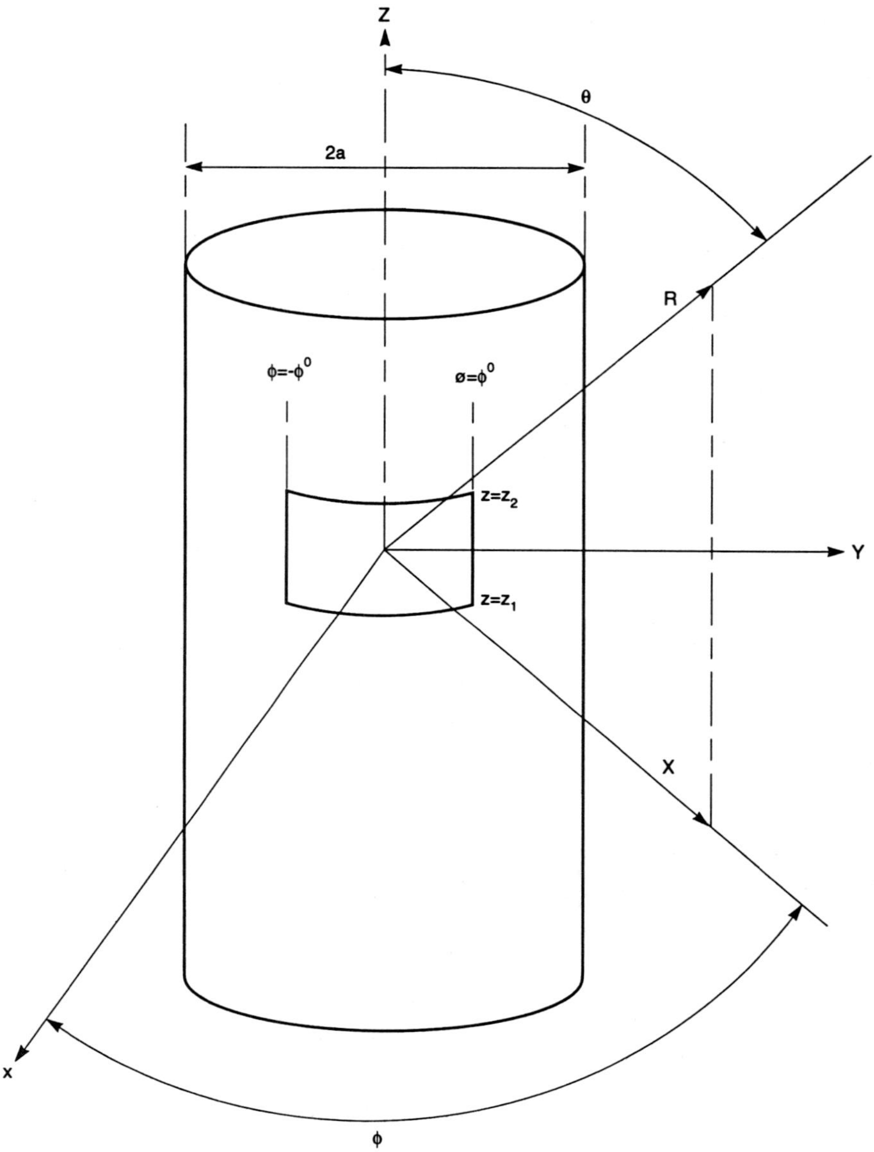

Figure 5.15 *Co-ordinate system for slotted cylinders*

where $H_n(Z)$ is the Hankel function of the second kind of order n and argument Z.

$$C_n = 1 \text{ for } n = 0 \text{ and } C_n = 2 \text{ for } n \neq 0$$

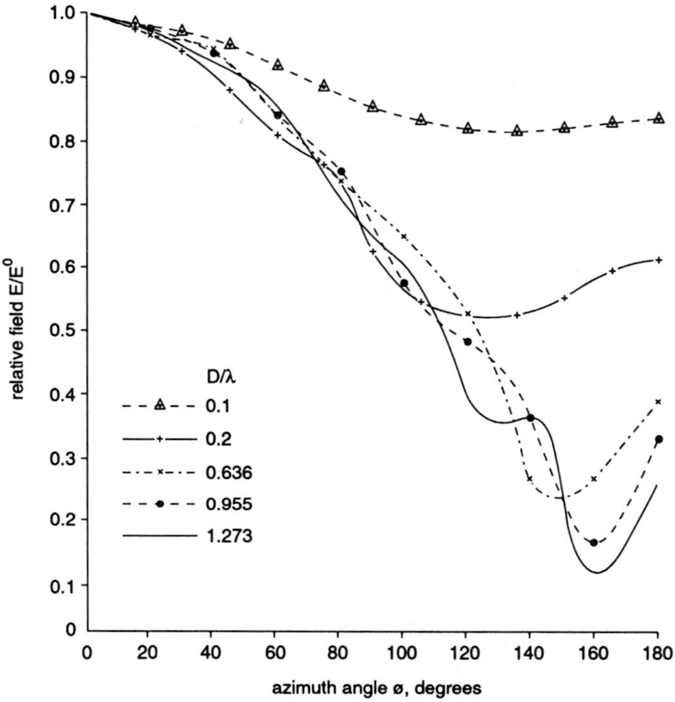

Figure 5.16 *Azimuth patterns of axial slots in vertical cylinders as function of diameter*

5.4.2 *The axial slot*

From the same reference, for the half-wave slot,

$$E_\theta = 0$$

$$E_\phi = \frac{\lambda_0}{2\pi^3} \frac{V_0}{a} \frac{\exp(-jk_0R)}{R} \left[\frac{\cos\left(\dfrac{\pi}{2}\cos\theta\right)}{\sin\theta} \right] \times \sum_{n=-\infty}^{\infty} \frac{j^n C_n \cos n\phi}{\sin\theta H_n^1(k_0 a \sin\theta)} \frac{\sin n\phi_0}{n\phi_0}$$

If the slot is thin, the last term $\sin n\phi_0 / n\phi_0 = 1$

It will be noted that the term $\cos(\pi/2 \cos\theta)/\sin\theta$ is simply the term for the E_Φ pattern of a horizontal half-wave dipole.

Numerous writers have calculated patterns of axial slots on cylinders. They include Sinclair [18], Papas and King [16], Wait [19] and Bosse [4]. A number of patterns are shown in Fig. 5.16. It will be seen that for diameters less than $0{\cdot}1\lambda$ the pattern is almost circular and only changes comparatively slowly as the diameter is increased. While this gives some freedom in cylinder diameter for

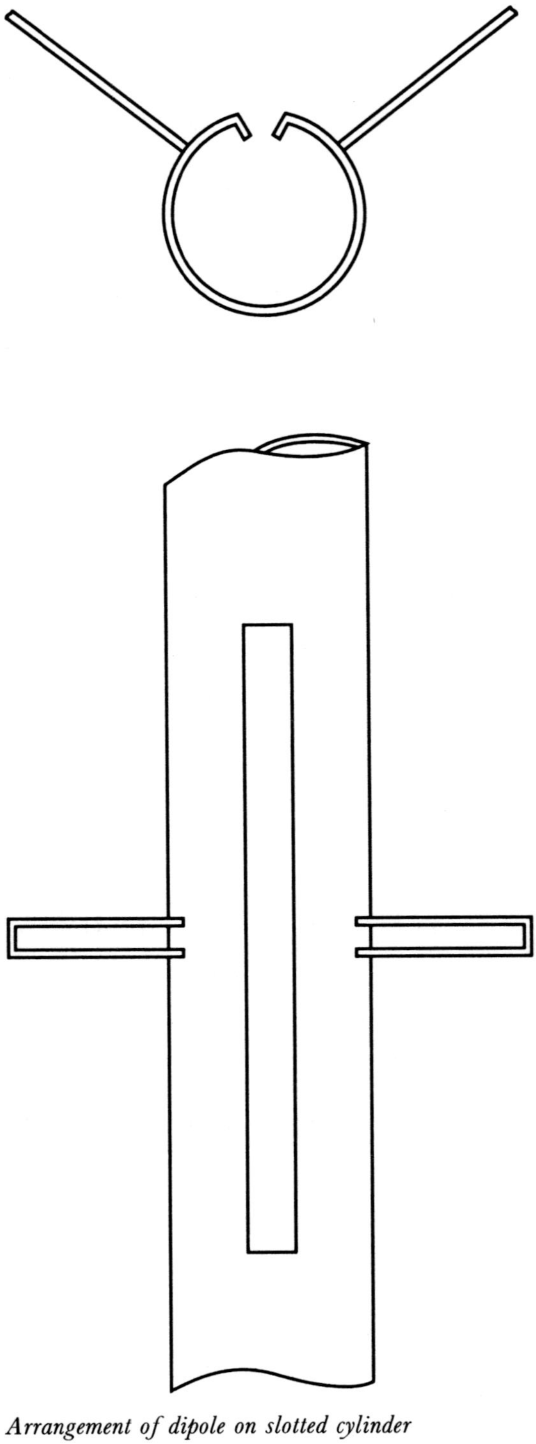

Figure 5.17 *Arrangement of dipole on slotted cylinder*

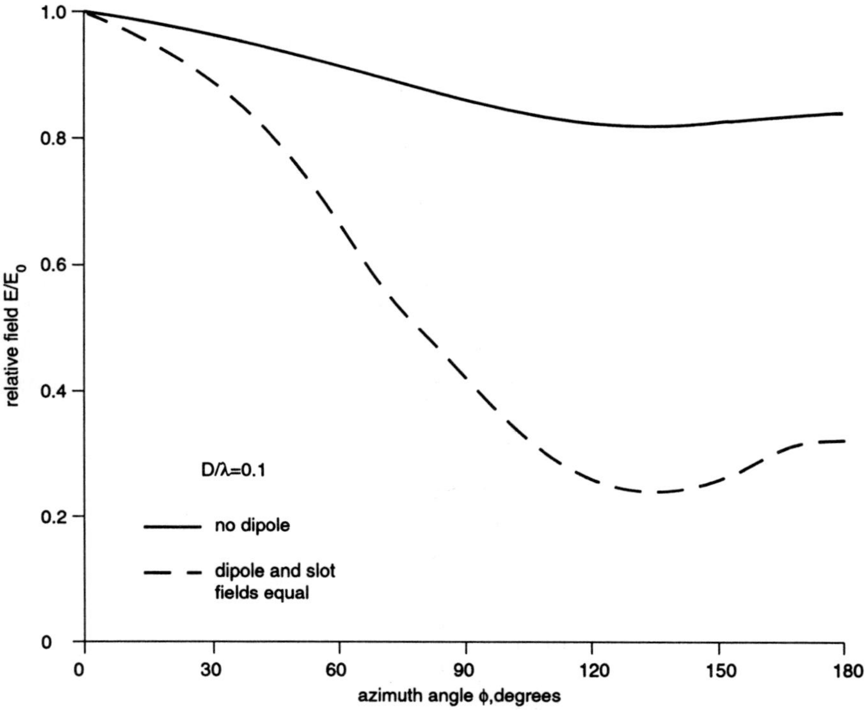

Reference 4

Figure 5.18 *Calculated pattern of slotted cylinder with dipole*

omnidirectional coverage it means that inconveniently large diameters are necessary for a reasonable directivity. Bosse [4] showed that by combining a slotted cylinder with a dipole much improved directivity can be achieved. He found that in order to get the best agreement in phase between the slot and dipole fields the dipole arms should be offset about 30° as shown in Fig. 5.17. With this arrangement patterns were calculated for various ratios of dipole field to slot field (Fig. 5.18). Patterns for equal fields with a cylinder diameter of $0{\cdot}1\lambda$ show the marked improvement possible. Theory has been confirmed by measurements on a pylon antenna for the VHF broadcast band 87–100 MHz for which the following details are given:

Frequency	87–100 MHz
Cylinder diameter	$0{\cdot}144\lambda$
Dipole arms	0·6 m
Slot length	5·3 m
Forward gain	5·4
Rear gain	0·6
Mismatch, *m*	1·3
Z_0	not quoted

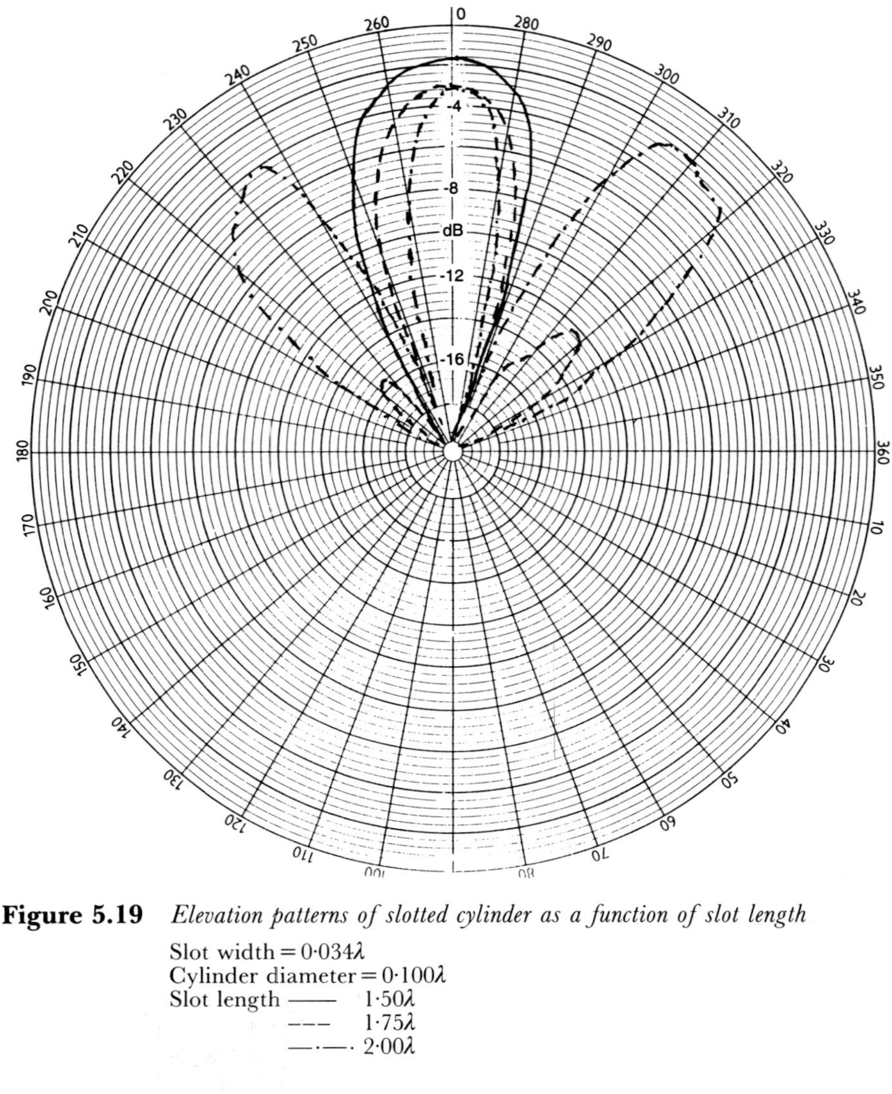

Figure 5.19 *Elevation patterns of slotted cylinder as a function of slot length*

Slot width = 0·034λ
Cylinder diameter = 0·100λ
Slot length ——— 1·50λ
--- 1·75λ
—·—· 2·00λ

It will be noted that this is a long slot (1·5λ at 87 MHz) using techniques developed by the Lorenz company (see Bosse [3]). From the photograph in Ref. 4 it appears that four dipoles were attached to this long slot, presumably at the points where they had most influence. The relative fields were adjusted by altering the lengths of the dipole arms.

5.4.2.1 *The long slot antenna*

The axially-slotted cylinder can be considered as a balanced transmission line loaded by close-spaced loops which provide a distributed shunt inductive reactance. The phase velocity of propagation along such a line is greater than the velocity of light and is controlled by the internal cross-sectional area *A* of the

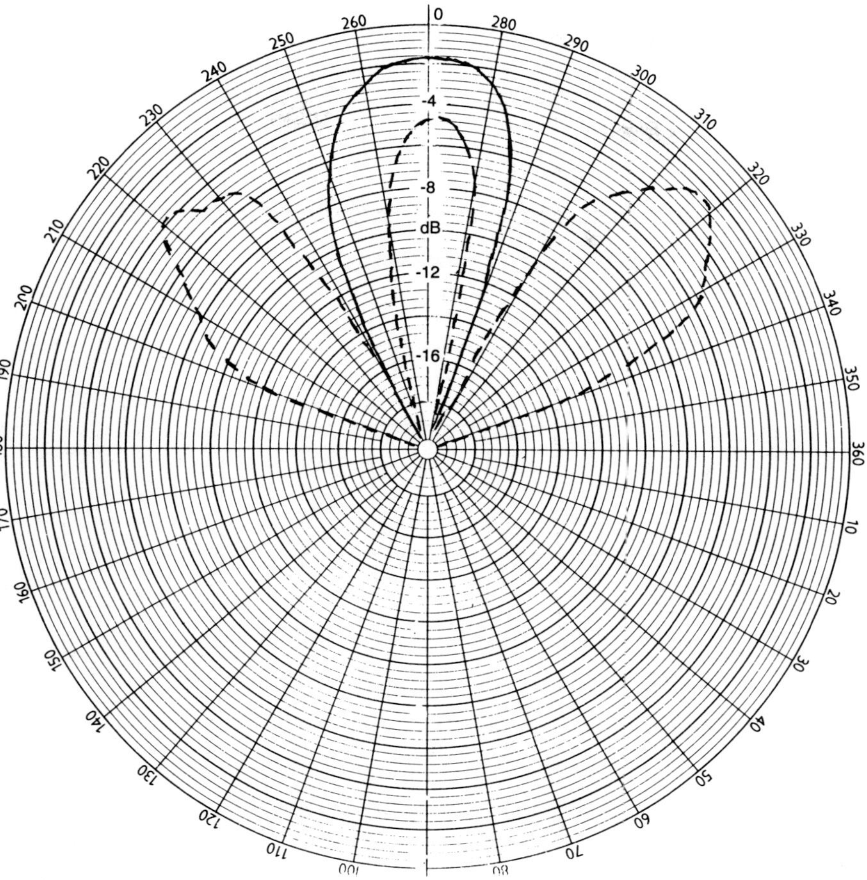

Figure 5.20 *Elevation patterns of slotted cylinder as a function of cylinder diameter*

Slot width = 0·034λ
Slot length = 1·5λ
Cylinder diameter
—— 0·100λ
– – – 0·150λ

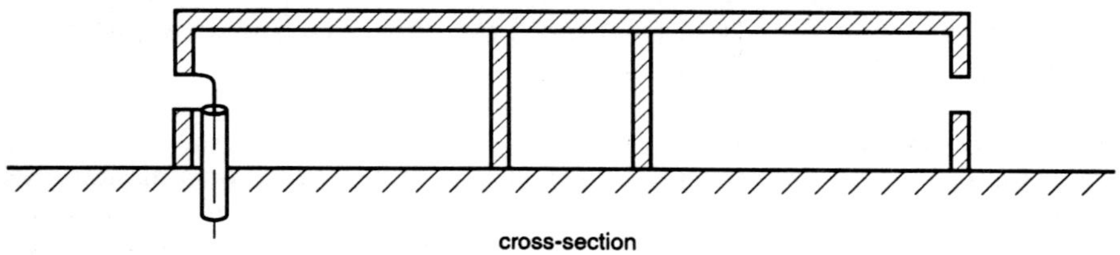

Figure 5.21 *Pillbox annular slot*

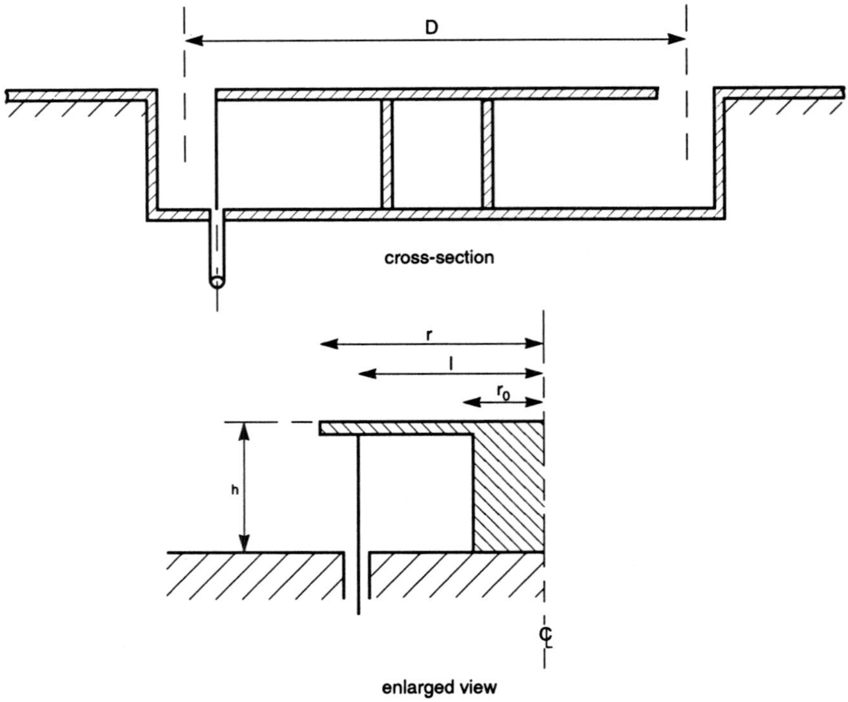

Figure 5.22 *Flush mounted annular slot*

cylinder. As A is decreased, the velocity increases to about four times that of light. Further decreases in A make the propagation along the line exponential. There is a combination of slot and cross-sectional area for which the phase velocity and amplitude remain substantially constant, typically when the slot is 2λ long and the cylinder is $0{\cdot}14\lambda$ internal diameter. These dimensions assume that the cylinder is thin-walled; increasing the thickness at the slot edge increases the capacitance loading and hence alters the velocity in the slot. Details and patterns are given by Alford [1] and further data on phase velocity as a function of slot and cylinder dimensions are given by Jordan and Miller [10]. The cylinder need not be circular in cross-section; in one design for a satellite tracking station at 138 MHz the cavity was a square section aluminium tube 4·4 m long, 0·3 m side.

Fig. 5.19 shows elevation plane patterns for various slot lengths for a long slot $0{\cdot}034\lambda$ wide in a cylinder $0{\cdot}100\lambda$ diameter. Fig. 5.20 shows patterns in the same plane for cylinder diameters of $0{\cdot}1\lambda$ and $0{\cdot}15\lambda$. These suggest that very large bandwidths are not to be expected.

Alford [1] shows that the azimuth plane pattern can be modified by fixing wings along the slot edges so that the slot is effectively at the apex of a corner reflector. The wings may be bent back to widen the pattern or forward to narrow it. In the latter case the wings have some effect on the phase velocity along the slot and this has to be offset by increasing the capacitance across the slot.

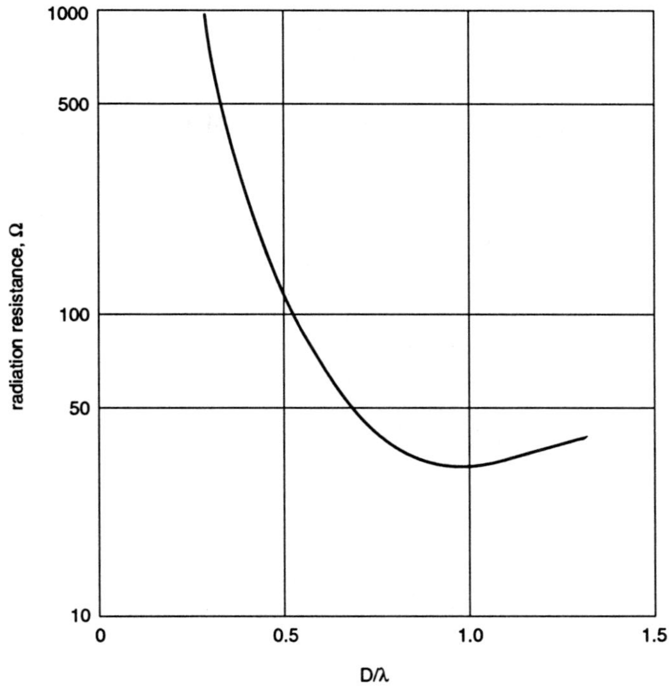

Figure 5.23 *Resistance of annular slot as a function of diameter*

The impedance of a long slot radiator depends on four factors:

- Ratio of phase velocity to velocity in free space
- Slot length
- Capacitive loading
- Shape of any wings.

5.5 The annular slot antenna

Fig. 5.1*b* shows the annular slot as the analogue of the loop antenna. From this analogy it is clear that a small diameter annular slot in a horizontal sheet will give an omniazimuth vertically polarised field. In the elevation plane the field is proportional to $J_1(Ka \sin\theta)$ where a is the mean radius of the annular slot, θ is the angle from the vertical and $J_1(x)$ is the first order Bessel function. To keep the field a maximum in the plane of the slot, Ka should not exceed 2.

Most of the published papers assume that the slot lies in the ground plane and this is probably the form most used. It was at one time used widely in aircraft to reduce drag but wiser counsels have prevailed as it is now realised that the weight and complexity of a flush mounted annular slot are not cost effective compared with the equivalent external monopole which can have better bandwidth, much lower weight and cost, and negligible drag. One form of annular slot which has been much neglected is the pillbox shown in Fig. 5.21. Unless flush mounting is the over-riding requirement the pillbox antenna has all the advantages:

- Does not require structural changes to the mounting surface
- Size not limited by constraints inside a vehicle, hence wider bandwidth possible
- Very simple to construct.

It is important to appreciate that there is probably little to be gained in making the antenna circular. For most applications a square antenna will function equally well and should be cheaper to construct. Whilst Fig. 5.21 shows a slot width less than the cavity depth, the antenna is simplified by making the slot width equal to the cavity depth. It then reduces to two parallel plates separated by a conducting box at their centres and fed from a coaxial cable across the slot. A single point feed is adequate for small diameter slots, the size criterion for omnidirectional patterns in azimuth being the same as for a horizontal loop, somewhat less than $0{\cdot}3\lambda$ in circumference. It is therefore perfectly possible to construct an antenna of two square sheets of aluminium kitchen foil separated by a thick sheet of expanded polystyrene heat insulation with the central earthing provided by a tobacco tin, for example. Clear plastic adhesive tape round the edges of the antenna will provide good resistance to moisture, at least in the short term. This is an extreme example but it should be obvious that this type of antenna need not be highly expensive to build.

Either form of antenna can be considered as a slot, radiating on one side only, and backed by a short-circuited radial transmission line. The following analysis is due to Johnson and Rothe [9], using the parameters shown in Fig. 5.22. The radiation resistance R_s of an annular slot radiating on one side is obtained by Babinet's principle from the resistance R_L of a loop of the same size:

$$R_s = \frac{(120\pi)^2}{2R_L}$$

Values of R_L are given by Moullin [14]:

D/λ	0·3	0·5	0·7	0·9	1·0	1·25	1·5
$R_L/30\pi^2$	0·5	1·4	3·0	3·8	3·7	3·0	4·0

from which the curve of Fig. 5.23 has been constructed. R_s is shunted by a capacitance C given approximately by

$$C = 0{\cdot}017816\text{D} \ln D/h \text{ picofarads}$$

where D and h are in mm.

The impedance $Z(r)$ of the short-circuited radial transmission line is given by

$$Z(r) = \frac{-j60h}{r}\,\frac{J_0(kr)N_0(kr_0) - J_0(kr_0)N_0(kr)}{J_1(kr)N_0(kr_0) - J_0(kr_0)N_1(kr)}$$

where J_0 and J_1 and N_0 and N_1 are Bessel functions at the first and second kind respectively of orders 0 and 1.

$Z(r)$ is chosen to resonate with the slot capacitance and, at resonance, the equivalent series resistance R_s' becomes

$$R_s' = \frac{X_s^2}{R_s} = \frac{Z(r)^2}{R_s}$$

Considering now the loop formed by the inner radius of the cavity, the feed line and the top and bottom surfaces of the cavity, the mutual impedance of this loop is

$$M = \frac{\mu h \lambda}{4\pi^2 r} \frac{N_0(kr_0)J_0(kl) - J_0(kr_0)N_0(kl)}{N_0(kr_0)J_1(kr) - J_0(kr_0)N_1(kr)}$$

μ is the permeability, normally unity.

Since $(\omega M)^2 = RR_s'$ where R is the input impedance at the feed,

$$R = R_s \left[\frac{N_0(kr_0)J_0(kl) - J_0(kr_0)N_0(kl)}{N_0(kr_0)J_0(kr) - J_0(kr_0)N_0(kr)} \right]^2$$

This neglects the inductance of the feed wire for which series capacitance compensation may be necessary.

Some useful design curves are given by Cumming and Cormier [7].

For broadband operation D/λ needs to be of the order of 0·8 which would not give a good omnidirectional pattern. This can be overcome by using four feeds at 90° spacing round the slot and combining these in parallel.

One other advantage of the pillbox antenna is that two can be stacked to cover different frequency bands. Using this arrangement the author was able to produce a dual-band antenna covering the VHF and UHF airborne communication bands 118–138 MHz and 225–400 MHz in a package only 115 mm high. The feed from the higher UHF antenna was taken through the electrically-dead centre of the lower, VHF, antenna. This arrangement would be useful for fitting on top of land vehicles or under comparatively slow aircraft with low ground clearance such as helicopters.

5.6 References

1 ALFORD, A.: 'Long slot antennas', Proc. National Electronics Conf. USA, 1946, Vol. 2, pp. 143–155

2 BOOKER, H.G.: 'Slot aerials and their relation to complementary wire aerials (Babinet's principle)', *J. IEE*, 1946, **93** Pt. IIIA, p. 620

3 BOSSE, H.: 'Breitband-Rohrschlitzantenne', *VDE-Fachberichte*, 1951

4 BOSSE, H.: 'Rohrschlitzantennen mit horizontaler Richtung', *FTZ*, 1953, Pt. 3, pp. 123–127

5 CARY, R.H.J.: 'The slot aerial and its application to aircraft', *J. IEE*, 1952, **99** Pt. III, pp. 187–196

6 COLLIN, R.E., and ZUCKER, F. (Eds.): 'Antenna theory' (McGraw-Hill Book Co., New York, 1969) chap. 14

7 CUMMING, W.A., and CORMIER, M.: 'Design data for small annular slot antennas', *IEEE Trans.*, 1958, **AP–6**, pp. 210–211

8 FROOD, D.G., and WAIT, J.R.: 'An investigation of slot radiators in rectangular metal plates', *Proc. IEE*, 1956, **103** Pt. B, pp. 103–109

9 JOHNSON, W.A., and ROTHE, P.: 'A wide-band circular slot radiator'. RAE Tech. Note RAD453, 1949

10 JORDAN, E.C., and MILLER, W.E.: 'Slotted cylinder antenna', *Electronics*, Feb. 1947, pp. 90–93
11 KRAUS, J.D.: 'Antennas' (McGraw-Hill Book Co., 1950) p. 370
12 LAZARUS, D.: 'Slot antenna development at Radio Research Laboratory'. RRL Harvard Univ. Report 411–263, 17 Nov. 1945
13 LOPEZ, M.V.: 'A technique for fore-shortening a cavity-backed slot antenna'. 15th Annual Symposium on USAF Antenna Research and Development, University of Illinois, 14 Oct. 1965
14 MOULLIN, E.B.: 'Radiation from large circular loops', *J. IEE*, 1946, **93** Pt. III, pp. 345–351
15 NEWMAN, E.H., and THIELE, G.A.: 'Some important parameters on the design of T-bar fed slot antennas', *IEEE Trans.*, 1975, **AP–23**, pp. 97–100
16 PAPAS, C.H., and KING, R.: 'Currents on the surface of an infinite cylinder excited by an axial slot', *Quart. Appl. Math*, 1949, **7**, pp. 175–182
17 RADIO RESEARCH LAB. STAFF: 'Very high frequency techniques' (McGraw-Hill Book Co., New York, 1947) chap. 7
18 SINCLAIR, G.: 'The patterns of slotted cylinder antennas', *Proc. IRE*, 1948, **36**, pp. 1487–1493
19 WAIT, J.R.: 'Radiation characteristics of axial slots on a conducting cylinder', *Wireless Engineer*, Dec. 1955, pp. 316–323

Chapter 6

The notch antenna

This type of antenna has been used for forty years but there are few references to it in the literature. Its most widely known use is probably for HF communication on high performance aircraft but in fact its first use was for VHF telemetry and command on aircraft and missiles, Johnson [3]. These early antennas were all short, narrow-band devices; the self-resonant notch which is physically practical in the VHF and UHF bands is the main topic of this chapter.

6.1 Principle of the notch antenna

If a metallic rod is in an electromagnetic field whose electric vector lies in the same plane as the rod, an electric current will be induced in the rod. The phase and magnitude of the current will be determined by the length of the rod. For small diameter rods the current will be uniform in cross-section and will be zero at the ends. As the rod diameter becomes a significant fraction of the wavelength the current at the ends becomes finite and the current distribution around the circumference of the rod ceases to be uniform. If the rod cross-section is altered to highly-elliptical or even to a flat strip then the current is concentrated in the regions of sharpest curvature. Fig. 6.1 shows a typical current distribution on an aerofoil, Granger and Morita [2].

A short cut into one edge of the sheet will provide a means of abstracting energy from the incident field. Such a cut is known as a notch and typical notch antennas are shown in Fig. 6.2. Johnson [3] derives radiation resistance and reactance from the formulae for a monopole mounted on the edge of a semi-infinite sheet and lying in the plane of the sheet, Fig. 6.3. The notch is the slot analogue of this arrangement and its radiation resistance can then be determined by Babinet's principle. The radiation resistance at the mouth of the notch is then

$$R_F = \frac{2 \times 120\pi}{3[C(kh) - \cot kh \times S(kh)]^2}$$

where C and S are Fresnel integrals.

When h is less than $0{\cdot}1\lambda$ a reasonable approximation is

$$R_F = \frac{3}{8}\, 120\pi\, \frac{\lambda}{h}$$

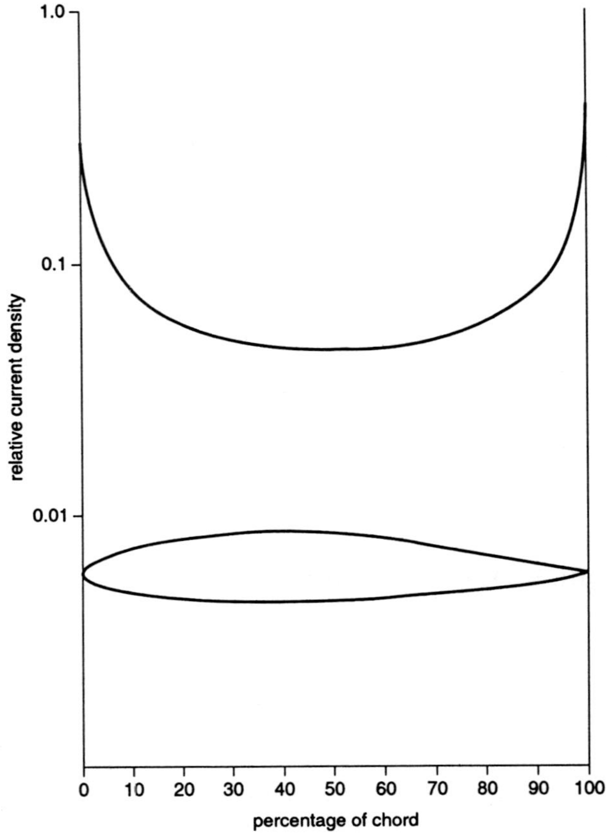

Reference 2

Figure 6.1 *Current distribution on an aerofoil section*

A full derivation is given by Wolff [4] but it should be noted that both Johnson and Wolff have inconsistencies in their formulae. Fig. 6.4 shows values of R_F against h/λ.

The reactance in shunt across the mouth of the notch can be determined by considering it as a short-circuited transmission line of average characteristic impedance

$$Z_{av} = \frac{60\pi^2}{\log_e\left(\frac{16h}{w} - 1\right)}$$

where W is the width of the notch.

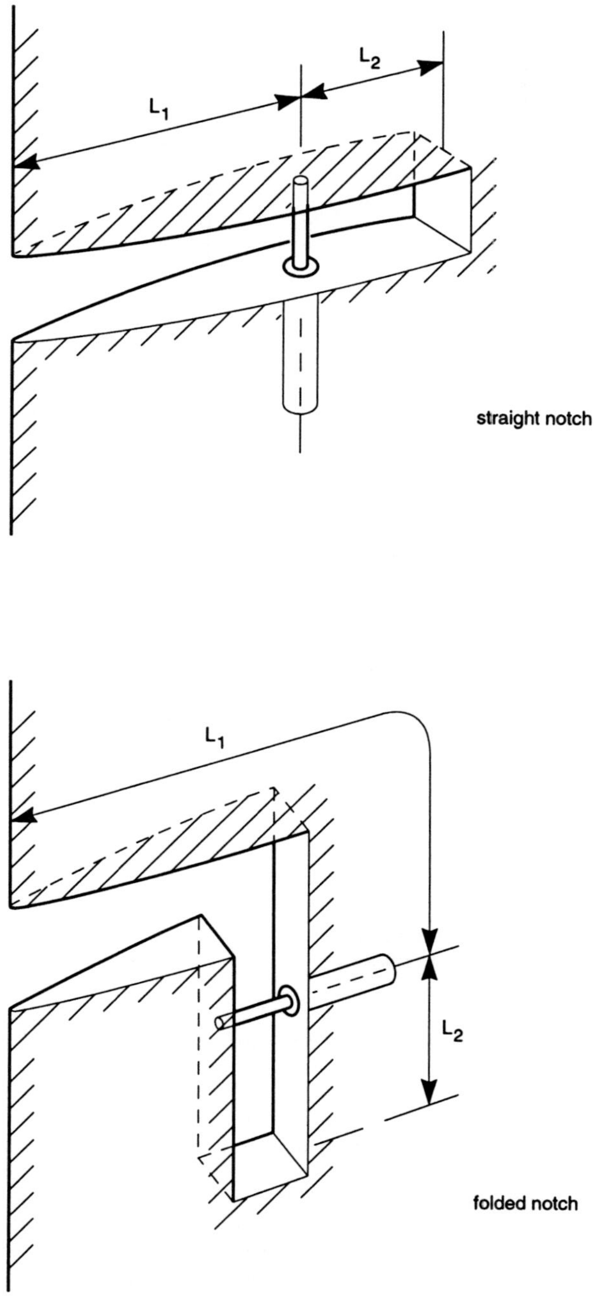

Figure 6.2 *Typical notch antennas*

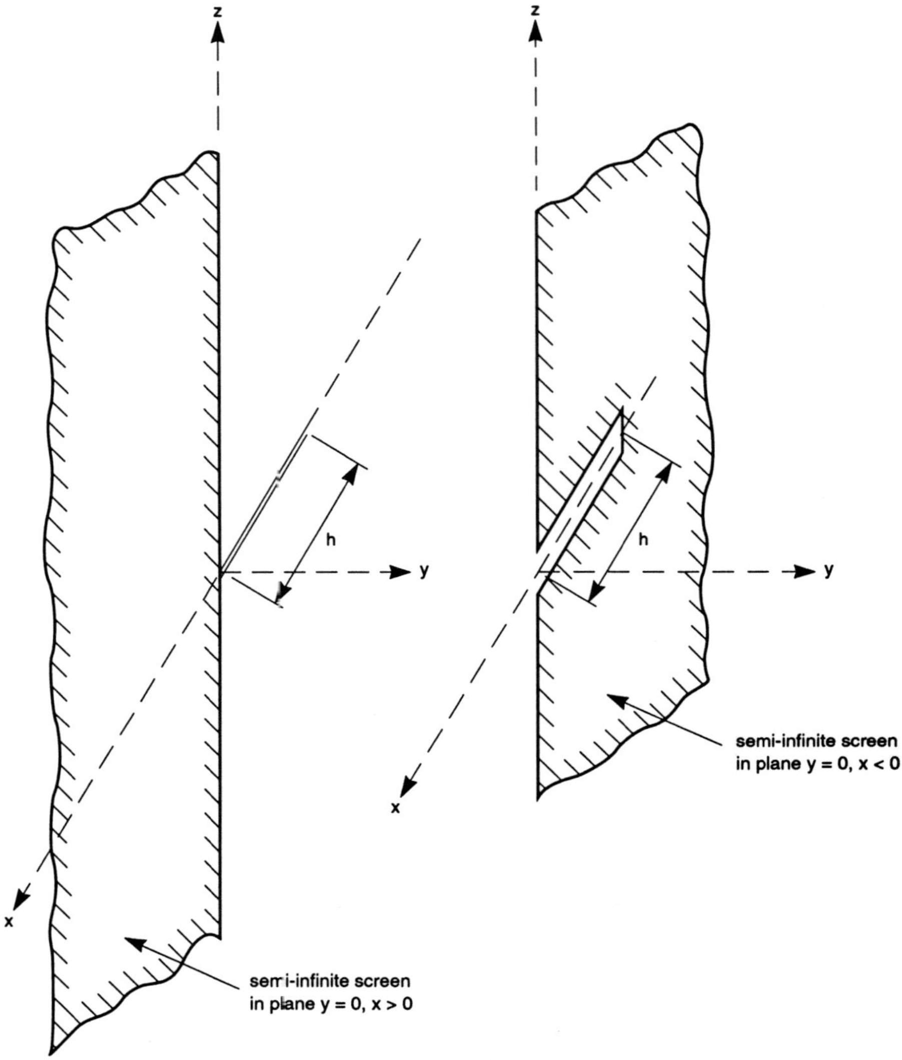

Figure 6.3 *Monopole on edge of sheet and notch*

6.2 Practical antennas

For a notch approximately a quarter-wavelength the radiation resistance at the mouth will be about 400 ohms. Matching to a 50 ohm transmission line can easily be achieved at a single frequency by tapping along the notch at a point where the conductance = 1/50 mhos and adjusting the length of notch between tapping point and short circuit to compensate the shunt reactance. Fig. 6.5 shows the equivalent circuit of a notch and Fig. 6.6 is a graphical representation of the matching process, assuming a notch Z_0 of 100 ohms.

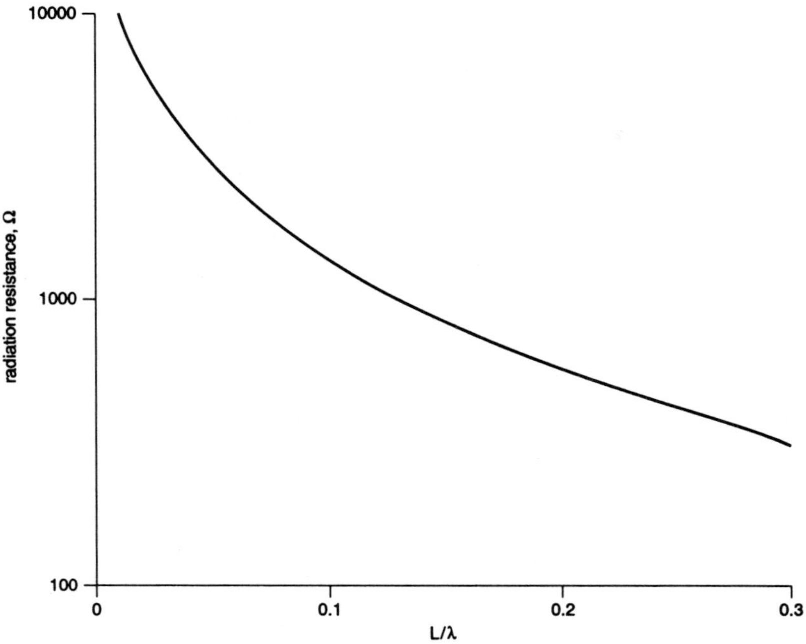

Figure 6.4 *Radiation resistance at mouth of notch*

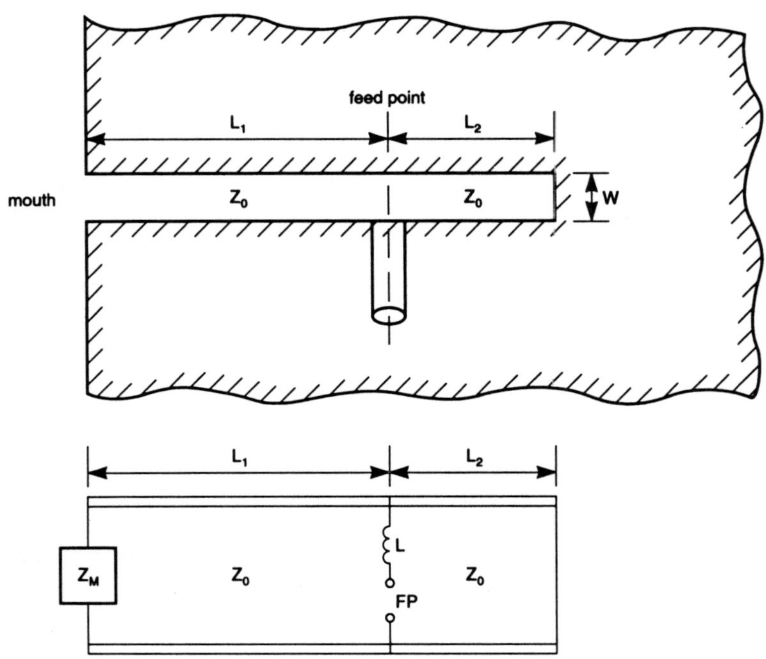

Figure 6.5 *Equivalent circuit of notch*

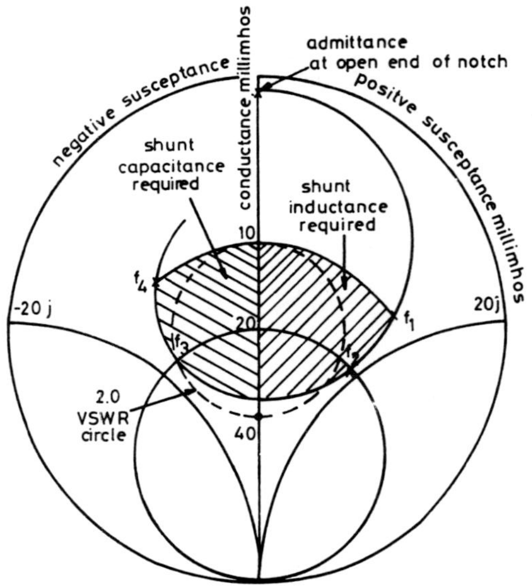

Figure 6.6 *Graphical representation of matching*

f_1–f_4 typical feed point admittances

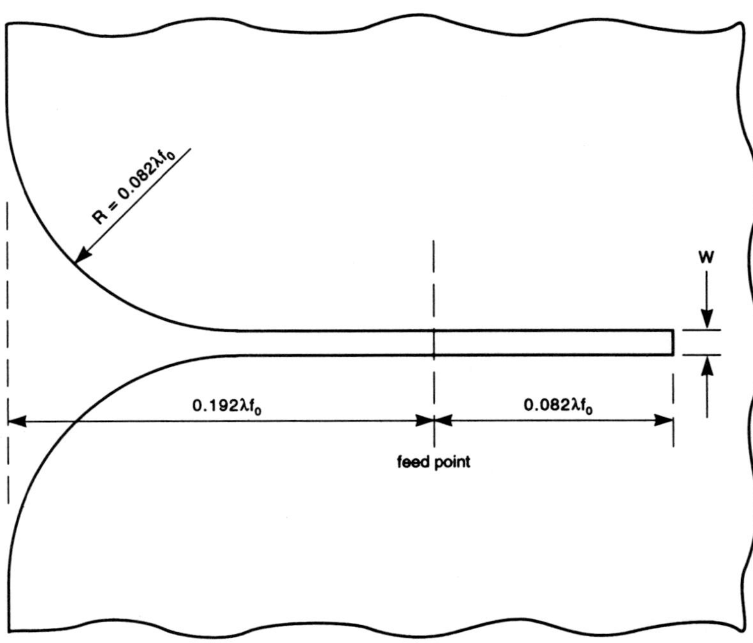

Figure 6.7 *Dimensions of broadband planar notch*

T = sheet thickness
W/T = 7·5
f_0 = lowest operating frequency

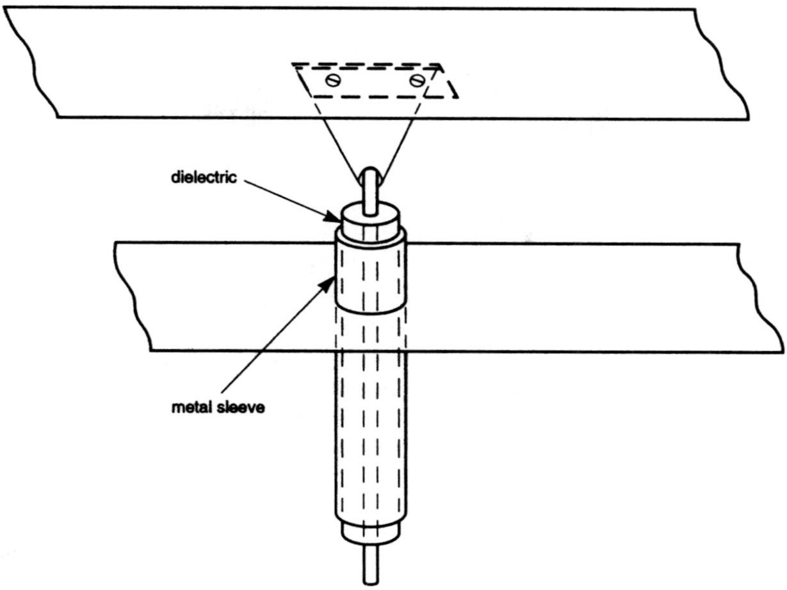

Figure 6.8 *Low inductance feed for solid notch*

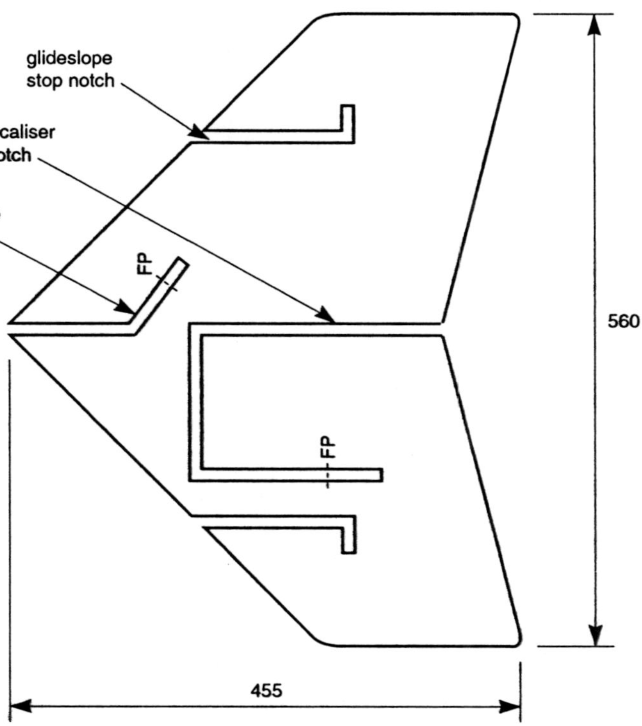

Figure 6.9 *Two-band notch antenna on printed board*

Dimensions in mm
FP = feed-point

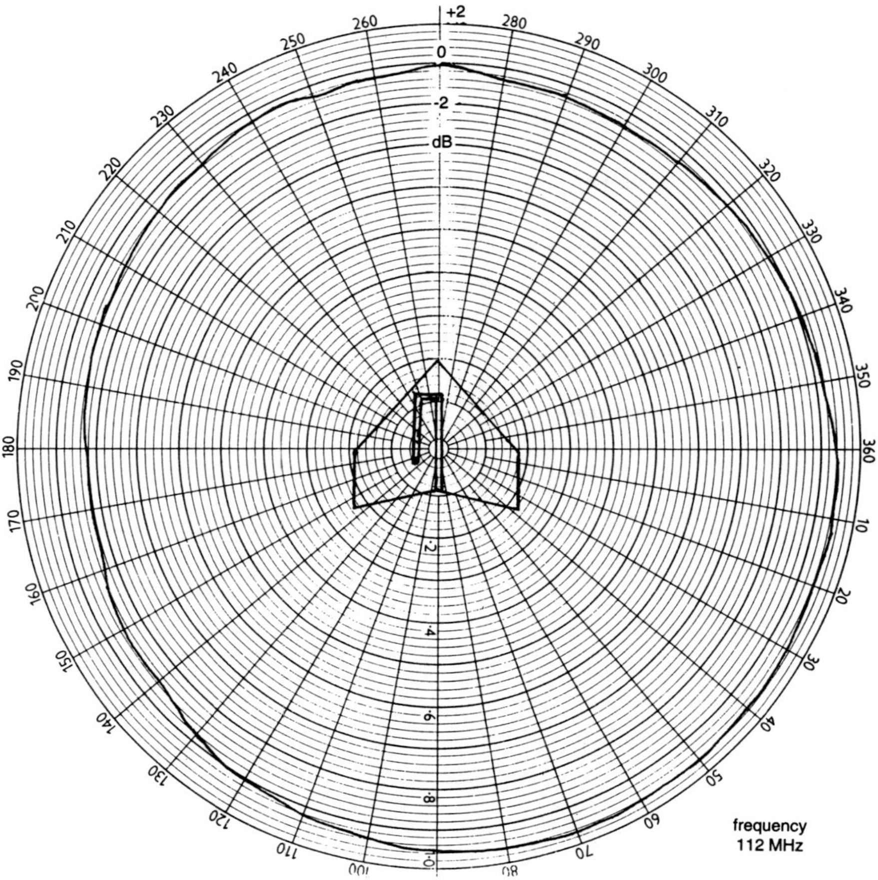

Figure 6.10 *Azimuth radiation pattern of printed notch ILS localiser*

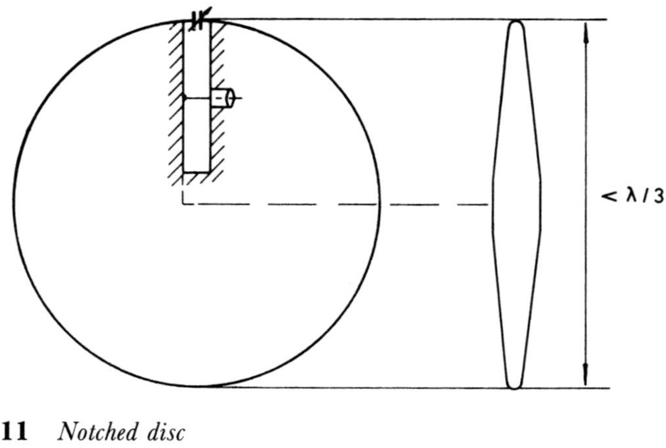

Figure 6.11 *Notched disc*

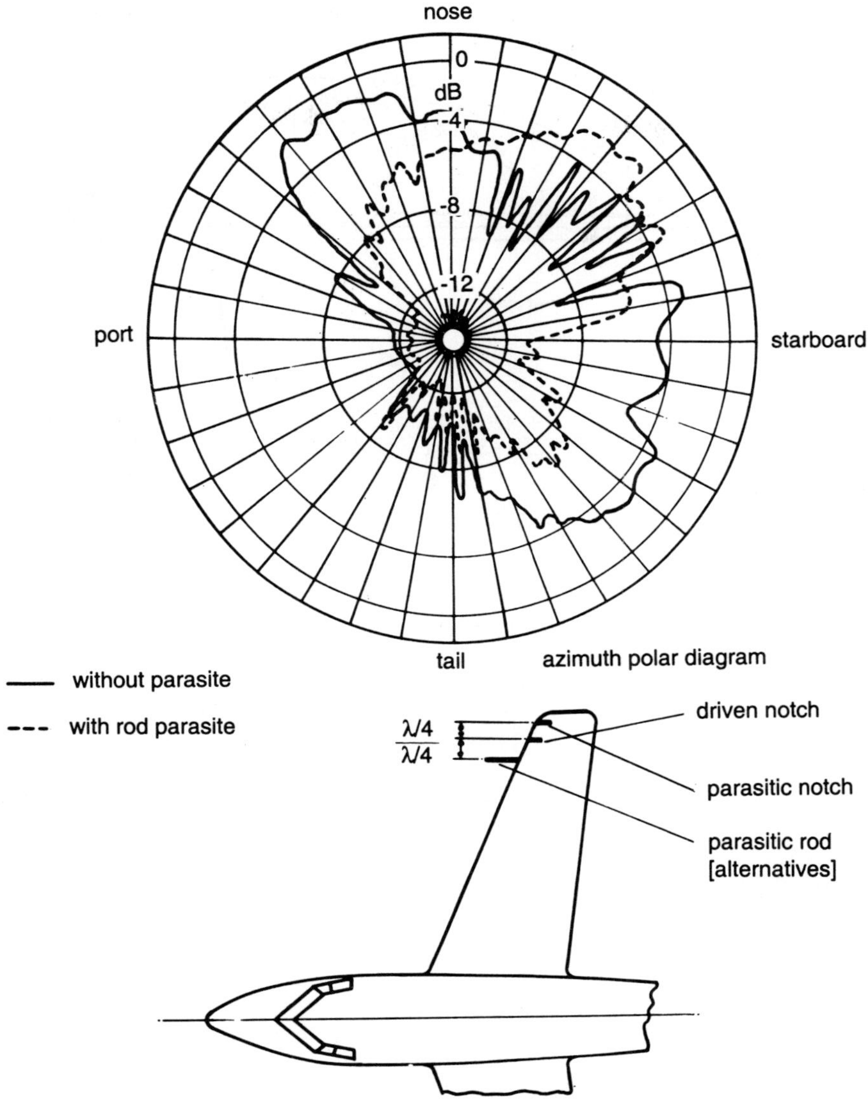

Figure 6.12 *Notch in long metal edge*

As Fig. 6.2 indicates, the notch does not have to be in a flat sheet and many aircraft applications have an aerofoil cross-section. This provides in a parallel-sided notch a continuously varying characteristic impedance which can be an advantage in broadband matching. The equivalent arrangement in a planar notch is a taper from mouth to feed-point. A good approximation is obtained by using a constant radius instead of an exponential taper, Fig. 6.7. In the UHF band it may be necessary to take into account the series reactance of the feed

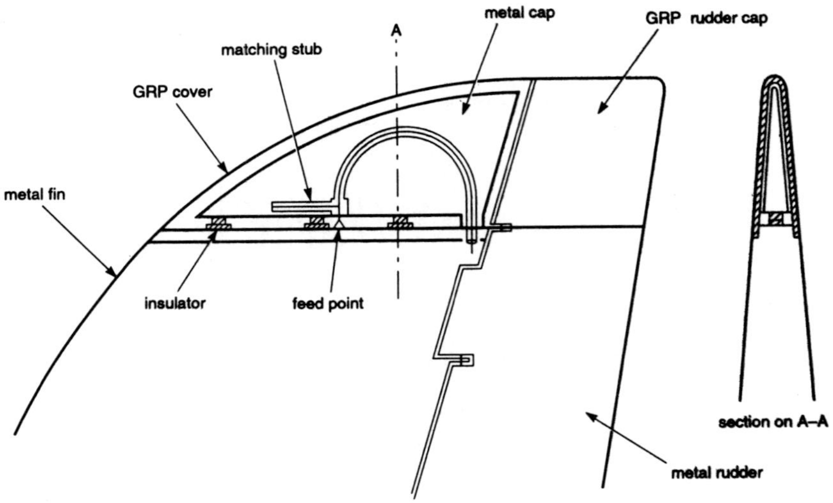

Figure 6.13 *Notch-fed fincap VHF antenna*

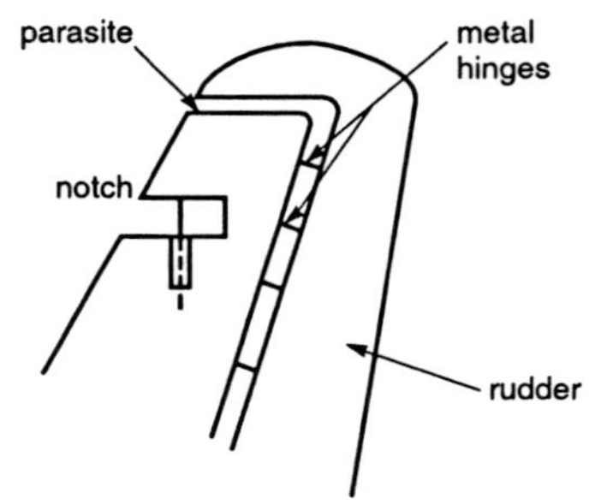

Figure 6.14 *Redesigned fincap*

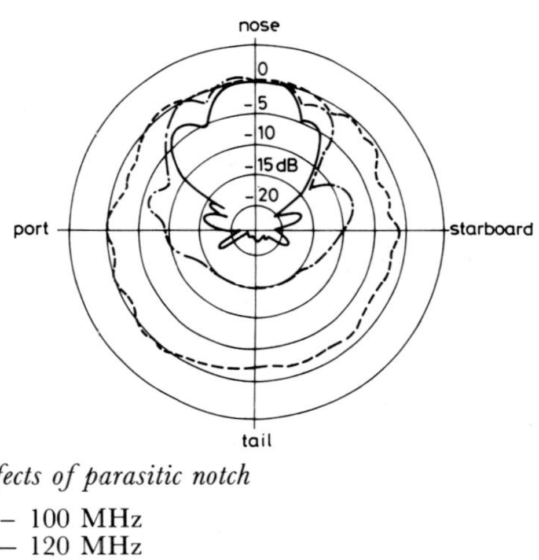

Figure 6.15 *Effects of parasitic notch*

--- 100 MHz
—— 120 MHz
–·–·– 140 MHz

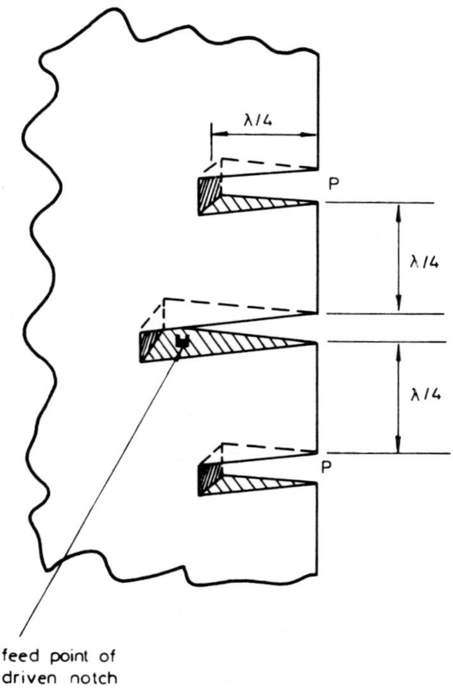

Figure 6.16 *Triple notch antenna*
P = parasitic notch

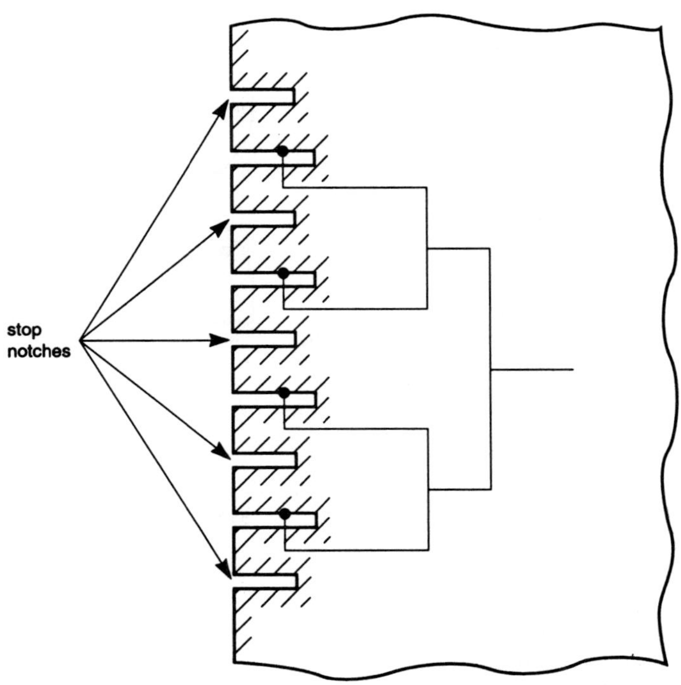

Figure 6.17 *Stacked notches*

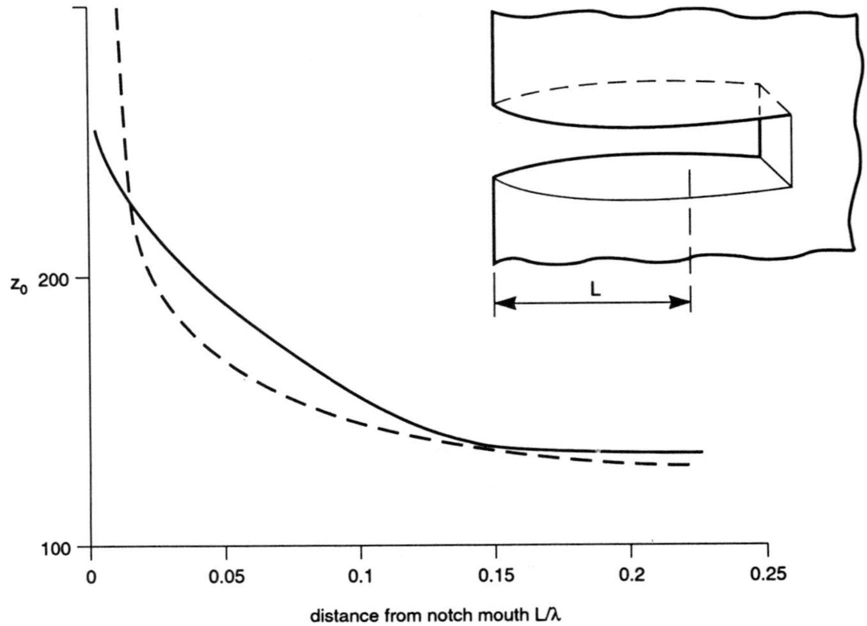

Figure 6.18 *Z_0 of aerofoil notches*

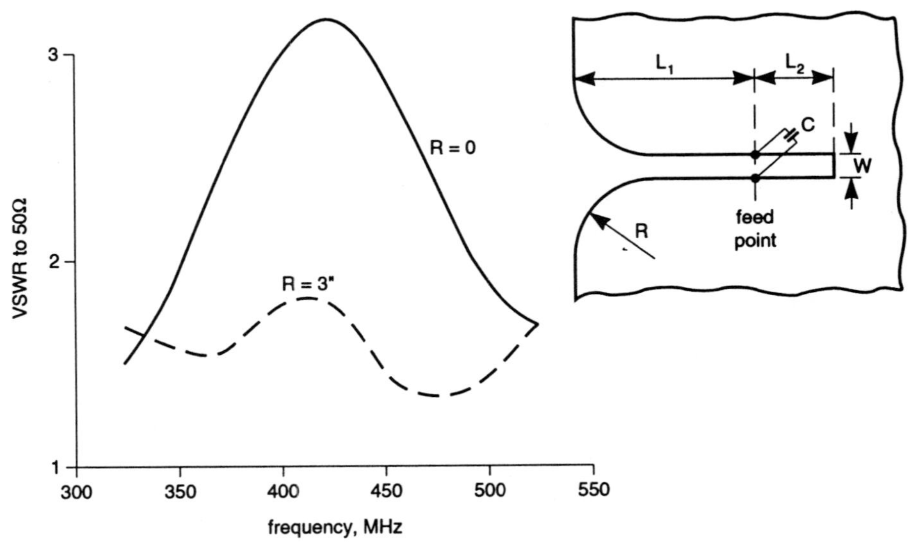

Figure 6.19 *Effect of radius on bandwidth*

$L_1 = 7''$; $L_2 = 3''$; $W = \frac{1}{4}''$; C = 3·3 pF
Sheet thickness = $\frac{1}{32}''$

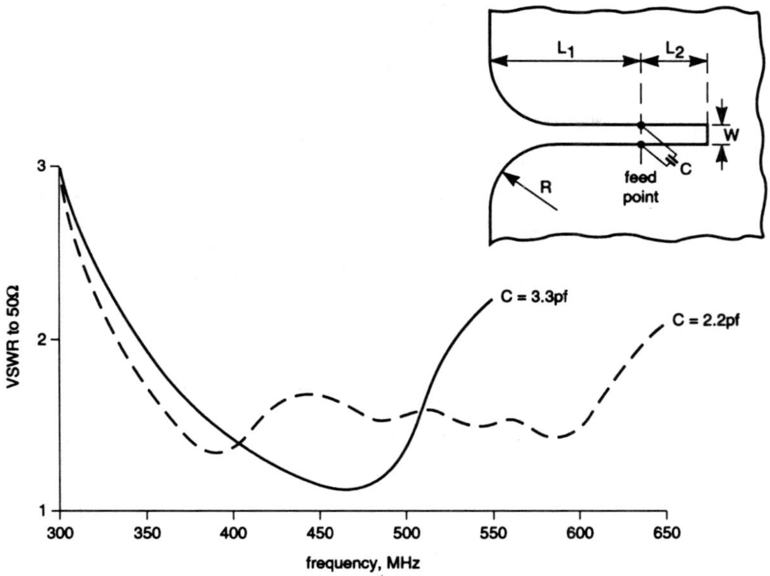

Figure 6.20 *VSWR of broadband notch*

$L_1 = 7''$; $L_2 = 3''$; $R = 3''$; $W = \frac{1}{4}''$
Reduced series inductance
Sheet thickness $\frac{1}{32}''$

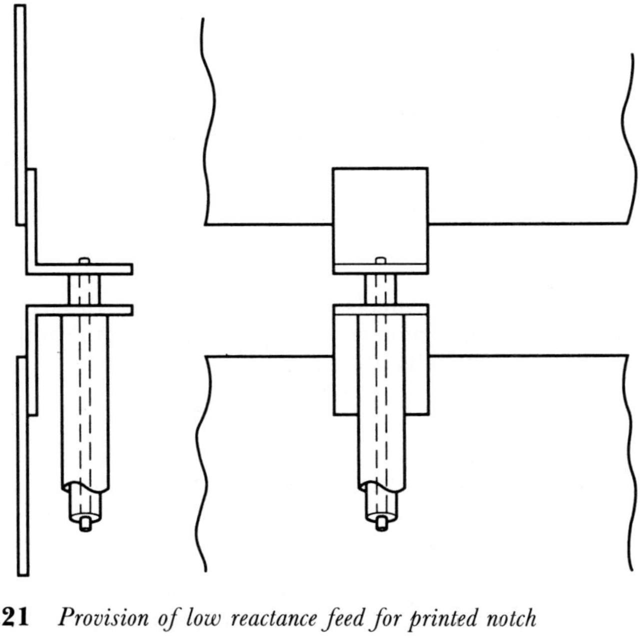

Figure 6.21 *Provision of low reactance feed for printed notch*

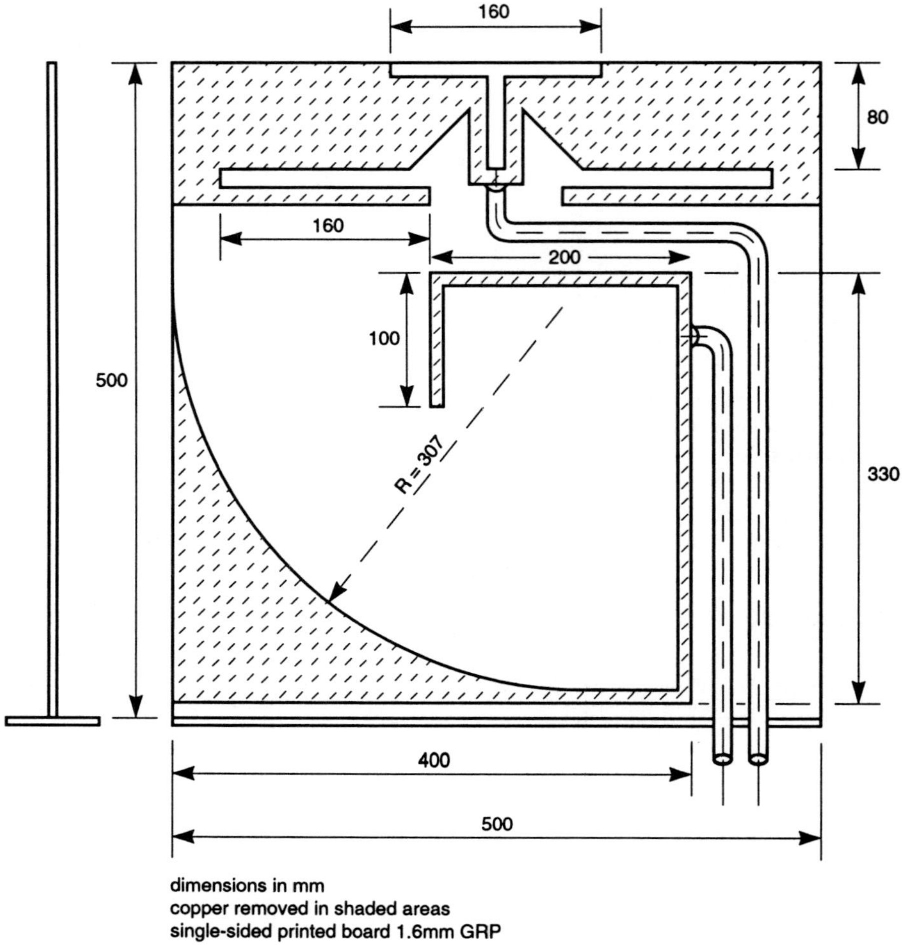

Figure 6.22 *Dual band printed antenna*

system. Typically the notch antenna is fed by coaxial cable whose inner extends across the notch; the series inductance may become significant and some method of reduction may be advisable, Fig. 6.8.

6.3 Radiation patterns

One advantage of the notch antenna over the dipole is that it can be fed from a coaxial cable without the need for any balancing system. This makes it possible to create an array from a series of notches in a single metallic sheet, a very simple structure resulting. However, the radiation pattern of a notch in a long sheet will be the same as that of a loop having the same perimeter. In a small sheet, therefore, the radiation pattern in the plane of the sheet will be nearly circular. Since the current is concentrated close to the sheet edges it is possible to fold the notch without having much effect on either its impedance or radiation pattern. Fig. 6.9 shows the localiser portion of an airborne ILS antenna which has the near-circular radiation pattern of Fig. 6.10 in the plane of the antenna. Another application to an omnidirectional horizontally-polarised antenna is the notched disc of Fig. 6.11 which can be pole-mounted and may be a physically preferable alternative to a Vee-dipole or a loop. The example shown used a loaded short notch as only small bandwidth was required but a coiled notch could equally well be used.

The radiation characteristic of a long wire is shown by Fig. 6.12 which is the azimuth pattern of a notch in the leading edge of an aircraft wing. The reflection from the forward fuselage gives rise to a number of narrow lobes on the starboard side. When the current flow along the wing is reduced by the use of a parasitic radiator, a typical dipole pattern is obtained. The use of parasites to control the radiation patterns of notches in large surfaces arose from a chance discovery by the author in 1956 during the development of a fin-cap antenna for a fighter aircraft. In the original aircraft design the VHF antenna was a notch-fed plate at the top of the fin, Fig. 6.13. This operated over the range 100–156 MHz with a VSWR of 2·5 to 50 ohms. A photograph is shown in Burberry [1]. The fin was redesigned as Fig. 6.14 and the notch moved down the fin. As the radiation patterns show, the gap between fin-top and rudder is resonant within the wanted band with disastrous effects on the radiation patterns, Fig. 6.15. Subsequent examination has shown that such parasitic notches may occur in many places on an aircraft and doubtless on other structures as well.

The deliberate employment of parasitic notches produced the antenna shown in Fig. 6.16. This operates in the 1 GHz band with a 28% VSWR bandwidth to 2:1. As shown the antenna was bolted onto the trailing edge of a helicopter fin, or into the tail of a fighter aircraft, but it has equally well been constructed as a printed antenna in a dielectric cover. Mounted to give vertical polarisation the antenna has a dipole pattern in elevation; azimuth coverage extends to 120° either side of the antenna axis, so two such antennas would give omniazimuth cover. This antenna is unaffected by the structure outside the parasites and is equally effective as a free standing item.

Null steering systems designed to combat interference which may be deliberate or unintentional often use circular arrays. It has been shown that the use of directional elements improves system performance and one antenna arrangement envisages the use of a number of flat sheets radiating from a common axis. Each sheet, either of metal or as a printed board, carries a stack of notches separated by parasites; the number of elements in the stack is determined by the elevation pattern requirements and the gain needed. A single notch with parasites has a gain of about 2 dB above a half-wave dipole. Fig. 6.17 shows a stack of notches; it should be noted that the parasites reduce the mutual coupling between elements so matching is comparatively simple. Another useful property of the notch is that if several need to be connected in parallel it is easy to find a tapping point to give a radiation resistance higher than the final output impedance of the array.

6.4 Calculation of notch parameters

From experimental work it appears to be desirable that the characteristic impedance of the notch should be high at the mouth, of the order 300–400 ohms, decreasing to a value around 140 ohms at the short circuit. This arrangement provides the broadest bandwidth; for narrow band applications a constant Z_0 will be satisfactory. The only limitation then will be that of voltage breakdown across the transmission line if high power is used.

Using the ratio D/W for a parallel plate line where D is the distance between plates and W their width, characteristic impedances have been plotted for two typical notches in aircraft wings. These are shown in Fig. 6.18 for an arbitrary frequency.

For the planar notch this method is inappropriate. A simple technique has been developed for determining the Z_0 of the section beyond the feed-point. At a constant frequency the conductance transformed from the mouth to the feed-point remains constant regardless of the distance between the feed point and the short circuit, which adds a series susceptance varying with length and of value $Y_0 \cot kl$. Provided that any series inductance in the feed system is accounted for, a series of admittance plots at constant frequency will all lie on the same conductance curve and the value of Y_0 can be calculated. For air-spaced notches free space velocity can be assumed, but for printed antennas it will be necessary to calculate the relative velocity factor as well. This can readily be done by making the short-circuited line rather longer than a half-wavelength in air and shortening it progressively until the admittance has moved through a complete circle on the Smith chart. By plotting susceptance against length on a Cartesian diagram a sufficiently accurate value of velocity factor should be obtainable. Using this method a 12·7 mm ($\frac{1}{2}$ in) wide notch in a metal sheet 1·22 mm (0·048 in) thick has a Z_0 of 196 whilst for a 6·35 mm ($\frac{1}{4}$ in) width the Z_0 is 140.

It should be noted that a dielectric cover over an aerofoil notch will modify the impedance characteristic significantly. In the examples shown in Fig. 6.18 where the notches were 101·6 mm (4 in) wide and the GRP cover only 6·35 mm ($\frac{1}{4}$ in) thick, the tolerances between individual covers were sufficient to cause significant differences in bandwidth. As would be expected, the effect is greatest at the mouth of the notch where Z_0 is highest.

6.5 Broadbanding

The advantage of tapering a planar notch was briefly mentioned in Section 6.2. The effect on VSWR is clearly seen in Fig. 6.19 in which two otherwise identical notches, one with and one without a radius at the mouth, are compared. It should be noted that these both incorporate a small capacitance of 3·3 pF across the feed. Further improvement can be obtained by reducing the series inductance. Fig. 6.20 shows VSWR for two such notches with differing shunt capacitance. One method of providing both the low series inductance and the shunt capacitance is shown in Fig. 6.21. If desired, a dielectric can be placed between the plates; a small piece of double-sided printed board would be a practical arrangement.

6.6 Notch-fed monopole

In some circumstances a grounded monopole may be needed as, for example, when two separate antennas are mounted in the same structure. Such an arrangement is shown in Fig. 6.22. The lower antenna operates between 94 and 105 MHz and the upper one around 190 MHz as a top-loaded sleeve monopole. The two notches below the upper antenna prevent current flow from it down the edges of the printed board. Such an antenna would be suitable for an aircraft fin-cap under a slim dielectric cover.

6.7 Short notches

If space does not permit a resonant notch the length can be reduced by capacitance loading across the mouth. Antennas of this type have much reduced bandwidth. Typically a notch of length $0{\cdot}064\lambda$ has a bandwidth of only 4% for a VSWR of 5:1 referred to 50 ohms. The high Q of such an antenna does have an advantage in reducing coupling from other systems. Johnson [3] indicates that two short notches $0{\cdot}033\lambda$ long spaced only 100 mm apart on the same surface could be used for transmission and reception on separate frequencies; this compares with a spacing of 1·8 m needed with quarter-wave monopoles. The gain of such a short notch was within 0·5 dB of the monopole.

6.8 References

1 BURBERRY, R.A.: 'Aerial systems for aircraft', *J. Royal Aeronautical Society*, 1956, pp. 101–113

2 GRANGER, J.V.N., and MORITA, T.: Radio frequency current distributions on aircraft structures, *Proc. IRE*, 1951, **39**, pp. 932–938

3 JOHNSON, W.A.: The notch aerial and some applications to aircraft radio installations, *Proc. IEE*, 1955, **102** Pt. B, pp. 211–218

4 WOLFF, E.A.: '*Antenna analysis*' (John Wiley and Sons, New York, 1966)

Chapter 7

Directional antennas

Introduction

This chapter is concerned mainly with antennas of modest gain in the range 5 to 15 dBi. It does not consider shaped reflector or 'dish' antennas. Certain other directional antennas, i.e. slot, notch and log periodic types, are discussed in Chapters 5, 6 and 8 respectively.

The main classes discussed here are:

- Aperiodic reflectors
- Parasitic elements
- Travelling wave antennas
- Short backfire
- Helix
- Arrays of elements
- Steerable arrays

7.1 Aperiodic reflectors

7.1.1 *Plane sheet*

One of the simplest directional antennas consists of a centre-fed dipole in front of a plane conducting sheet. Since this is the basis for many more complex arrangements it is important to consider in some detail the effects of a finite sheet on antenna performance.

Consider the arrangement of Fig. 7.1. In the *xz* plane where the sheet edges AB and CD are orthogonal to the dipole it might be supposed that there was little effect due to sheet size. Fig. 7.2 shows that this is not entirely true although the effect of these edges is far less than those of the edges BC and AD. These results are from some early work on diffraction by Booker [5] which may be considered as part of the foundation of Geometrical Theory of Diffraction (GTD) discussed in Chapter 14. These patterns indicate that if the sheet extends $\lambda/2$ beyond the dipole centre no advantage will accrue from further extension. In fact as we shall see in the section on corner reflectors a height of $0{\cdot}6\lambda$ is usually considered adequate with half-wave dipoles.

In the *xy* plane the edges have much more effect as Fig. 7.3 demonstrates. Moullin [17] shows that for $S/\lambda = 0{\cdot}25$ it is worthwhile making the distance to the edge, L, about one wavelength to increase the front-to-back ratio but hardly worthwhile extending it further. For smaller S/λ, smaller values of L may be

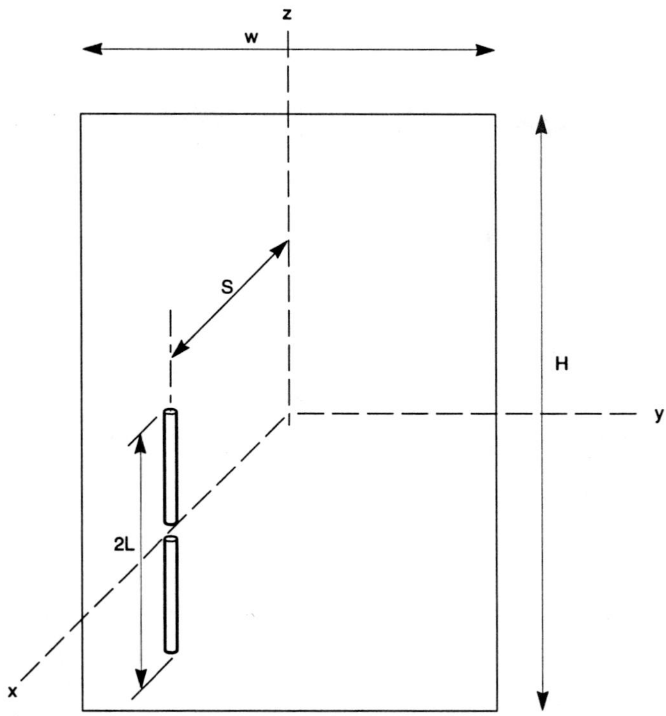

Figure 7.1 *Dipole and flat rectangular sheet*

acceptable. Fig. 7.4 shows front-to-back ratio as a function of sheet size in this plane.

The gain of a dipole and flat sheet combination oscillates with spacing, falling from a peak of about 7 dB above a dipole at small spacings to zero at $S=0{\cdot}5\lambda$ and rising to about 6 dB for spacings of an odd number of quarter-wavelengths. At small spacings, however, resistive losses may reduce the gain and, when the spacing is large, diffraction will lead to larger back lobes and reduced gain. Some measured and calculated figures are given in Table 7.1. It appears that for small sheets the gain may be enhanced in some instances.

The radiation resistance does not appear to be greatly affected by the sheet size provided the latter is large enough to give reasonable radiation patterns. Fig. 7.5 shows how the resistance oscillates with spacing. This is for an infinitely thin $\lambda/2$ dipole in front of an infinite sheet.

The sheet does not have to be solid: a grid of wires can be used to reduce windage. In fact the wires perpendicular to the dipole do not contribute and an array of rods parallel to the dipole can be used. The thickness and spacing of the rods can be adjusted to allow a given amount of leakage using the nomograph by Mumford [18] which can also be found in the 'Microwave Engineers Technical and Buyers Guide', 1968. An approximate formula, Smith [24], for

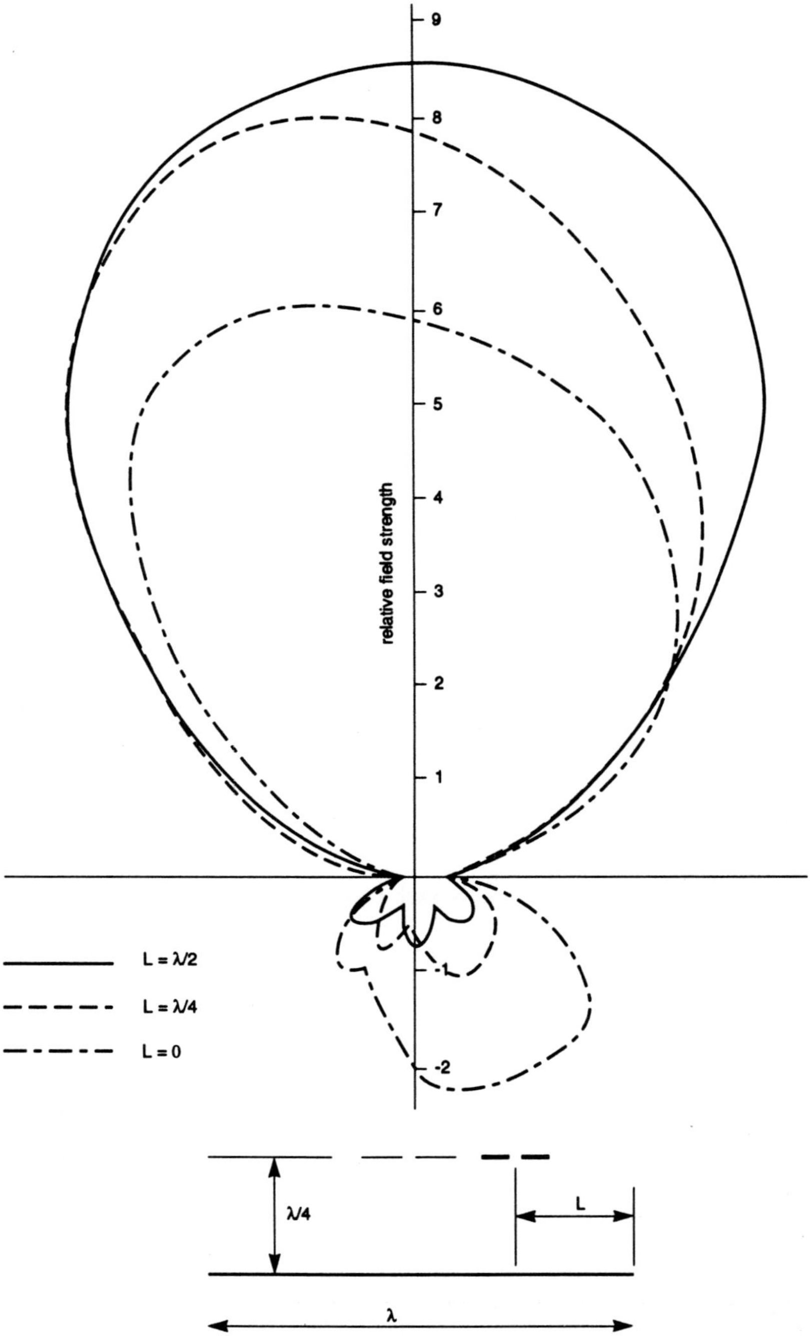

Reference 5

Figure 7.2 *Radiation patterns in xz plane as a function of distance from sheet edge*

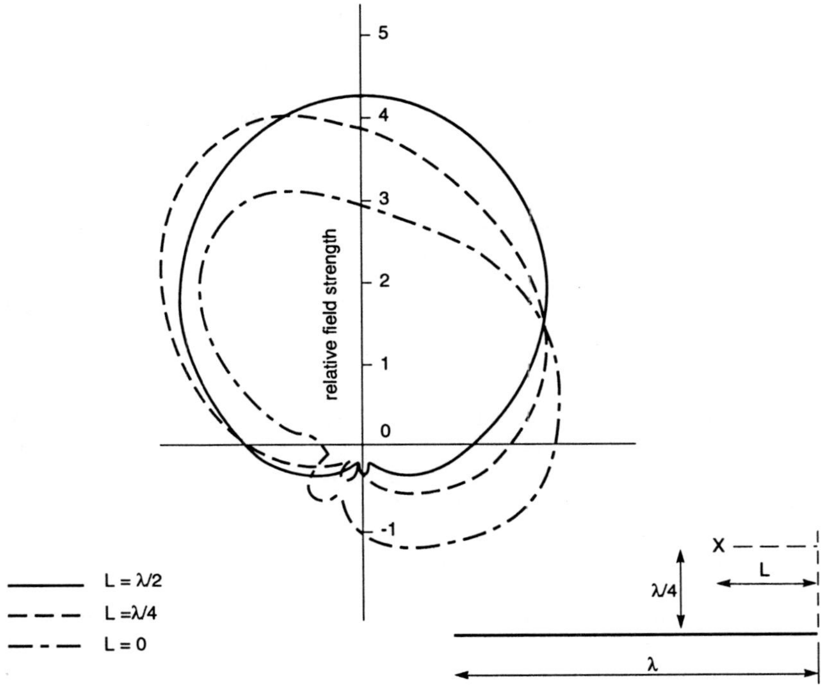

Reference 5

Figure 7.3 *Radiation patterns in xy plane as a function of distance from sheet edge*

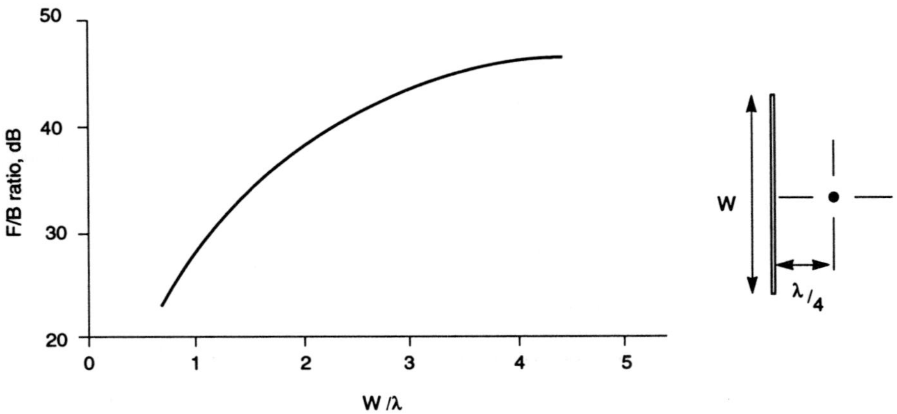

Figure 7.4 *Front-to-back ratio as a function of sheet width w*

Table 7.1 *Effects of finite sheet on antenna performance*

	Square of side L/λ	Spacing of dipole S/λ	Gain w.r.t. dipole dB
Measured	1	0·2	5·2
Calculated	Large	0·215	5·8
Measured	1·5	0·775	7·0

the transmission coefficient of a screen of wires of radius a spaced d apart is:

$$\gamma = (2d/\lambda)\ \ln\ (d/2\pi a)$$

As a rough guide, the screening ratio will exceed 0·9 if the spacing d is less than $\lambda/8$ and if a is greater than $0{\cdot}01\lambda$.

Using stacks or bays of dipoles in front of a plane sheet leads to matching problems if several elements are to be fed in parallel. An alternative adopted in Germany was to use fat full-wave dipoles in front of a screen of rods (Fig. 7.6). This arrangement had a number of practical advantages:

(i) The dipole thickness could be chosen so that the dipole and screen combination gave the best impedance for combining a number of elements.
(ii) The dipole arms could be metallically connected at their mid-points to the structure supporting the screen.
(iii) The use of fat dipoles (length/diameter $\simeq$10) meant that the length was only $0{\cdot}7\lambda$ at mid-band.
(iv) De-icing systems could be installed without having to bridge insulators.

The antenna system is discussed in detail in Stohr [25] and Bosse [6]. A VSWR of 1·3 to 60 ohms was achieved over the band 41–68 MHz.

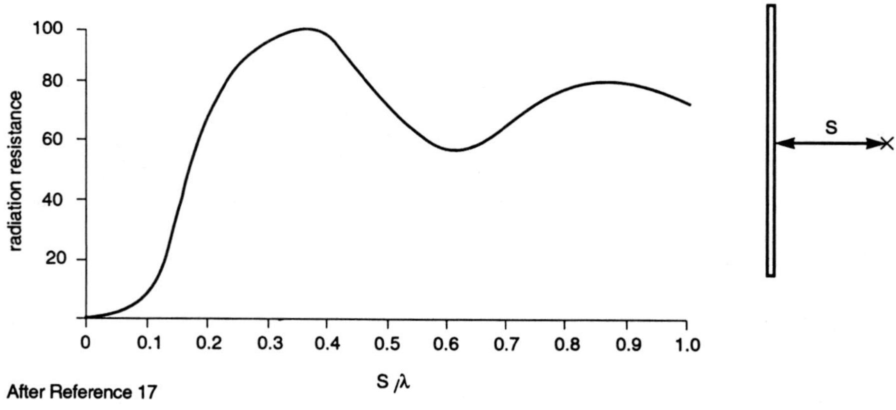

After Reference 17

Figure 7.5 *Radiation resistance of half-wave dipole and flat sheet*

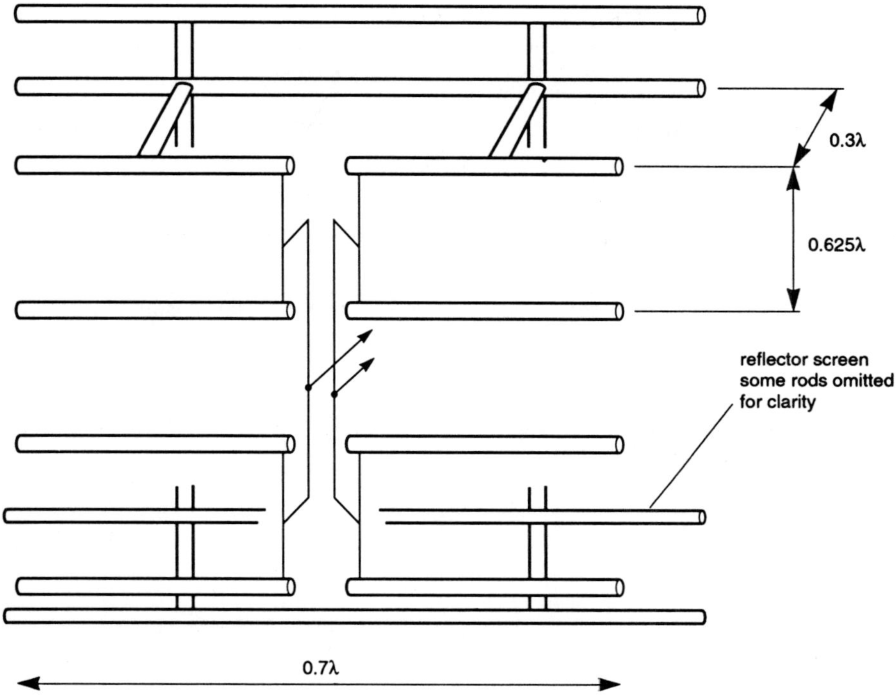

After Reference 6

Figure 7.6 *Stack of four full-wave dipoles and reflector screen*

7.1.2 *Corner reflectors*

These antennas have been covered so thoroughly by Moullin, Kraus [14] and Harris [11] that there is no point in reproducing their work.

One point that needs to be made with corner antennas of finite side length is that the length of the side must be sufficient that rays reflected near the end are radiated outward and not back along the axis. The point at which axial reflection occurs is when the side length $L = S\sqrt{2}$ on a 90° corner so it is usual to make $L = 2S$. Similar considerations apply to corners of smaller angles so that the side length needs to be proportionately longer. Since the resistance changes more rapidly with frequency for these corners of smaller angle it may be better mechanically and electrically to use a stack of two 90° corners than one 60° corner. For example, with a spacing of 0·6λ a pair of 90° corner antennas fed in parallel would have a resistance of about 65 ohms. The array would have sides of 1·2λ and a height of 1·2λ since the sides of a single corner need be only 0·6λ high.

For a corner angle of 270°, Harris [11] shows some interesting patterns. With a spacing of about 0·3λ almost uniform coverage obtains over 180° with a rapid

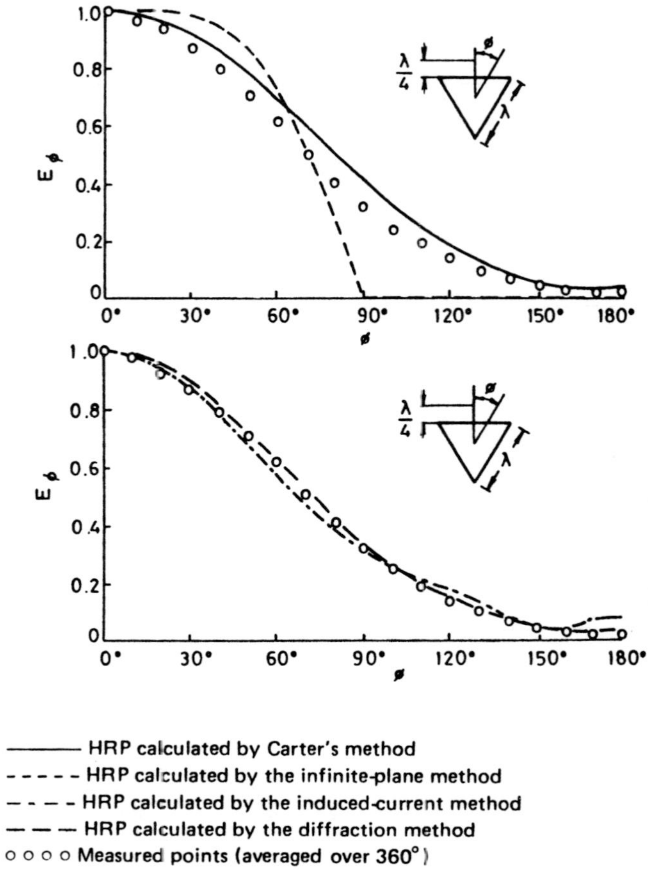

Figure 7.7 *Azimuth patterns of a vertical dipole on a triangular mast*

decrease thereafter. This offers the possibility of putting a dipole on the corner of a building whose walls have been covered with wire mesh or metal foil, that is, if you happen to want radiation in that direction.

7.1.3 *Triangular mast*

A similar situation is the mounting of a dipole on a horizontal mast of large side. This was explored by Knight [13] and patterns for both horizontal and vertical dipoles are shown in Figs. 7.7 and 7.8. The need to consider edge diffraction accounts for the failure of the infinite-plane calculation with the vertical antenna whereas the horizontal antenna gives better agreement with measurement. This agrees with the findings in Section 7.1.1 with plane reflectors. Some experimental results for square and triangular masts are given by Andrews [3].

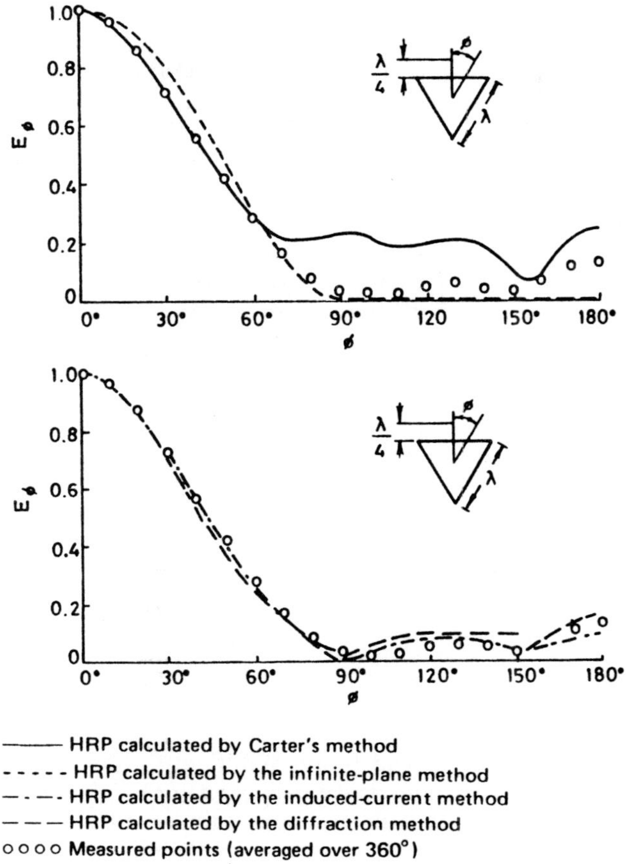

Figure 7.8 *Azimuth patterns of a horizontal dipole on a triangular mast*

7.1.4 *Elliptical cylinders*

Patterns for elements close to elliptical cylinders have been considered by Sinclair [23]. Circular cylinders were dealt with by Carter [7] in a classical paper. One result giving a useful directional pattern is shown in Fig. 7.9.

7.2 Parasitic elements

The simplest arrangement consists of a single driven element parallel to an unfed element whose length and spacing have been chosen either to enhance the radiation in the direction of the parasite (director) or to reduce it (reflector). Whilst many if not most antennas use dipoles or monopoles and rod parasitic elements there is in principle no reason why loops, slots or notches should not be used similarly. The use of parasitic notches has been reported in Chapter 6 but has not been explored very extensively. Parasitic slots have been used to modify the patterns of single slots but, because they require a backing cavity, are most suited to use on conductive bodies such as aircraft or missiles. There does not appear to be much in the literature on parasitic loops in this frequency range unless the cubical quad, to be discussed later, could be so described.

Walkinshaw [28] calculates radiation pattern, gain and impedance for dipoles with up to four parasites in terms of parasite spacing and self-reactance. The latter is a function of element thickness and length in terms of wavelength. Walkinshaw's paper shows that for a single parasite a director at spacings of about 0·1λ should be used or a reflector at larger spacings. The best dimensions for maximum forward gain are not necessarily those giving the best front/back ratio. Typical gains of 5·5 dB w.r.t. a dipole for a director or 5·2 dB for a reflector can be achieved. At a spacing of 0·1λ the radiation resistance of a half-wave dipole will only be 10–15 ohms. A folded dipole is often used to improve matching. For bandwidths up to 20%, a dipole plus reflector at 0·15–0·20λ is to be preferred, with slightly reduced gain but better impedance characteristics. For wider bandwidths a combination of reflector and single director has been used, the dimensions being chosen so that the effective bands of the two pairs, dipole and director and dipole and reflector, overlapped. The director is effective at the higher frequencies, the reflector at the lower. A bandwidth of 36% has been achieved in this way.

7.2.1 *Yagi–Uda antennas*

In view of all that has been written about these antennas since Uda's original paper in Japanese in 1926, it is not proposed to discuss them in detail. The first paper in English was by Yagi [29]. Other useful papers are by Smith [24], Uda and Mushiake [27] and Fishenden and Wiblin [10]. Smith gives an approximate formula for the maximum power gain of a long Yagi antenna as $G = 9{\cdot}2l/\lambda$ referred to isotropic. Yagis with optimum gain usually have rather narrow bandwidths. Considerable improvement in bandwidth can be achieved at the sacrifice of gain. As an example, two 5-element Yagis (one reflector, driven

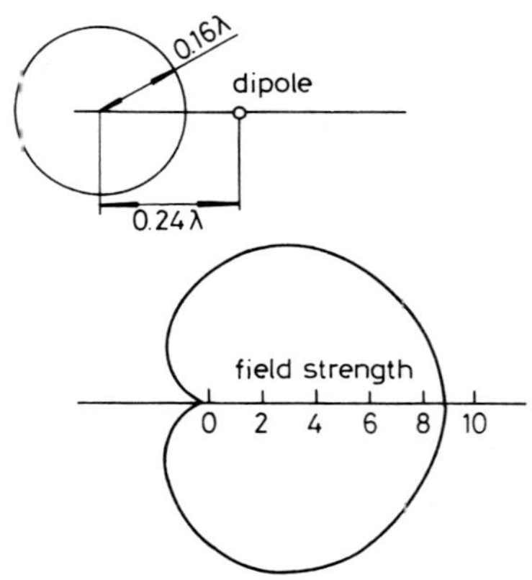

Figure 7.9 *Pattern of a dipole close to and parallel to a circular cylinder*

element and three directors) had gains respectively of 8 dB over a 9% band for the optimised antenna and 4 dB over a 60% band for a broadband version.

Very long antennas pose structural problems and it may be better to use arrays of Yagis of say 10 dB gain rather than a very long antenna. Smith recommends a spacing of $1{\cdot}5\lambda$ between two antennas in the same plane and 3λ for antennas parallel to one another to reduce interaction. These spacings may be too large and give interference lobes within the desired beamwidth in which case some compromise will be necessary.

A problem which has not been mentioned in the main references but is nevertheless important in some areas is the effect of ice loading. Snow has a very low dielectric constant so its effect will be small but ice, with a dielectric constant of about 3 in the VHF and UHF bands, can cause serious effects. Experiments in Germany showed that it was possible to reverse the direction of the main beam by ice loading. The effects for a given thickness of ice were more severe on the thinner elements of a 500 MHz Yagi than on one at 220 MHz. The effects can be simulated by putting close fitting tubes of, say, glass fibre over each element the thickness of the tubes being that of the predicted ice thickness in the desired location of the antenna.

7.2.1.1 *Some Yagi variants*

(a) Gain optimised elements

Landstorfer [16] has experimented with gain-optimised dipoles of length $1{\cdot}5\lambda$. With similar parasitic elements, a three-element array as in Fig. 7.10 gave a gain of 11·5 dB (isotropic). The best gain shown by Walkinshaw for a three element array is 8·7 dB so a significant improvement is possible. Such a high gain cannot be achieved without sacrifice of bandwidth: in the example given the bandwidth was 2·5% for 11 dB gain and 8% for 8·5 dB gain. Good sidelobe level and front/back ratio are claimed.

(b) Delta loops

Dipole elements do not on the whole lead to broadband Yagis. An alternative approach by Tsukiji and Tou [26] using twin-delta loops appears to offer some promise. One arrangement is shown in Fig. 7.11. The parasitic element consists of two equilateral triangles joined at their apices. The perimeter of each triangle is C_2. Similarly the driven element consists of two equilateral triangles of perimeter C_1, truncated close to their apices and fed by twin line. The diameter of the tubing of which each loop is constructed is $2a$. In the example given this is small compared with the wavelength; its influence on performance is not given.

The parasite operates as a reflector for $\lambda \leqslant C_2 \leqslant 2\lambda$ and as a director for $0{\cdot}5\ \lambda \leqslant C_2 \leqslant 0{\cdot}9\ \lambda$. The maximum gain of 9·8 dBi was achieved for the parasite as a reflector with $C_2/C_1 = 0{\cdot}9$ and spacing $d/C_1 = 0{\cdot}1$. For a gain of 7 dBi nearly an octave bandwidth was achieved with $C_2/C_1 = 1{\cdot}1$. This is rather better than the performance with a director.

(c) Cubical quad

Whether the inventors of the twin-delta loop were familiar with the cubical quad is not clear. This antenna has long been used by amateurs but appears to be little known outside their ranks. Its derivation from a stack of two horizontal Yagis is shown in Fig. 7.12. The typical 'quarter-wave' quad is shown in Fig. 7.13*a*. Means for tuning are shown in Fig. 7.13*b* in the form of a short-circuited and open-circuited stub on the reflector and director, respectively. The quoted gain is 7–7·5 dB. The antenna is said to be broadband and dimensionally non-critical. It is attractive because of its compact form. A half-wave quad (sides approximately $\lambda/2$) can have a gain of 12·5 dB.

7.3 Backfire antennas

The original antenna as proposed by Ehrenspeck [9] would now be classified as a 'long backfire'. In essence this is a surface wave structure between a large and a small reflector as shown in Fig. 7.14. Bach Andersen [4] states that no simple theory is available. Since then, however, Kumar [15] has proposed a theoretical analysis which gives good agreement with measured results. His paper includes results for a plane reflector with rim and a stepped reflector with rim. The latter gives lower sidelobe levels. Ehrenspeck showed the gain was a function of reflector diameter up to about 4λ for a 4λ long backfire antenna. Here the gain reached about 16 dBd (18 dBi) and oscillated thereafter. This was for a plane reflector: with the stepped reflector of Fig. 7.14 the gain could be increased to 20 dB for $D=6\lambda$.

This is rather a large antenna for all except the upper end of the UHF band and is rather a clumsy structure. Later, however, Ehrenspeck found that the

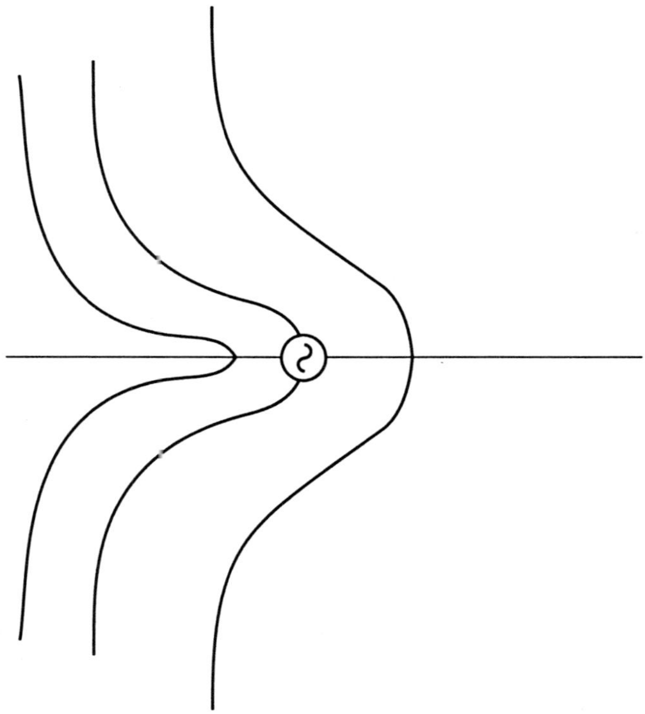

Figure 7.10 *Short Yagi with gain-optimised elements*

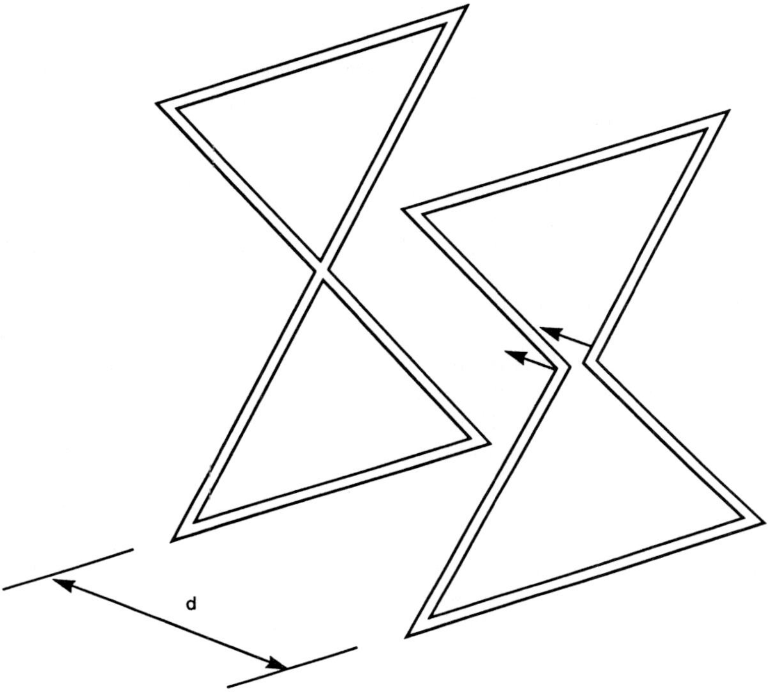

Figure 7.11 *Short Yagi with twin-delta loops*

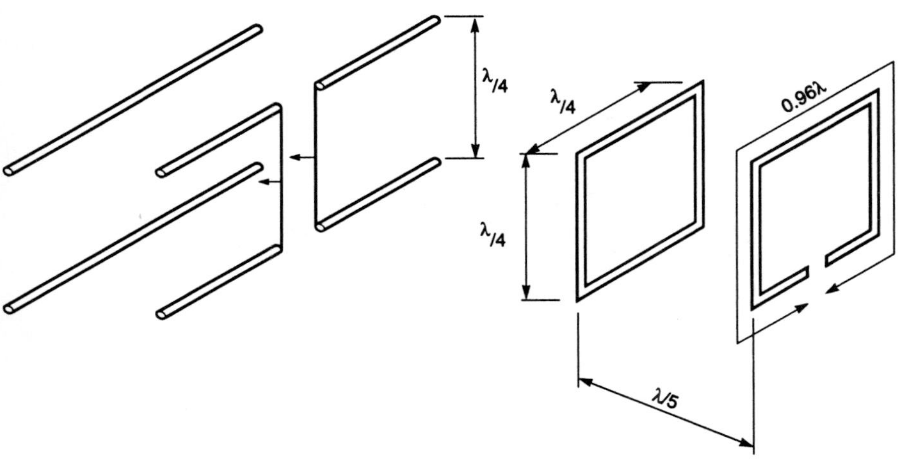

Figure 7.12 *Quad antenna derived from pair of dipoles and reflectors*

much smaller and simpler short backfire could give almost as much gain. Narrow-band gains of the order of 18 dBi have been claimed. The author has built a number of these short backfires and believes that a broadband gain of the order of 12 dB would be realistic. Fig. 7.15 shows a typical arrangement in which d would be 2–2·5λ, $h_1 = h_2 = 0{\cdot}25\lambda$, $2r = 0{\cdot}5\lambda$, w up to 0·6λ. For linear polarisation a dipole feed could be used: a practical arrangement is the slotted feed of Fig. 2.9 with the slots extended up to the reflector, Fig. 7.16. The reflector may be a circular disc or a $\lambda/2$ strip. For circular polarisation crossed dipoles would be suitable. Support of the small reflector would be difficult in this case and it is probably best attached to a dielectric plate fixed to the rim and thus enclosing the antenna.

The rim is important in reducing sidelobes and increasing gain but the mechanism is not clear. It is possibly the position of the edge providing another radiation component at a different point in space that matters; in this case a closed loop above a plane reflector might have the same effect. There is some evidence that rim corrugations Fig. 7.17 improve the sidelobe levels as Fig. 7.18 shows.

The antenna is simple to construct and not very critical dimensionally. VSWR bandwidth could be as much as 2:1 with appropriate broadband dipoles. It is probable that for linear polarisation the large reflector need not be circular: a rectangular reflector with greater width in the E plane than the H plane should be worth trying. The rim need then only be on the E plane sides so the reflector becomes a simple channel.

7.4 Helical antennas

A helix fed at one end can radiate in several modes depending on its diameter D and pitch S as shown in Fig. 7.19. When D/λ is small the antenna resembles a monopole. There is a range of $C/\lambda = \pi D/\lambda$ between 0·75 and 1·33 in which the radiation is unidirectional and this is the axial mode, the only one of importance as a directional antenna.

The topic has been so thoroughly covered by Kraus [14] that there is little point in doing more here than summarise the salient features.

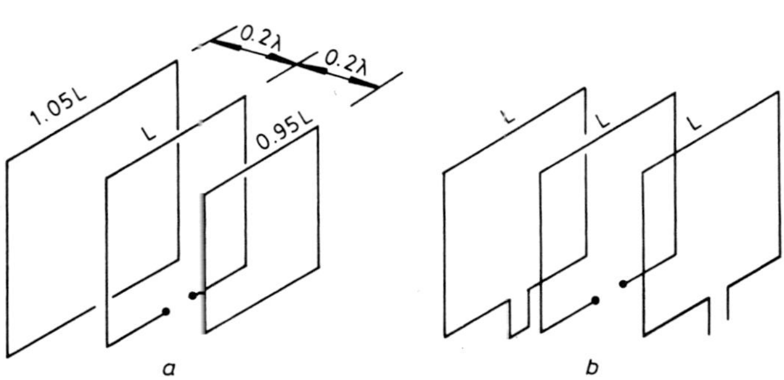

Figure 7.13 *Two forms of 3-element quad*

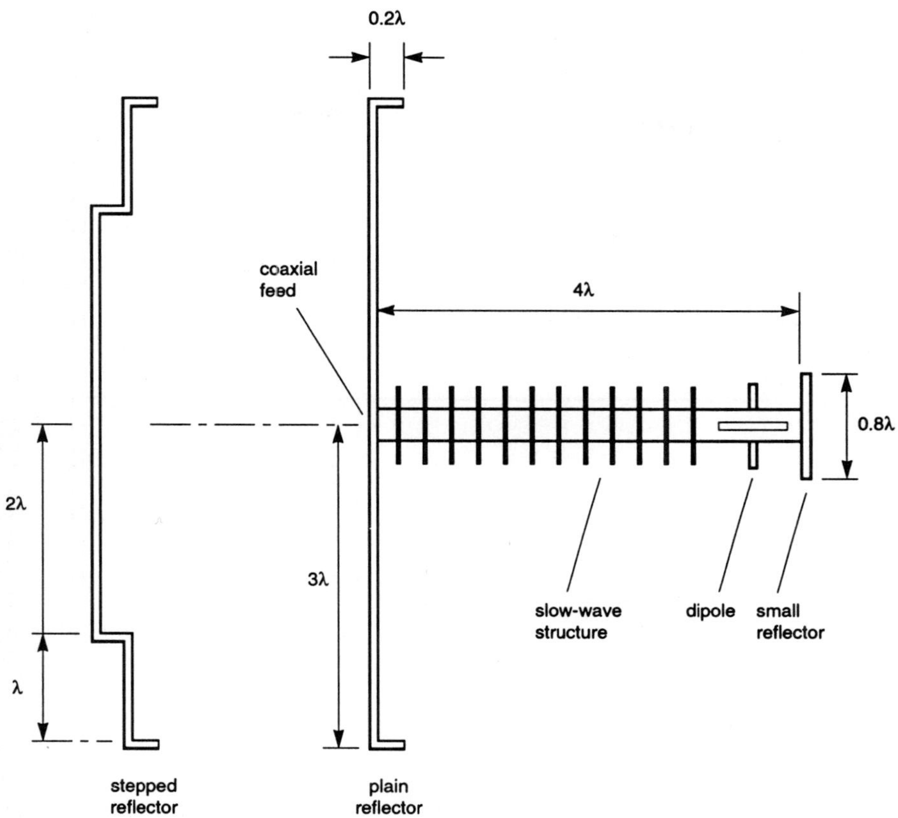

Figure 7.14 *Long backfire antenna, dipole fed*

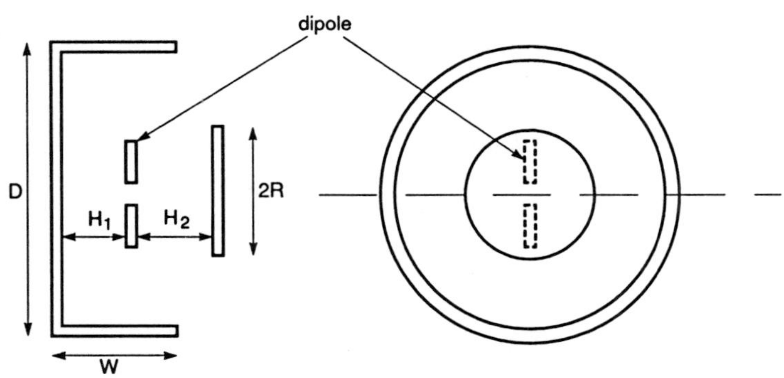

Figure 7.15 *Short backfire antenna*

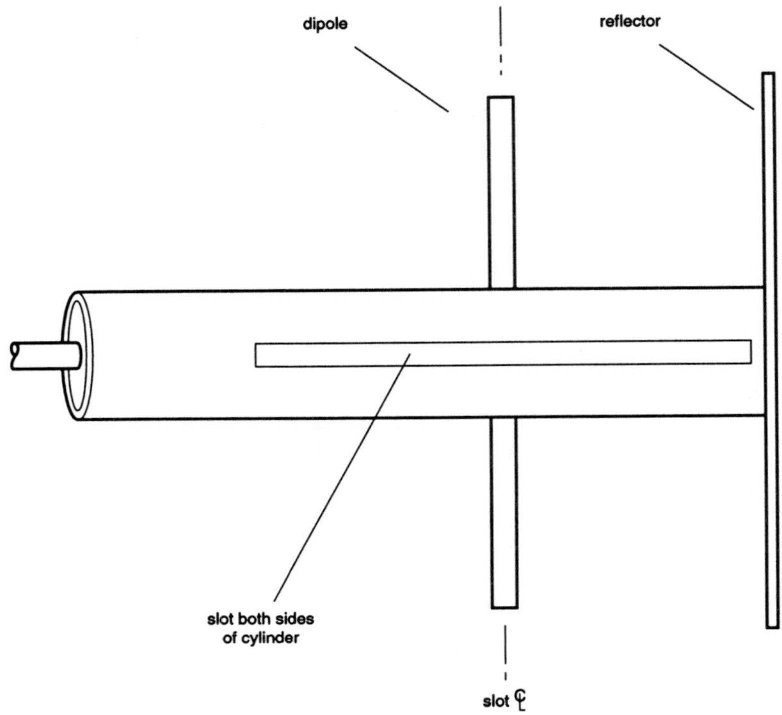

Figure 7.16 *Slotted fed dipole and reflector*

- The ground plane diameter G should be at least 0·5λ.
- The pitch angle α should be between 12° and 14°.
- The number of turns n should be greater than 3; shorter helices produce distorted patterns because of wave reflection from the open end of the helix.

Then,

Radiation pattern $$E = \sin\left(\frac{90^\circ}{n}\right)\frac{\sin(n\psi/2)}{\sin(\psi/2)}\cos\phi$$

Half-power beamwidth $$B_1 = \frac{52}{C_\lambda\sqrt{nS_\lambda}} \text{ degrees}$$

Beamwidth to first nulls $$B_2 = \frac{115}{C_\lambda\sqrt{nS_\lambda}} \text{ degrees}$$

Directivity $$D = 15C_\lambda^2 nS_\lambda$$

Input resistance $$R = 140C_\lambda$$

Axial ratio $$AR = \frac{2n+1}{2n}$$

Where C_λ = circumference in free-space wavelengths
S_λ = spacing between turns in free-space wavelengths

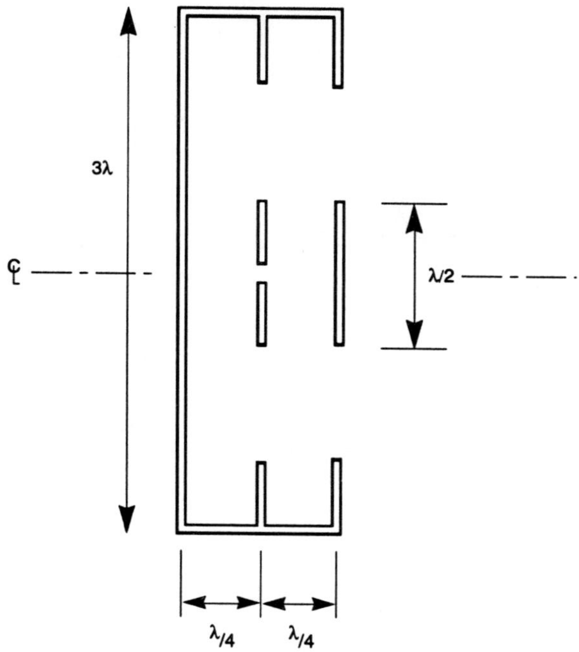

Figure 7.17 *Short backfire with rim corrugations*

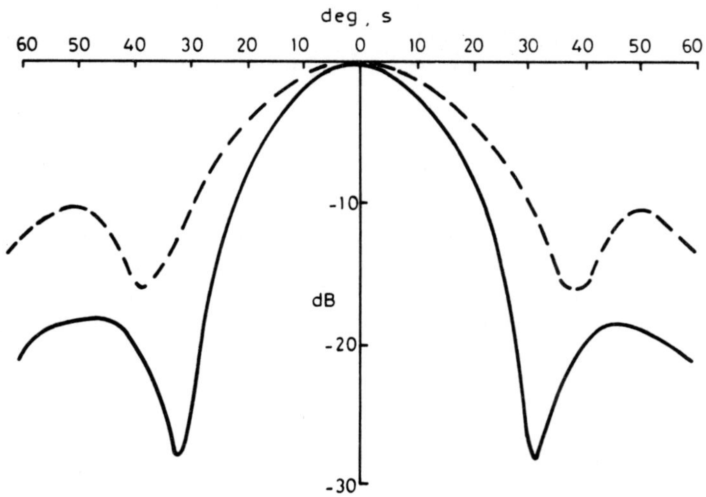

Figure 7.18 *E plane patterns of short backfire showing effect of rim corrugations*

—— with rim corrugations
– – – without rim corrugations

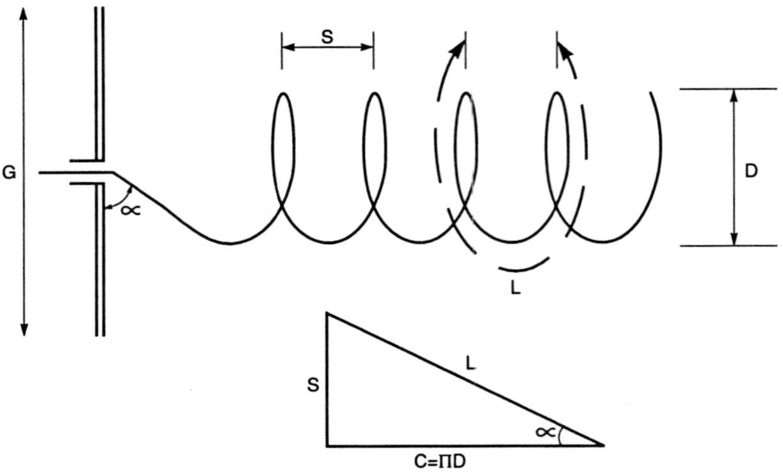

Figure 7.19 *Helical antenna*

ϕ = angle from helix axis
ψ = phase shift per turn

ψ is dependent on the phase velocity along the helix; see Kraus for details.

Typical gains for practical antennas are between 10 and 15 dBi. Sidelobe levels are not exceptional being typically −10 dB. King and Wong [12] have published pattern and gain data for antennas up to 8λ long showing gains up to 17·5 dB for 35 turns.

The problem with long helical antennas is supporting them without the supports affecting the antenna. Fig. 7.20 shows two possible methods. Trying to wind a helix on a dielectric tube is likely to prove disastrous

(*a*) Because of losses in the dielectric which is in the most dense part of the electric field
(*b*) Because of changes in velocity due to the dielectric constant of the dielectric.

In general the low-loss and low-density materials which would be electrically acceptable do not provide adequate mechanical support. This will be particularly true at frequencies above about 2 GHz and it may be desirable to enclose the antenna in a thin dielectric cover of diameter perhaps twice that of the helix in order to reduce mechanical loads on the antenna.

7.4.1 *Array of helices*

Apart from the statement that the reflector should be more than 0·5λ diameter, Kraus makes no reference to its effect. This is probably because with the long helices generally used there is negligible reflected wave. Shiokawa *et al.* [22] showed that the gain and axial ratio were very much affected by reflector size when the helix consisted of as few as two or four turns. It looks as if the characteristics of a backfire antenna are being superimposed on those of a

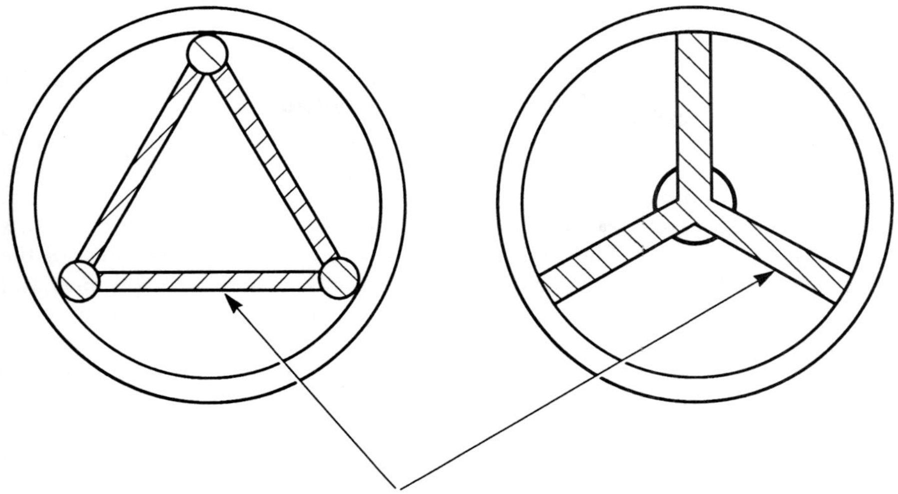

Figure 7.20 *Supporting an axial mode helix*

helical antenna. With a two-turn helix a gain of 9 dB was obtained with a reflector of λ diameter. Four such elements were mounted on a $1{\cdot}7\lambda$ ground plane and fed in phase. To reduce mutual coupling each element was surrounded by a $0{\cdot}7\lambda$ diameter cylinder $0{\cdot}25\lambda$ high. The resulting array had a gain of 13 dB for a height of only $0{\cdot}425\lambda$. The rim was shown to improve axial ratio, a figure of about 1 dB being achieved with the height shown.

7.4.2 *Multiwire helix*

Using a number of interlaced helices with feed-points offset from the axis allows for various feed options. The quadrifilar helix with four helices fed with relative phases of 0°, 90°, 180°, 270°, has been comprehensively reported by Adams *et al.* [2]. There is some evidence that ground plane size is more important than for the single helix. One of the improvements offered by the quadrifilar helix is a much increased frequency range: the forward axial mode can exist for $0{\cdot}4 < C_\lambda < 2$, a range of 5:1 compared with the single helix 1·7:1.

A good theoretical study of unifilar and multifilar helices is given by Bach Andersen [4].

7.4.3 *Zig-zag antenna*

This is a planar form of the helix which may be either unbalanced or balanced, Fig. 7.21. The balanced form is claimed to have much larger bandwidth, ±25% as against ±5% for the unbalanced form. Two good references are Cumming [8] and Sengupta [21].

7.5 Sandwich-wire antenna

This type of antenna is also known as the snake line, serpent line or meander line. It consists essentially of a three-line transmission line of which the two outer elements are earthed. In a finite array the centre element is fed at one end and terminated at the other. If the centre line is periodically distorted from being parallel to the outer lines it will radiate a beam polarised transversely to the array axis. The shape of the centre line can take a number of different forms as shown in Fig. 7.22. The radiating currents from adjacent crossings of the array longitudinal axis are in opposing directions: by choosing the physical and electrical spacing the beam can be endfire, backfire, broadside or anywhere in between. The wave velocity on the centre strip is assumed to be that of light so the phase velocity depends on the length of the strip between crossing points. If the strip is carried on a thin dielectric substrate the phase velocity will be slightly reduced.

It is usual to back the array with a cavity approximately $\lambda/4$ deep so that radiation is on one side only. The outer wires can be extended into strips or flanges but since the electric vector is normal to the edges their effect will be secondary on the radiation pattern.

This type of antenna was introduced by Rotman and Karas [19][20]. A good description is given by Zucker [30] and a more recent analysis is by Aboul-Atta and Shafai [1]. Obviously a similar arrangement would function with an undulating slot. This type of antenna is comparatively simple to construct. For frequencies above about 1 GHz it would probably be convenient to print the centre element on a thin dielectric sheet: a secondary, spaced, dielectric cover is recommended to minimise the effect of rain or snow on the antenna. For lower frequencies the centre element could be made stiff enough to support either on dielectric pillars or on bars across the cavity.

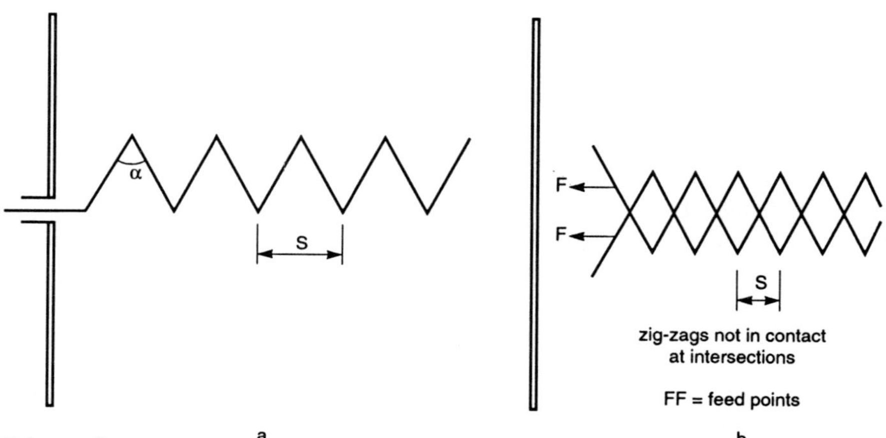

Reference 8

Figure 7.21 *Zig-zag antennas*

a Unbalanced
b Balanced

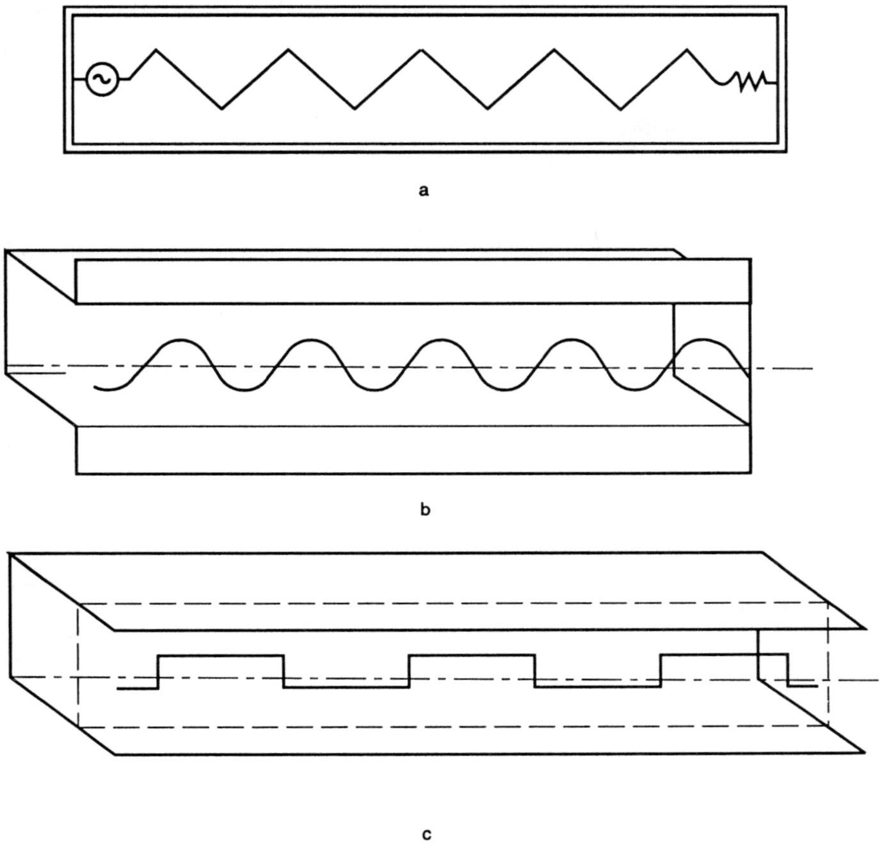

Figure 7.22 *Sandwich-wire antennas*

a Zig-zag
b Sinusoid
c Square wave

7.6 References

1 ABOUL-ATTA, O., and SHAFAI, L.: 'Hemispherically radiating meander-line planar array antenna'. IEE Conf. Publ. 219, April 1983, pp. 141–144

2 ADAMS, A.T., GREENOUGH, R.K., WALLENBERG, R.F., MENDELOVICZ, A., and LUMJIAK, C.: 'The quadrifilar helix antenna', *IEEE Trans.*, 1974, **AP–22**, pp. 173–178

3 ANDREWS, A.H.: 'Performance of UHF/VHF antenna adjacent to metal structures', *Electronic Technology (GB)*, 1979, **13**, pp. 210–214

4 BACH ANDERSEN, J.: 'Low and medium-gain microwave antennas' *in* RUDGE, A.W. *et al.* (Eds.): '*Handbook of antenna design*' (Peter Peregrinus, 1982) chap. 7

5 BOOKER, H.G.: 'Diffraction by aeroplane wings and aperiodic reflectors'. Telecommunications Research Establishment, UK, March 1941

6 BOSSE, H.: 'UKW-Breitband-Richtantenne', *FTZ*, 1952, (10), pp. 437–439

7 CARTER, P.S.: 'Antenna arrays around cylinders', *Proc. IRE*, Dec. 1943, pp. 671–692

8 CUMMING, W.A.: 'A non-resonant endfire array for VHF and UHF' *IRE Trans.*, 1955, **AP–3**, p. 52
9 EHRENSPECK, H.W.: 'The backfire antenna, a new type of directional line source', *Proc. IRE*, 1960, **48**, pp. 109–110
10 FISHENDEN, R.M., and WIBLIN, E.R.: 'Design of Yagi aerials', *Proc. IRE*, 1949, **96** Pt. III, p. 5
11 HARRIS, E.F.: 'Corner-reflector antennas' *in* JASIK, H. (Ed.): 'Antenna engineering handbook' (McGraw Hill Book Co., 1961)
12 KING, H.E., and WONG, J.L.: 'Characteristics of 1 to 8 wavelength uniform helical antennas', *IEEE Trans.* 1980, **AP–28**, pp. 291–296
13 KNIGHT, P.: 'Methods of calculating the horizontal radiation pattern of dipole arrays around a support mast', *Proc. IEE*, 1958, **105** Pt. B, pp. 548–554
14 KRAUS, J.: 'Antennas (McGraw-Hill Book Co., 1950) chap. 7
15 KUMAR, A.: 'Theoretical analysis of a dipole-fed long backfire antenna'. IEE Conf. Publ. 219, April 1983, pp. 136–140
16 LANDSTORFER, F.M.: 'New developments in VHF/UHF antennas'. IEE Conf. Publ. 169, Nov. 1978, pp. 132–141
17 MOULLIN, E.B.: 'Radio aerials', (Oxford University Press, 1949) chap. 5
18 MUMFORD, W.W.: 'Some technical aspects of microwave radiation hazards', *Proc. IRE*, 1961, **49**, pp. 427–447
19 ROTMAN, W., and KARAS, N.: 'The sandwich-wire antenna: a new type of microwave line source radiator'. IRE Natl. Conv. Record Pt. 1, 1957, p. 166
20 ROTMAN, W., and KARAS, N.: 'Printed circuit radiators: the sandwich wire antenna', *Microwave J.*, 1959, **2**, p. 29
21 SENGUPTA, D.L.: 'The radiation characteristics of a zig-zag antenna' *IRE Trans.*, 1958, **AP–6**, p. 191
22 SHIOKAWA, T., KARASAWA, Y., and YOKOI, H.: 'A ship-borne helical array antenna for maritime satellite communication'. IEE Conf. Publ. 195, April 1981, pp. 303–307
23 SINCLAIR, G.: 'The patterns of antennas located near cylinders of elliptical cross-section', *Proc. IRE*, June 1951, pp. 660–668
24 SMITH, R.A.: 'Aerials for metre and decimetre wavelengths' (Cambridge University Press, 1949) p. 79
25 STÖHR, W.: 'Breitband antenne für Richtfunkverbindungen', *Radio Mentor*, 1951, **17**
26 TSUKIJI, T., and TOU, S.: 'High gain and broadband Yagi-Uda antenna composed of twin-delta loops'. IEE Conf. Publ. 195, April 1981, pp. 438–441
27 UDA, S., and MUSHIAKE, Y.: 'Yagi–Uda antenna' (Maruzen Co. Ltd., Tokyo, 1954) (in English)
28 WALKINSHAW, W.: 'Theoretical treatment of short Yagi aerials', *J. IEE*, 1946, **93** Pt. IIIA, pp. 598–610
29 YAGI, H.: 'Beam transmission of ultra short waves', *Proc. IRE*, 1928, **16**, pp. 715–741
30 ZUCKER, F.J.: 'Surface and leaky-wave antennas' *in* JASIK, H. (Ed.): 'Antenna engineering handbook' (McGraw-Hill Book Co. 1961) chap. 16

Chapter 8

Broadband antennas

What constitutes a broadband antenna is subjective, which is why some antennas so described by their inventors may not be considered so by potential users. An essential feature is that certain characteristics — usually including radiation pattern coverage and VSWR — should remain within specified limits over a frequency range usually at least an octave. The limits will vary according to the specification; thus an antenna whose VSWR is less than 5:1 over an octave might be a satisfactory broadband receiving antenna for electronic surveillance monitoring (ESM) but could be unsuitable for a transmitting role in an electronic countermeasures system (ECM). It should be noted that the main bandwidth limitation is likely to be radiation pattern coverage rather than VSWR.

8.1 Omnidirectional antennas

8.1.1 *Discone*

The discone, Fig. 8.1, is not identical to an inverted cone because the size of the top plate is critical to the functioning of the antenna whilst the cone's ground plane is not defined. The significant paper on the subject is by Nail [15] who examined the effect of the various parameters. He showed that optimum values of D and S in Fig. 8.1 are independent of L and C_{min}. For most purposes choosing $S=0{\cdot}3\ C_{min}$ and $D=0{\cdot}7\ C_{max}$ will give satisfactory results. The bandwidth for a given VSWR is inversely proportional to C_{min}.

The lowest operating frequency tends to f_0 as ϕ increases, where f_0 is the frequency for which L is a quarter-wavelength. For a VSWR to 50 ohms not exceeding 2:1, the lowest operating frequency is related to ϕ as follows:

ϕ^0	f/f_0
25	1·27
35	1·16
60	1·14
70	1·22
90	1·34

Below the operating frequency the VSWR rises rapidly showing in effect a low frequency cut-off.

Discones for the lower end of the VHF band are usually constructed with radial elements for both the disc and the cone as in Fig. 8.2. This is necessary to reduce weight and windage and it also permits a portable arrangement in which the elements are screwed or clamped into a central boss which can be mounted

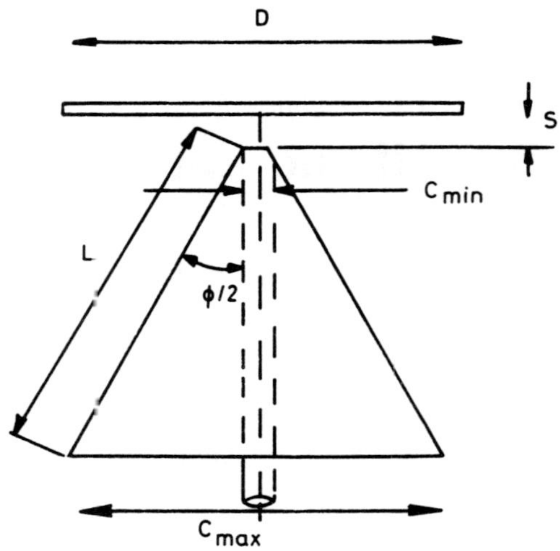

Reference 15

Figure 8.1 *Discone parameters*

on a support mast. Fig. 8.3 shows the VSWR characteristic of an antenna which was designed for the range 90–250 MHz. This was measured on an automatic impedance equipment sweeping over a wide range. Some of the peaks on the curve almost certainly arise from resonance on the elements. It has been found beneficial to join the radials, both of the disc and the cone, with conducting rings. These should be closer together near the centre of the disc and the apex of the cone than at the extremities. This will not only smooth the VSWR curve but will help the elevation pattern of the antenna.

At the upper end of the operating frequency range the peak of the radiation pattern tends to tilt increasingly downward until the signal level in the azimuth plane becomes undesirably low. This is demonstrated by patterns at 2 GHz and 3·5 GHz for an antenna which had a good VSWR performance from 500 MHz to 5·5 GHz. The patterns are shown in Fig. 8.4.

As a general rule a frequency range of 4 or 4·5:1 probably represents the best that can be achieved with consistent VSWR and pattern performance. Using the guide-lines described above an antenna was constructed with a solid disc and a cone of 16 radials. The range 250–1000 MHz was achieved with a VSWR better than 2·5:1 to 50 ohms.

8.1.1.1 *Grounded discone*

Kandoian [10] pointed out that if the base of the cone was grounded to a conducting surface the cone was effectively extended and the lower cut-off frequency lowered. This has been confirmed experimentally by the author with a discone mounted on the roof of a metal van: the lowering in frequency is obviously a function of ground plane size and probably of cone angle.

Kandoian also claimed that neither the cone nor the disc need be circular but could be highly elliptical thus allowing the antenna to be installed in an aircraft fin cap. This could provide a single antenna for the communication bands between 100 and 400 MHz but the antenna would be taller than a top-loaded sleeve monopole (see Chapter 3) which would serve the same purpose.

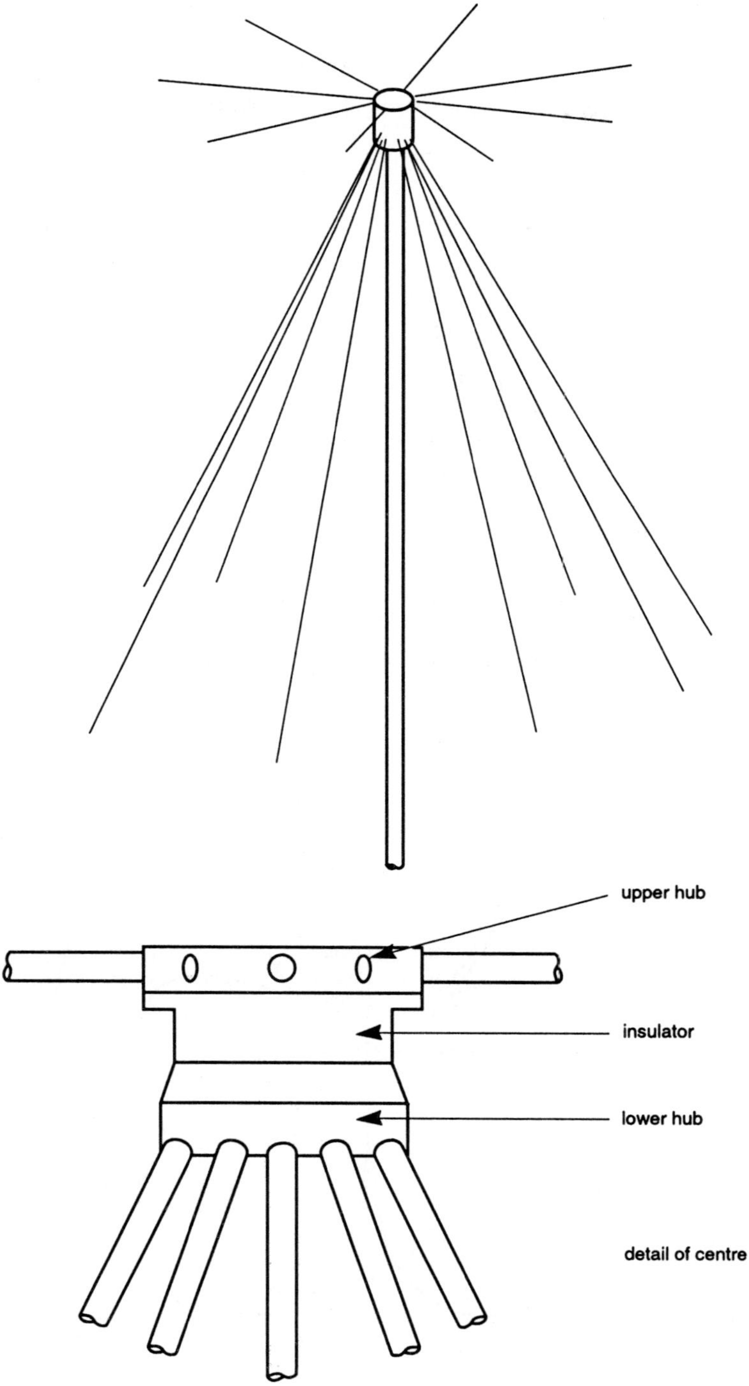

Figure 8.2 *Discone with radial elements for VHF operation*

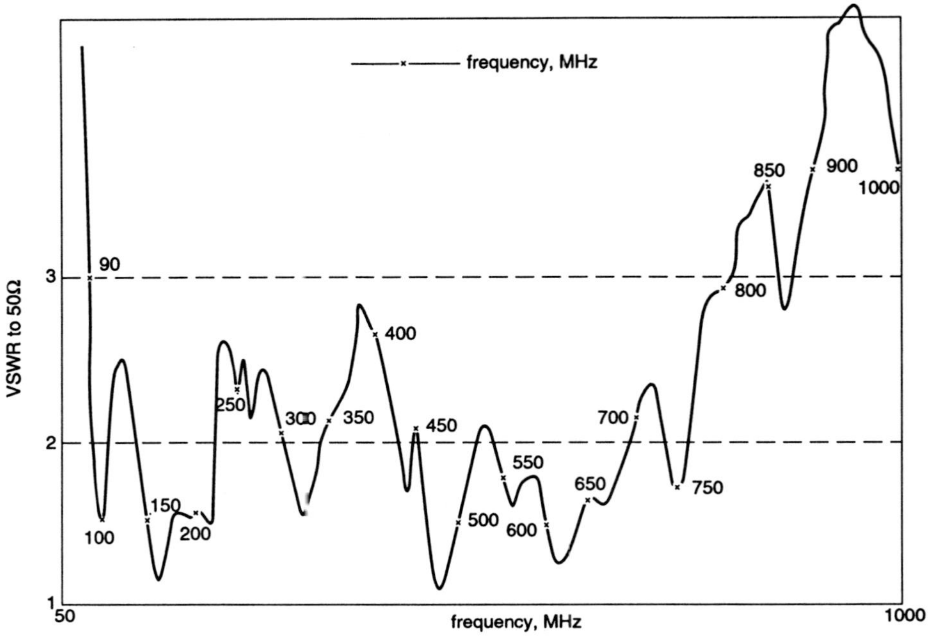

Figure 8.3 *VSWR performance of discone with eight radials*

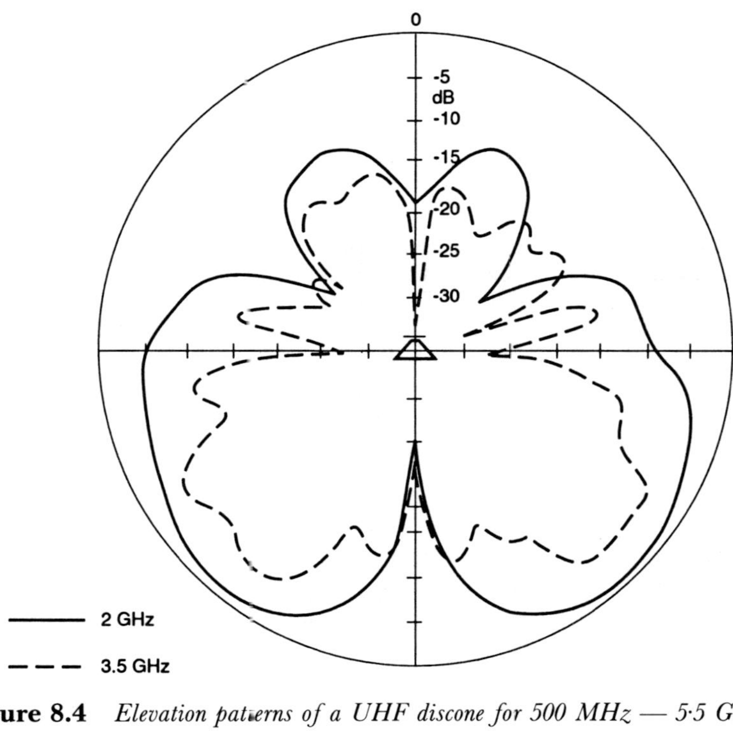

Figure 8.4 *Elevation patterns of a UHF discone for 500 MHz — 5·5 GHz*

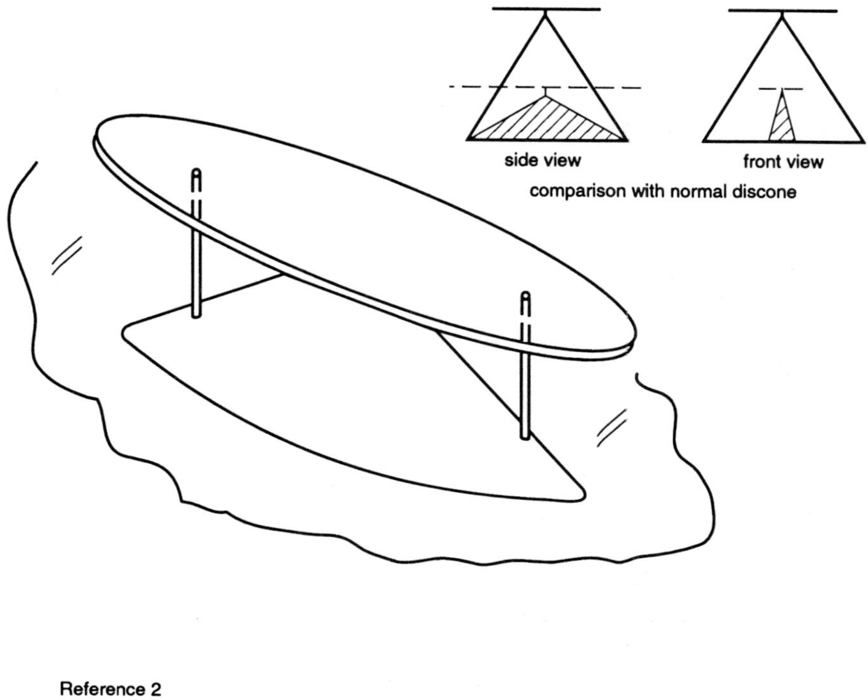

Figure 8.5 *Aerodiscone antenna*

8.1.1.2 *Aerodiscone*

Kandoian's work was built on by Barbano [2] who developed a non-circular discone which he called the Aerodiscone. The cone is either of aerofoil or highly-elliptical cross-section and the top plate, of similar section, is larger than for a conventional discone. The height is much reduced to typically $\lambda/10$ to $\lambda/12$ at the lowest frequency. Fig. 8.5 shows a typical antenna and the side and frontal views emphasise the reduction in height. An important feature is the addition of two loading pins. These provide a conductive connection between the disc and the cone and may be considered as operating rather as a tee-match dipole. Each pin will have a self-inductance which is seen at the end of a transmission line Z_0 which decreases towards the feed-point. Since the antenna will tend to have a capacitive reactance at its lower frequencies the shorted transmission line, which will be less than $\lambda/4$ at these frequencies, adds an inductive shunt compensation. Fig. 8.6 shows the improvement in VSWR on an antenna whose dimensions were $C_1 = 152{\cdot}4$ mm, $C_2 = 28{\cdot}6$ mm, $D_1 = 168{\cdot}3$ mm, $D_2 = 50{\cdot}8$ mm, $H = 38{\cdot}1$ mm, $L = 25{\cdot}4$ mm.

From the measured radiation patterns it appears that the loading pins have little effect. As the frequency increases the elevation patterns in the plane through the major axis tend to have more energy at high angles than at 0°. The results must depend to some extent on the size of the ground plane but it appears that a 3:1 frequency range can be achieved normally.

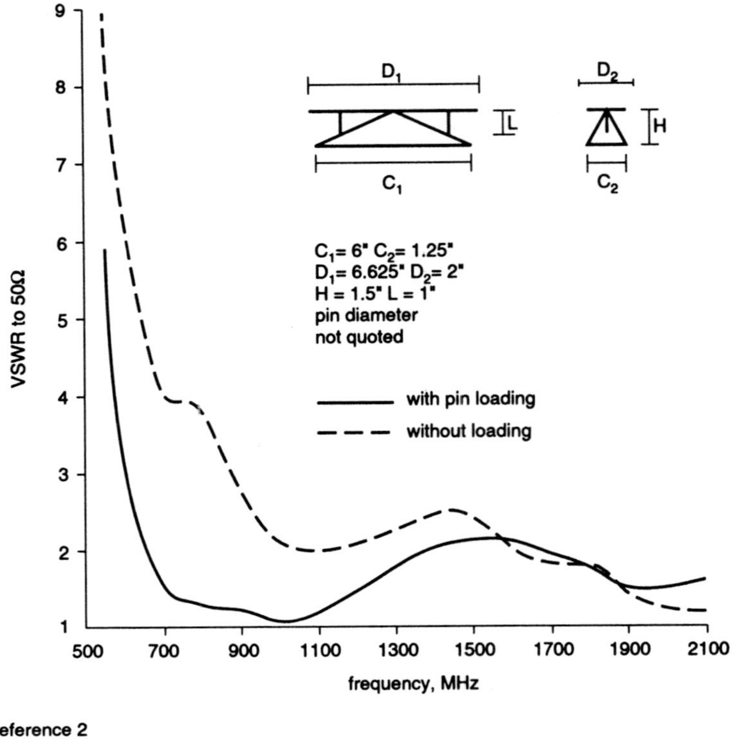

Figure 8.6 *VSWR of aerodiscone showing effect of pin-loading*

Barbano also describes what he calls an 'inverted discone'. This is no more than a highly elliptical cone above a ground plane with pin loading providing support as well as compensation. In the ultimate the antenna becomes a wide-angled triangle (Fig. 8.7). Fig. 8.8 shows the VSWR for such an antenna.

It should be obvious that the compensation due to pin loading can be altered either by altering the length of the pins or their diameter. In any given antenna experimentation will be necessary to optimise the bandwidth.

8.1.1.3 *Double discone*

Barbano's use of pin-loading was utilised by the author and colleagues in producing an omnidirectional vertically polarised antenna covering the range 25–300 MHz. One discone was mounted above the other, the disc of the lower frequency antenna serving as a ground plane for the upper antenna. It would have been possible to use a cone as the upper antenna but the discone gave better control of elevation pattern at the higher frequencies since maximum gain in the azimuth plane was required. A frequency range of about 3·5:1 for each antenna meant that VSWR was not a critical item and the design could be adjusted for optimum elevation pattern. The coaxial cable to the upper antenna was taken through one of the hollow loading pins so that it did not affect the operation of the lower antenna. The arrangement is shown schematically in Fig. 8.9.

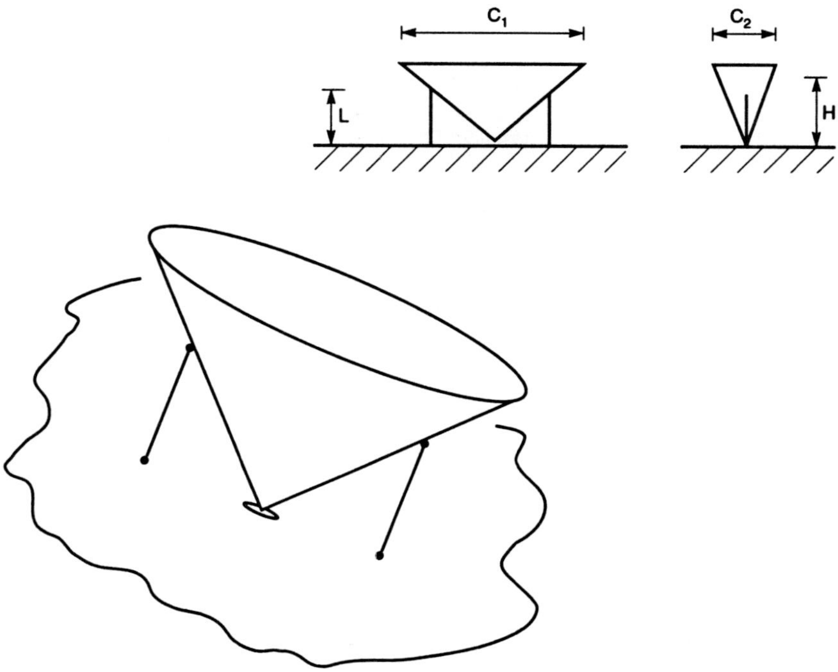

Reference 2

Figure 8.7 *Inverted discone*

8.1.2 *Wide-band bent sleeve dipole*

Some experimental work was carried out by the author in 1958 (i.e. before Barbano's work was published) to derive an antenna capable of covering both VHF and UHF aircraft communication bands in a form suitable for high performance aircraft. A scale model was used: this is shown in Fig. 8.10 but there is some doubt over some of the dimensions. Fig. 8.11 shows measurements of VSWR for an antenna mounted on a 2·5 m diameter fuselage. It was found that the rod length and the sleeve height should be roughly equal: reducing the rod length gave a lower VSWR but over a reduced bandwidth.

8.1.3 *Horizontally polarised antenna*

There does not appear to be any single antenna which will give a good omniazimuth pattern with horizontal polarisation over a wide frequency range and at the same time have a good VSWR characteristic. The terminated loop does have a broadband pattern characteristic but its impedance is anything but broadband. Lamberts [12] describes what appears to be an Alford loop consisting of four broadband elements each $\lambda/2$ long forming the sides of a square. This is shown to have horizontal radiation patterns with a maximum variation of 1·56 (4 dB) over a 1·66:1 frequency range. The VSWR is not particularly good, rising to 2·9 at one end of the band and 3·5 at the other.

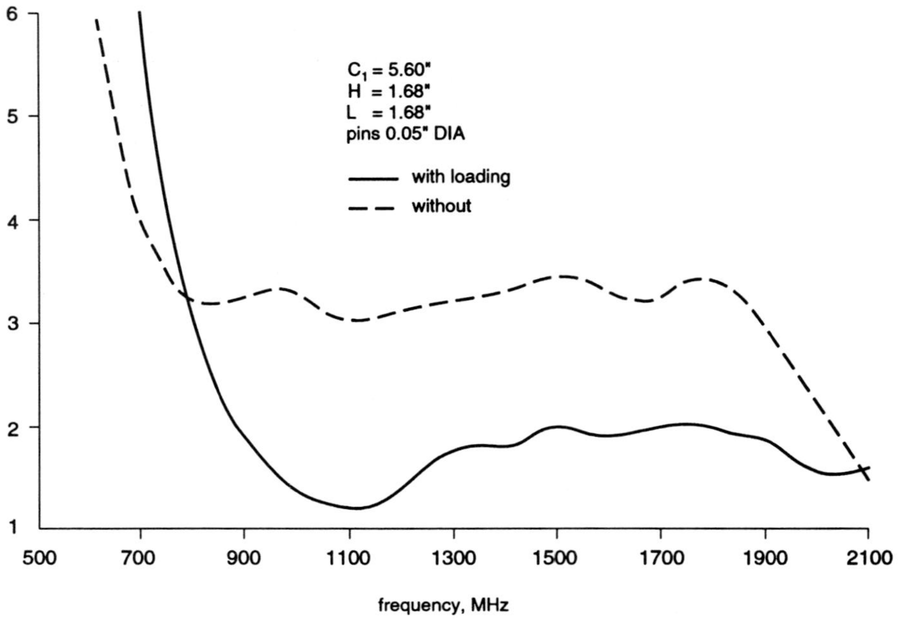

Reference 2

Figure 8.8 *VSWR of inverted discone in planar form*

8.2 Directional antennas

8.2.1 *Frequency independent antennas*

If, using any scaling factor an antenna can be transformed into a structure equal to the unscaled one then it can be classed as frequency independent. Its form can then be defined solely in terms of angles. One member of this class is the equiangular spiral in which scaling merely has the effect of rotating the antenna. The three-dimensional form is the conical spiral. Whilst these two types of antenna have the same performance over their frequency range, another class, the log-periodic antenna has a performance which repeats for each multiplication by a fixed scaling factor τ. In practice the performance of the antenna remains substantially constant over a small frequency band so by careful design a continuously-consistent performance can be obtained. The whole topic is well covered by Rumsey [17].

8.2.2 *Equiangular spiral antenna*

Fig. 8.12 shows what is effectively a dipole with triangular arms (bow-tie antenna) curled up in an equiangular spiral. In theory the arms should extend from 0 to infinity but in practice the antenna can be truncated at both ends, the

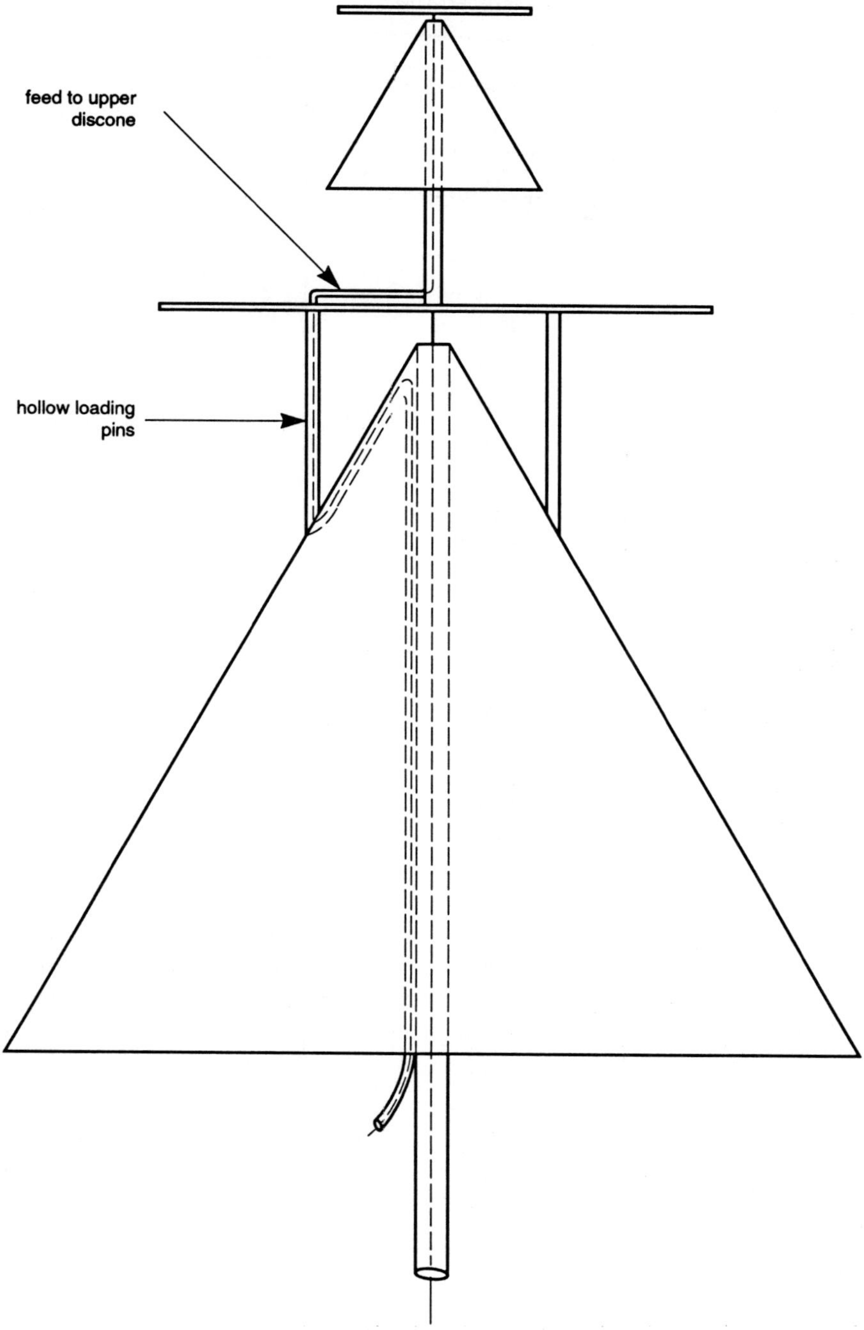

Figure 8.9 *Double discone*

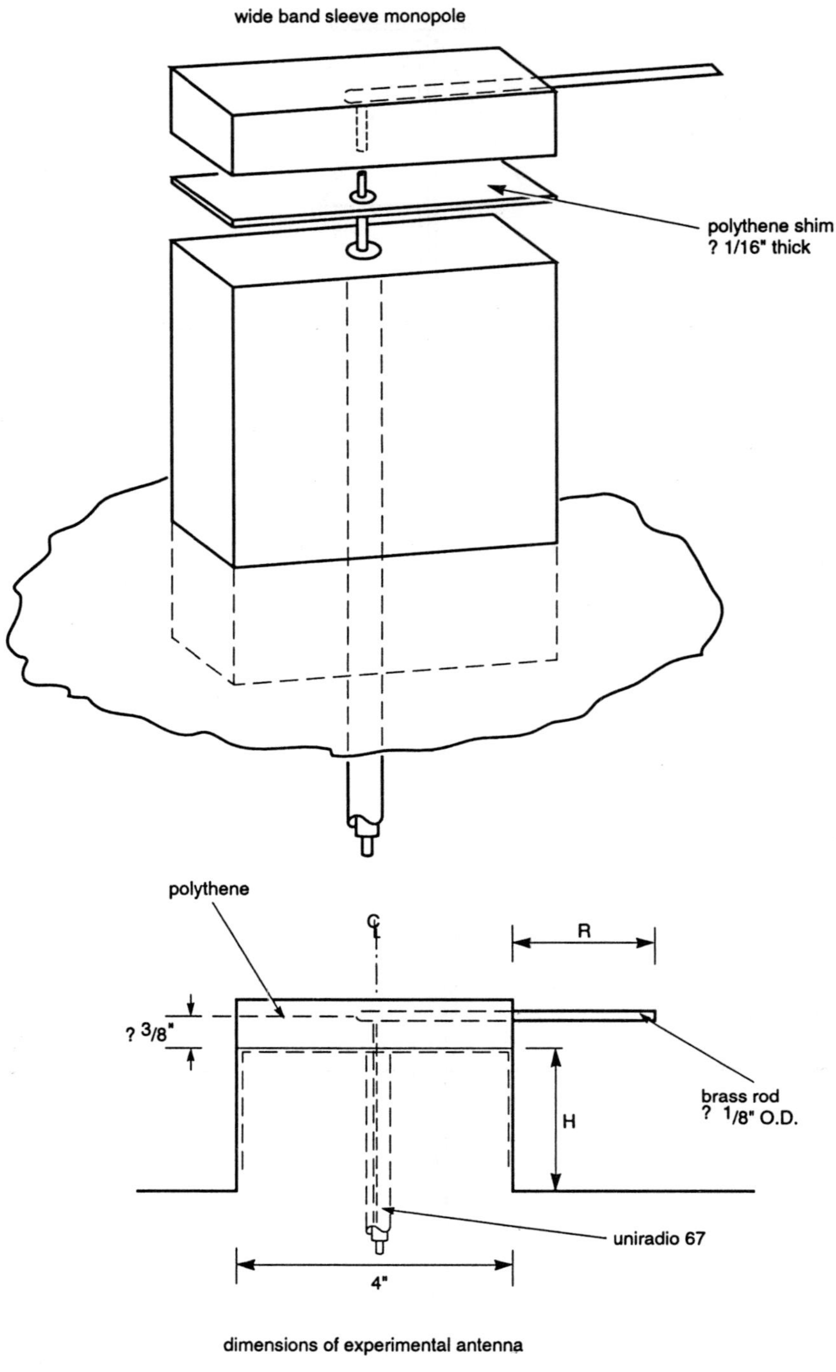

Figure 8.10 *Wideband bent sleeve monopole*

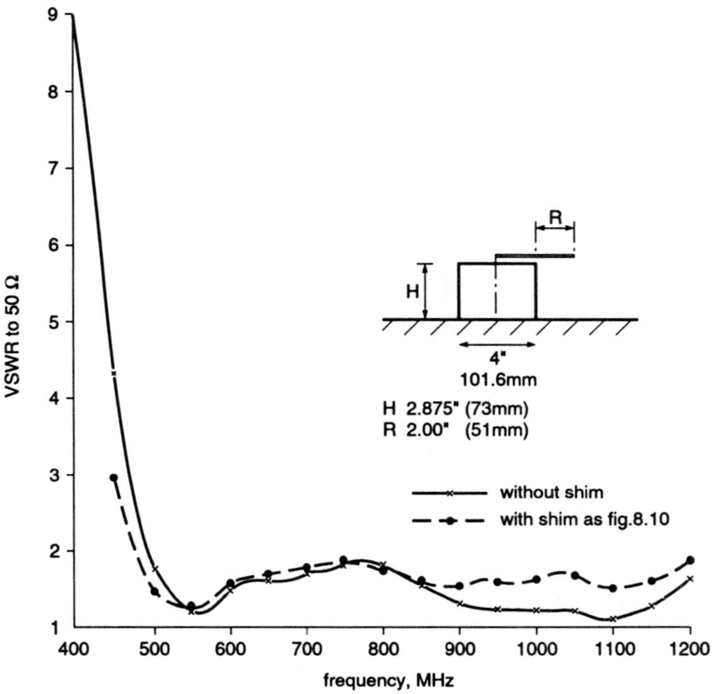

Figure 8.11 *VSWR curves for antenna of Fig. 8.10*

frequency being accordingly limited. The upper frequency cut-off f_1 is determined by the radius r_0 of the feed region which should be less than $\lambda_1/8$. The lower frequency cut-off f_2 is a function of the attenuation along the spiral. If there is insufficient attenuation then there will be reflection at the outer end of the arms and the antenna performance will be frequency dependent. If the attenuation is adequate for the required bandwidth then the termination at the outer radius is not important. The possibility therefore exists of making the antenna in a metal sheet as in Fig. 8.13. Here it becomes clear that the antenna is either a spiral dipole or a spiral slot. When β in Fig. 8.12 is 90° then the slot and the dipole are self-complementary and the impedance is 60π ohms.

The attenuation along the arms is a function of the rate of spiral $1/a$ and the angle ϕ' which is related to the radius ratio $C = \exp - a\phi'$. Jasik [9] gives a chart from which a two-arm antenna can be constructed.

It should be noted that the arms of the equiangular spiral can be defined as follows:

$$\textit{Arm 1} \quad r_1 = r_0 \exp a\phi$$

$$r_2 = r_0 \exp a(\phi - \phi') = Cr_1$$

$$\textit{Arm 2} \quad r_3 = r_0 \exp a(\phi - \pi)$$

$$r_4 = r_0 \exp a(\phi - \pi - \phi') = Cr_3$$

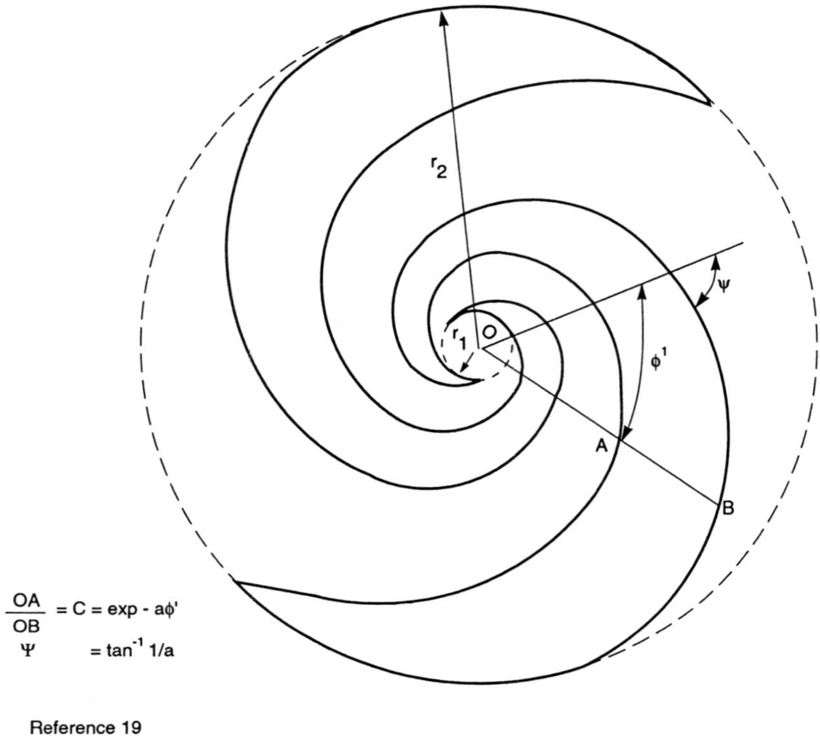

Figure 8.12 *Two-arm equiangular antenna*

Equiangular spirals produce bi-directional radiation patterns having maxima at right angles to the plane of the antenna, and essentially circular polarisation on axis. Axial ratios near unity can be achieved for bandwidths of at least 10:1.

8.2.2.1 *Feed system*
A balanced feed is required which must not disturb the current distribution on the antenna. When the latter is cut in a metal sheet as in Fig. 8.13, then a coaxial cable can be laid along one arm with its inner connected to the other arm. Obviously the width of the arm must be greater than the cable diameter at the feed-point but this is not a serious problem with the small diameter solid-jacketed cables now available. A dummy cable on the other arm ensures correct balance of arm currents.

The impedance of a self-complementary antenna is inconveniently high for a coaxial line. Experimental work reported by Wolff [19] shows that the spiral slot impedance can be reduced to a convenient figure by making the arms thin ($C \rightarrow 1$), Fig. 8.14. The bi-directional radiation from an equiangular spiral is not often required; the antenna is more usually used with a backing cavity. In this instance a broadband balun will usually be employed. Baluns covering

Figure 8.13 *Equiangular spiral antenna in metal sheet*

bandwidths up to 10:1 have been developed. The backing cavity will affect the impedance of the spiral and it is usual for broadband operation to put an absorbent material on the back face of the cavity to minimise its effect.

8.2.3 *Archimedean spiral*

The generating curve for the Archimedean spiral is

$$r = k\phi \exp j\sigma$$

where r is the radius at angle ϕ. The pitch is controlled by the constant k and for a two-arm spiral σ is either 0 or π.

The width of each arm remains constant for all values of ϕ unlike the equiangular spiral. It is claimed that the Archimedean form gives better control of circularity of radiated signal at low frequencies near the cut-off.

The radiation from a spiral occurs from the region in the aperture where the currents are in phase. This is a region of circumference one wavelength. This permits another method of minimising the effect of a backing cavity: an inverted cone, concentric with the spiral, is placed below it with the cone angle chosen so that the distance from the active region to the cone surface is $\lambda/4$. This has a

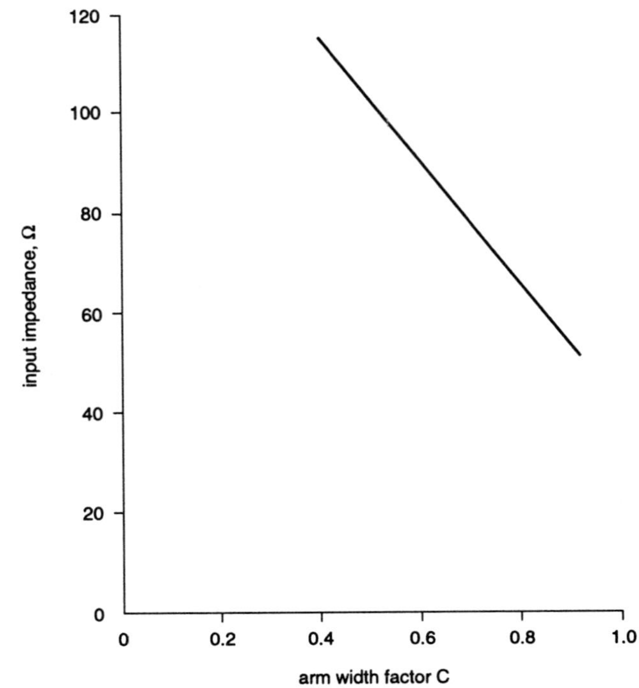

Figure 8.14 *Impedance of equiangular spiral slot antenna as a function of arm width*

secondary advantage in that a twin-line feed from the spiral can be taken through the truncated apex of the cone to the balun shielded by the cone surface.

8.2.4 *Conical spiral*

The flat spiral can be considered as a special case of the conical spiral in which the cone angle is 180°. Referring to Fig. 8.15 and using the same nomenclature as for the flat equiangular spiral

$$r_1' = r_0' \exp \phi\, a \sin \theta_0$$

$$r_2' = r_0' \exp a \sin \theta_0(\phi - \phi') = Cr_1'$$

where $C = \exp -a\phi' \sin \theta_0$
θ_0 is the half angle of the cone.

The radiation from the conical spiral is circularly polarised over a wide angle from the axis, $\theta = 0$, but the pattern is unidirectional, the back lobe decreasing as $\theta°$ decreases. The lowest usable frequency is determined by the base diameter which should be at least $3\lambda/8$. This equates to a circumference of approximately λ which was shown to be the active region of a flat spiral.

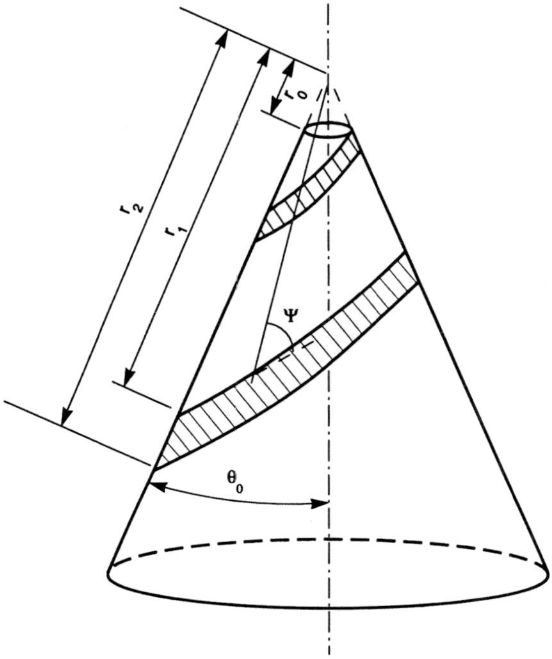

Figure 8.15 *Conical spiral*

Ramsdale and Crampton [16] have summarised the requirements for good frequency independent performance as follows:

For broad beamwidth

ψ: small, loosely wound spiral
ϕ': small, narrow arms
θ_0: optimum, cone angle 20–25°

For high front/back

θ_0: small
ϕ': optimum, typically $\geqslant 90°$

For circularity of azimuth pattern

- Improves with number of arms
- Improves with ϕ'

For extension of frequency-independent range

ψ: large to maximise track length
θ_0: large to maximise track length
ϕ': optimum arm width to maximise radiation per unit length of arm ($\geqslant 90°$)

Clearly a particular specification may require a compromise between the different factors.

So far we have only considered spirals with two arms spaced at 180°. It is also practicable to use four arms spaced at 90°. This allows different methods of feeding, enabling the antenna to radiate in different modes. Ramsdale and Crampton [16] discuss three methods and indicate the distance of the active region from the cone vertex for each mode, measured along a cone generator.

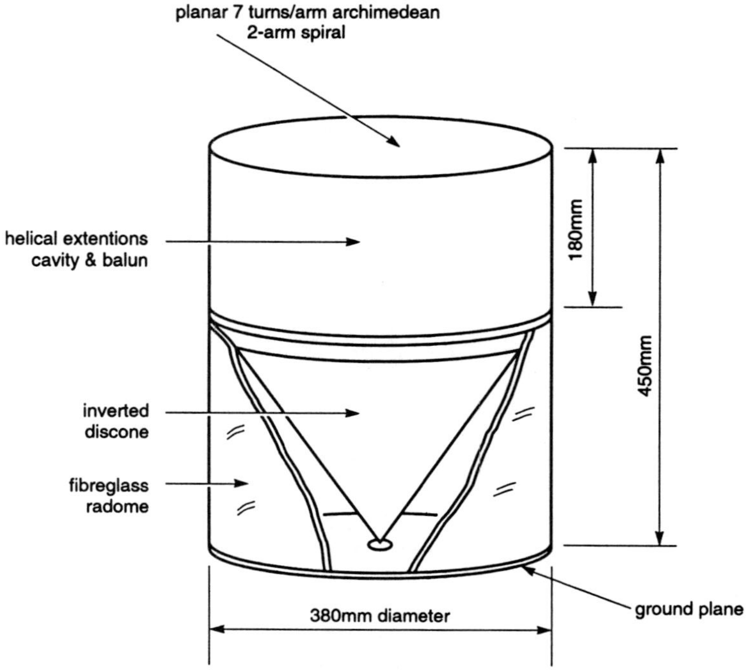

Figure 8.16 *Spiral and discone combination*

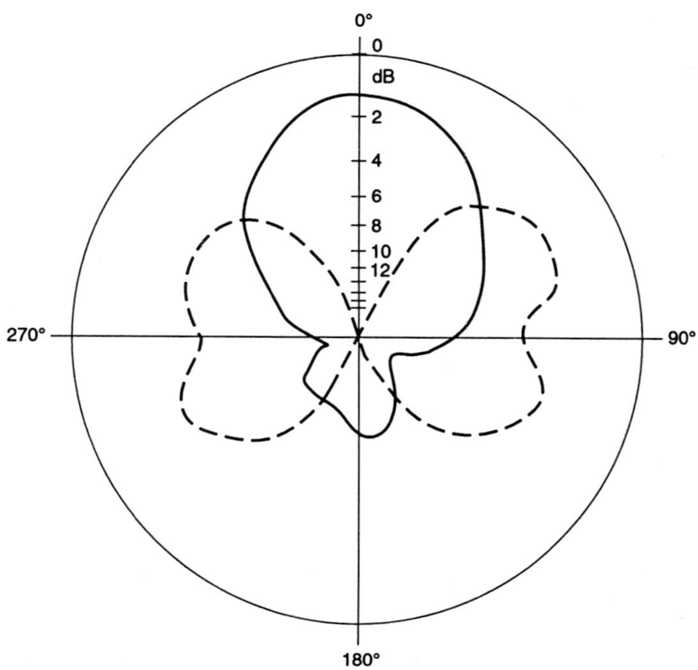

Figure 8.17 *Elevation pattern of spiral–discone combination*

Mode 1
Progressive 90° shift between arms
Active region from 0·3 λ
Mode 2
Progressive 180° shift between arms
Active region from 0·6 λ
Mode 3
As mode 1
Active region from 1·2 λ

Ancona [1] compares two- and four-arm conical spirals standing on a plane reflector and shows that in the azimuth plane the pattern of the four-arm is more circular and the ellipticity is low whereas the two-arm spiral has a high ellipticity in this plane. These antennas were both self-complementary. The four-arm was fed with a 90° progressive phase shift.

Ramsdale and Crampton experimented with end loading of the arms with resistance to reduce the lower frequency cut-off. They demonstrated that this was effective in both the two-arm and four-arm antennas in producing unidirectional patterns below the normal cut-off frequency. Impedance bandwidth was extended considerably, the 2:1 VSWR cut-off for mode 1 on a four-arm spiral being reduced to below 250 MHz for an unloaded antenna with cut-off above 600 MHz.

One point which has not been stressed so far is the critical effect of the feed region on the antenna impedance. This point is illustrated by Dyson [6].

8.3 Combination antennas

Restricted space has produced a number of ingenious antenna systems for broadband operation. One of these described by Woodman [20] combines a planar spiral and an inverted discone to give hemispherical coverage over the range 250–1500 MHz. The critical dimension in this case was diameter which could not exceed 380 mm, (about $\lambda/3$ at 250 MHz). This is too small for satisfactory operation of a planar spiral so its effective length was increased by adding helical extensions below the spiral. An Archimedean spiral was used so the arm width remains constant; the helical extensions were slightly narrower and were interleaved. For ease of manufacture they were printed on a very thin film which could be laid round a cylinder of non-conducting material and soldered to the arms of the spiral. A broadband balun was used with an impedance transformer to connect to a 50 ohm coaxial line.

A cavity of at least 300 mm would be required for a lowest frequency of 250 MHz. This space was not available so a shallower cavity filled with low-loss dielectric was used. A conical insert provides an optimum reflector, its slope being chosen so that the surface is $\lambda/4$ below the active portion of the spiral.

Below the spiral an inverted discone is mounted, the whole arrangement being as shown in Fig. 8.16. The antenna system was required to give linear polarisation at low angles and circular polarisation at high angles when mounted on a mast-head: the success of this solution is shown by the elevation radiation pattern of Fig. 8.17. The VSWR to 50 ohms for both antennas was below 2:1 from 250 to 1200 MHz rising to 3:1 up to 1600 MHz.

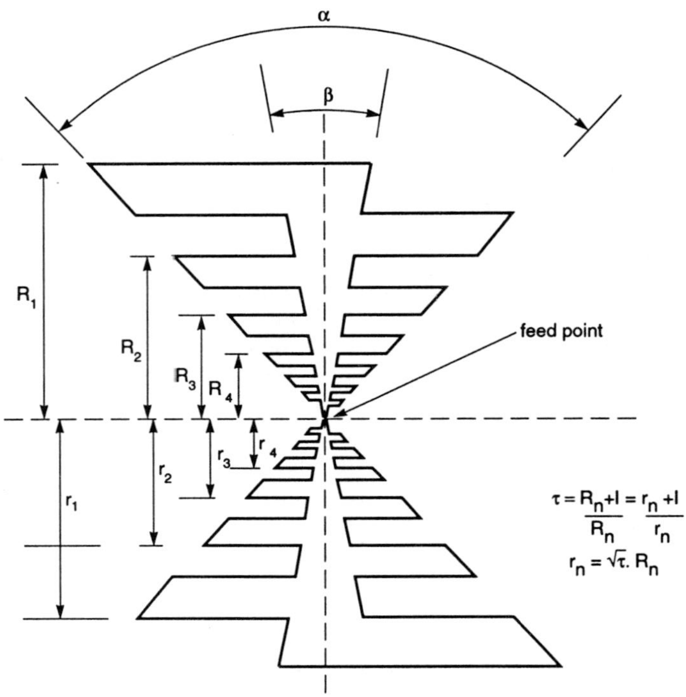

Reference 9

Figure 8.18 *Flat log-periodic antenna*

8.4 Log-periodic antennas

Unlike the spiral antennas, the log-periodic types have the same characteristics when the frequency is multiplied by a fixed factor τ. In the theoretical antenna extending from 0 to infinity there are no frequency limits but in practice the size of the elements fix upper and lower limits. Within these limits a few but not all of the possible physical arrangements retain their frequency independence. In these the dimensions are chosen so that there is little change in performance over a small range of frequency. By choosing τ appropriately the next small range of frequencies is sufficiently close to the previous set for the performance to vary only slightly in passing from one set to the next and hence over the operating range of the antenna.

Fig. 8.18 shows what might perhaps be described physically as a serrated bow-tie dipole. The two halves are fed in antiphase at their vertices. It will be seen that the antenna is defined by two angles which fix the shape of the tooth and its depth. The ratio of distances of each tooth from the plane through the vertices is $\tau = R_{n+1}/R_n$ for one side, $\tau = r_{n+1}/r_n$ for the other. The relationship between the two sides is $r_n = R_n\sqrt{\tau}$.

Some finite-length log periodic antennas have a low-frequency cut-off where the longest tooth is of the order of $\lambda/4$. Currents beyond this length of tooth die away rapidly so it is clear that the active region of the antenna decreases as the frequency increases, that is, the aperture remains substantially constant. A high-frequency cut-off occurs when the smallest tooth is of the order of $\lambda/4$.

In the example shown in Fig. 8.18 the two arms lie in a plane: the angle ψ between them is 180°. A whole range of antennas can be constructed by altering ψ: when the angle is less than about 60° a useful unidirectional beam can be obtained and such an antenna is useful for feeding a paraboloid or other shaped reflector. It must be pointed out that most of the radiation is due to the currents along the length of the teeth and the polarisation is therefore normal to the axis of the arm unlike the bow-tie antenna in which it would be parallel to the axis. The pattern in the E plane will be determined by α, β and τ whilst the pattern in the orthogonal plane will be determined by ψ.

8.4.1 *Log-periodic dipole arrays*

Folding the two arms of Fig. 8.18 so that the included angle ψ becomes zero leads to a unidirectional form of log-periodic known as the log-periodic dipole array (LPDA), Fig. 8.19. In this the teeth are reduced in width to wires and the angle β becomes zero. Because the teeth of one arm lie between adjacent teeth of the other arm it will be obvious that the alternate elements of the array require

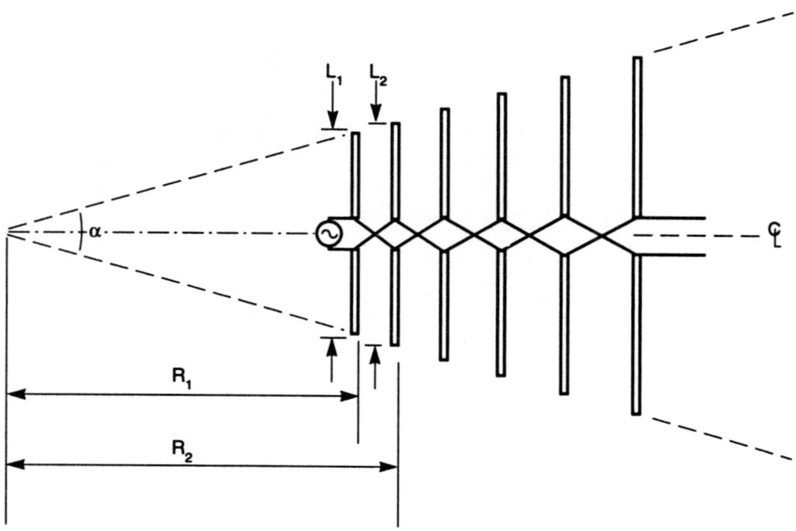

$$\tau = \frac{R_2}{R_1} = \frac{R_{n+1}}{R_n} = \frac{L_2}{L_1} = \frac{L_{n+1}}{L_n}$$

Reference 9

Figure 8.19 *Log-periodic dipole array*

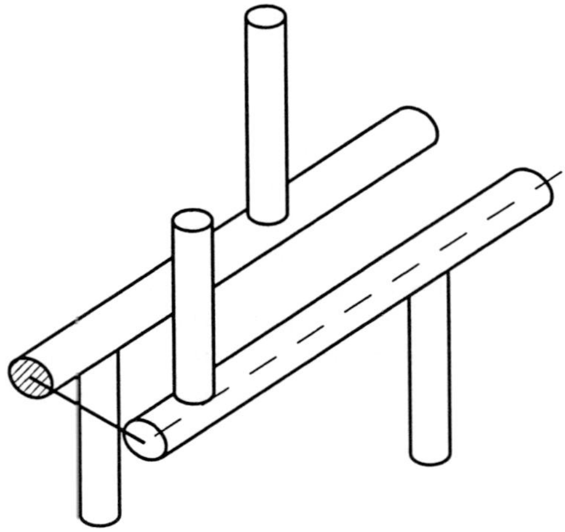

Figure 8.20 *Method of feeding LPDA*

180° phase shift between them. One convenient method of feeding is shown in Fig. 8.20. The dipole arms are attached to two parallel conductors which form a transmission line. A coaxial cable is taken through one conductor, its inner being joined to the opposite conductor at the high frequency end of the array. The twin line extends beyond the longest elements where, since there is virtually no current, the line can be short-circuited without affecting the performance of the array over its design frequency band.

This type of log-periodic antenna was first developed by Isbell [8] whose paper gives useful performance details. Smith [18] also gives many design details but one of the most useful papers is by Carrel [3]. The significant factors are τ, the ratio between adjacent element lengths and spacings, and σ, the spacing in wavelengths between one element and its next smaller neighbour.

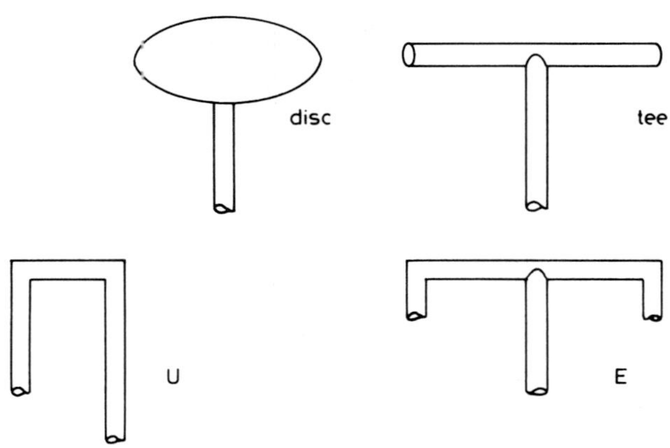

Figure 8.21 *End-loading for LPDA elements*

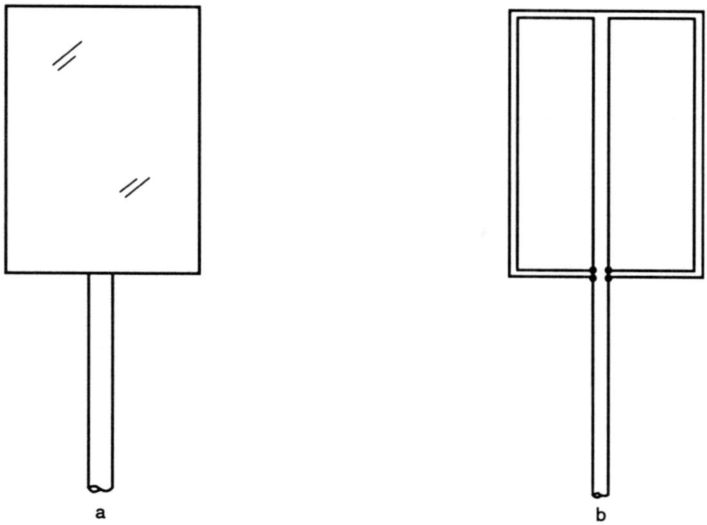

Reference 11

Figure 8.22 *Kuo's top-loaded elements*

a Printed element
b Wire element

The useful range of τ is shown to be between 0·76 and 0·98 for values of σ between 0·06 and 0·20. Gains are between 8 and 12 dB over a dipole, the latter only being achieved with half-wave elements for values of τ near the upper limit.

The mean resistance R_0 of the antenna is shown to be a function of the transmission-line characteristic impedance Z_0, which has little effect on the directivity. Since their impedance change with frequency is more rapid than with fatter elements, thinner elements give less directivity. VSWR referred to R_0 tends to rise with Z_0 and with decreasing σ. Carrel indicates that a VSWR of better than 1·4 should generally be possible over a wide frequency band.

Mayes and Carrel [14] indicate that higher gains can be achieved with arrays consisting of vee-dipoles of overall length $3\lambda/2$ or $5\lambda/2$. Straight dipoles of this length would give multi-lobed patterns. Gains of 12 and 18 dB, respectively, have been recorded. Apart from the width of such arrays the disparity between E and H plane patterns may make them less generally useful.

At the lower end of the VHF band even half-wave elements may be uncomfortably large and various attempts have been made to reduce array size. The problem is similar to that encountered with low frequency dipoles and monopoles and similar solutions have been tried by DiFonzo [5]. Forms of end-loading are shown in Fig. 8.21. Another approach is by Kuo [11] who uses loaded elements analogous to dumb-bell slots. There were tested in both solid (printed) form and as multiple wires, Fig. 8.22. Kuo reported loss in gain and directivity of about 1·5 dB, attributed to mutual coupling between adjacent

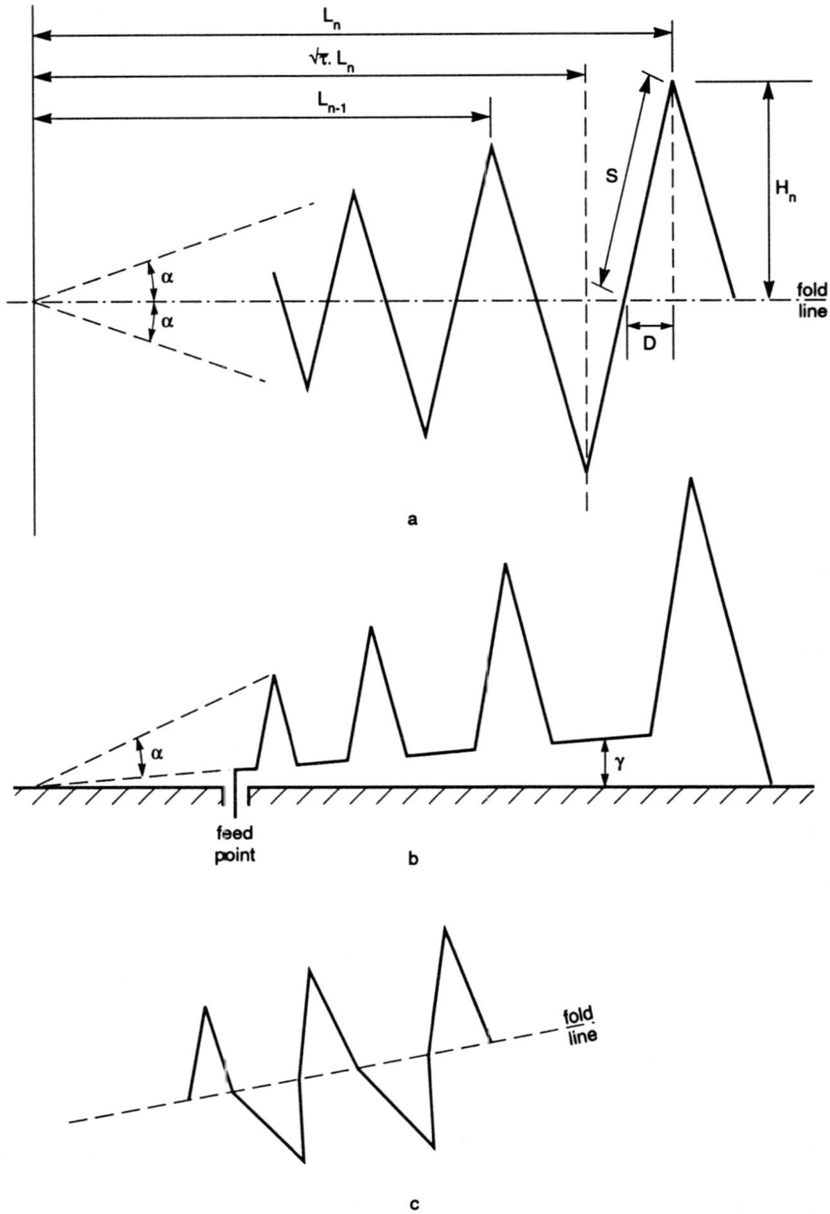

Reference 7

Figure 8.23 *Bent log-periodic zig-zag antenna*

a Basic zig-zag
b Side view
c Part perspective

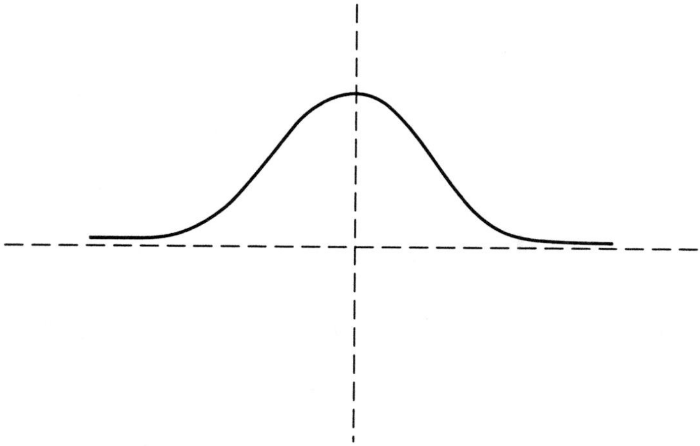

Figure 8.24 *3λ/2 dipole element of optimised shape*

elements. Size reductions of up to 35% were achieved and the frequency independent characteristics were maintained over at least a 2:1 frequency band with radiation patterns differing little from the full size antennas.

Since the only advantage in reducing the size of the dipoles is at the lower frequencies Difonzo and Kuo have both experimented with arrays with reduced-size low-frequency elements and full-size elements for the higher frequencies. Kuo shows that this results in a loss of only 1 dB compared with the full-sized antennas and the size reduction at the low frequencies is still maintained.

8.4.2 *Log-periodic monopole arrays*

Greiser and Mayes [7] developed a version of the dipole array which could be mounted over a ground plane to give a vertically polarised unidirectional array. This has been described as a bent log-periodic zig-zag (BLPZZ). The zig-zag of Fig. 8.23*a* is bent about the mid-line and one half is arranged horizontally, one vertically. A side view is shown in Fig. 8.23*b*. The horizontal portion does not radiate but provides the necessary phase shift between the vertical elements. Other methods of providing the phase shift were shown to be equally useful.

The original antennas were all constructed of uniform wire whereas strictly the wire diameter should be tapered. When this was tried there was a general improvement in the radiation patterns. Narrowing each tooth also improved the patterns.

The characteristic impedance of all the models described was of the order of 200 ohms. There is no information given on the controlling factors.

8.4.3 *LPDA with shaped dipoles*

It was noted earlier that higher gains could be achieved with vee-dipoles of overall length $3\lambda/2$ and $5\lambda/2$. An alternative approach by Landstorfer [13] uses a $3\lambda/2$ dipole element shaped as in Fig. 8.24. The spatial difference ensures that

Table 8.1 *Performance with shaped dipoles*

Frequency band (MHz)	τ	α	Gain (dB)	No. of elements
470–610	0·94	20	12·5	6
370–470	0·87	60	12·5	3

currents from the centre and outside $\lambda/2$ sections are in phase in the forward direction. As the bandwidth of a single element is smaller than that of a simple $\lambda/2$ dipole, the ratio τ has to be chosen carefully to ensure that uniform performance is achieved over a reasonable bandwidth. Nevertheless some respectable performances are quoted (Table 8.1).

If, instead of single wires, multiple wires are used for each element, higher gains are possible. Landstorfer quotes a gain of 16 dB for $\tau = 0{\cdot}94$, $\alpha = 40°$ for the band 470–610 MHz. No information is given on the impedance characteristics of these arrays.

Cheng and Liang [4] have published a method of optimising the gain of a shaped $1{\cdot}5\lambda$ element. A sinusoidal current is not assumed and the wire radius is taken into consideration. Gains of about 7 dB per element are calculated.

8.5 References

1 ANCONA, C.: 'A hemispherical coverage conical spiral antenna and its aerospace applications'. IEE Conf. Publ. 77, June 1971, pp. 43–48
2 BARBANO, N.: 'The aerodiscone antenna', *Microwave J.*, Nov. 1966, pp. 57–62
3 CARREL, R.L.: 'The design of log-periodic dipole antennas'. IRE International Conv. Lec., **9**, Pt. 1, 1961, pp. 61–75
4 CHENG, D.K., and LIANG, C.H.: 'Shaped wire antennas with maximum directivity', *Electronics Letters*, 1982, **18**, pp. 816–818
5 DIFONZO, D.F.: 'Reduced-size log-periodic antennas', *Microwave J.*, Dec. 1964, pp. 37–42
6 DYSON, J.D.: 'The characteristics and design of the conical spiral antenna', *IEEE Trans.*, 1965, **AP–13**, pp. 488–499
7 GREISER, J.W., and MAYES, P.E.: 'Vertically polarised log-periodic zig-zag antennas'. Proc. Nat. Electronics Conf., 1961, pp. 193–204
8 ISBELL, D.E.: 'Log-periodic dipole arrays' *IRE Trans.*, 1960, **AP–8**, pp. 260–267
9 JASIK, H. (Ed): 'Antenna engineering handbook' (McGraw-Hill Book Co., 1961)
10 KANDOIAN, A.G.: British Patent 578457
11 KUO, S.C.: 'Size-reduced log-periodic dipole array antenna', *Microwave J.*, Dec. 1972, pp. 27–33
12 LAMBERTS, K.: 'Rundstrahlantennen mit horizontaler Polarisationsrichtung für breite Frequenzbänder im Meter — und Dezimeterwellengebiet', *Frequenz*, 1951, **5**, pp. 177–185
13 LANDSTORFER, F.M.: 'New developments in VHF/UHF antennas', IEE Conf. Publ. 169, Nov. 1978, pp. 132–141
14 MAYES, P.E., and CARREL, R.L.: 'Logarithmically-periodic resonant-V arrays'. IRE Wescon Record., 1961
15 NAIL, J.J.: 'Designing discone antennas', *Electronics*, Aug. 1953, pp. 167–169
16 RAMSDALE, P.A., and CRAMPTON, P.W.: 'Low frequency performance of hemispherical coverage conical log-spiral antennas'. IEE Conf. Publ. 195, Apr. 1981, pp. 298–302

17 RUMSEY, V.: 'Frequency independent antennas' (Academic Press, London, 1966)
18 SMITH, C.E.: 'Log-periodic design handbook' (Smith Electronics Inc., Cleveland, Ohio, 1966)
19 WOLFF, E.A.: 'Antenna analysis' (John Wiley & Sons, 1966)
20 WOODMAN, K.F.: 'Hemi-isotropic broadband UHF antenna system'. IEE Conf. Publ. 169, Nov. 1978, pp. 142–146

Chapter 9

Electrically-small antennas

Few engineers would choose to use an electrically-small antenna on purely electrical grounds, except for the important characteristic of selectivity which has been touched on in Section 6.7 in connection with the short notch. There are, however, many reasons for using small antennas, some of which are listed in Table 9.1.

There is no consensus on the definition of 'electrically small': some workers adopt a maximum dimension from the feed point of $\lambda/30$ but Schelkunoff and Friis [9] suggest $\lambda/8$. Now it is perfectly possible to achieve large instantaneous bandwidths with a top-loaded $\lambda/8$ monopole or with Josephson's open folded monopole (see Chapter 3) so this seems rather too large an upper limit. The $\lambda/30$ criterion was probably chosen because it permits the use of two approximations: constant current in the small loop and linear current in the short monopole or dipole. In general the characteristic expected of an electrically-short antenna is low radiation resistance and large reactance; hence very small instantaneous bandwidth with respect to the impedance of normal radio equipment. The short notch (Section 6.7) has a high radiation resistance at the mouth but may nevertheless have a reasonably high Q.

Because of the normally-low radiation resistance of small antennas, losses which may be negligible in resonant antennas may dominate the total resistance of a small antenna. More care must therefore be taken in the choice of materials than would otherwise be necessary. However, the presence of some loss resistance may be an advantage if the antenna has to be well matched to a transmitter. Without this loss it could be impossible to achieve a wide enough bandwidth for satisfactory operation of the radio system.

Four basic types of small antenna are discussed in this chapter. They are the dipole, the monopole, the loop and the notch. There do not appear to be any practical applications for the short slot antenna so this will not be discussed.

Operators of radio equipment, particularly on land vehicles, have sometimes objected to the use of very short antennas where these do not project above local obstacles. This is to consider the obstacle in terms of the size of the antenna instead of in terms of wavelength which would be correct. It is not true that 'if you can't see the antenna you can't communicate'. Unless there is a significant difference in the current distribution due to two different antennas on the adjacent structure, there will be no difference in their radiation patterns. Measurements on a number of vehicles with both resonant and electrically-small antennas have demonstrated this conclusively but it may still be necessary to convince the user.

Table 9.1 *Applications for electrically small antennas*

Reason	*Application*
Mechanical strength, i.e. windage	High speed aircraft Warships
Obstacle clearance on mobiles	Road vehicles: { garages, tunnels } Trains: rail gauge Aircraft: { ground clearance, rotor blades }
Danger to servicing staff	Aircraft
Icing	Aircraft, ships
Concealment of vehicle	Military ground vehicles
Concealment of function	Police, security forces
Size of tracked species	Radio tracking of animals
Visibility	Sensitive environments

9.1 The short dipole

In our range of frequencies the main application of the short dipole is probably as a field probe. The main properties required are:

- Minimum perturbation of the field being measured
- Response to a single polarisation.

These are achieved by making the dipole as short as possible consistent with adequate reception and maintaining extremely good balance between the two arms. It is of course also possible to use the probe to create a known field strength.

9.1.1 *Radiation resistance*

The input resistance R_{in} for a centre-fed dipole which is infinitely thin is

$$R_{in} = 20k^2(2l_{\text{eff}})^2$$

where l = length of one arm

$$k = 2\pi/\lambda$$

For the short antenna, where $l < \lambda/30$, the current distribution is for all practical purposes triangular, hence $l_{eff} = \frac{1}{2}l$. Then

$$R_{in} = 20k^2l^2$$

If the antenna is top-loaded so that the current is uniform, $l_{eff} = l$ and

$$R_{in} = 80k^2l^2$$

The advantage of top-loading for increasing the radiation resistance is clearly seen. If the dipole has a small but finite radius the radiation resistance is reduced.

9.1.2 *Input reactance*

The input reactance of a cylindrical dipole of radius a is given by

$$X_{in} = -j120\ (\ln 2l/a - 1)\ \cot kl$$

For short dipoles the input impedance is therefore

$$Z_{in} = 20(kl)^2 - j120(kl)^{-1}(\ln 2l/a - 1)$$

9.1.3 *Loss resistance*

As the radiation resistance of a short dipole is so low, the ohmic losses which may be negligible in a half-wave dipole have to be taken into consideration. Dummer and Blackband [4] give a convenient formula for the RF resistance of a straight wire. Converted into SI units it becomes

$$R_L = \frac{63{\cdot}253 \times 10^{-6}\sqrt{f}\sqrt{\rho}}{d} + \frac{3175{\cdot}155 \times 10^{-6}}{d^2}\rho \text{ ohms/metre}$$

where d = diameter of wire, mm
f = frequency, Hz
ρ = resistivity, μ ohm cm

Some appropriate values for ρ are:

Copper	1·72
Pure aluminium	2·82
Aluminium alloys	4

Although the loss resistance per unit length remains constant for constant cross-section, the power loss will vary if the current is not constant. An equivalent total loss resistance R_l at the feed-point is required. For an electrically short antenna, $l/\lambda < 1/30$, the current distribution may be considered as triangular. From this,

$$R_l = \tfrac{2}{3}R_L l \text{ where } l \text{ is the length of one arm of the dipole.}$$

The full analysis is given in Appendix 1.

9.1.4 *Pick-up of a short dipole*

The effective area A of a perfectly conducting short dipole is given by $A = 3\lambda^2/8\pi$. If the dipole is matched to its load all the power incident on the antenna can be collected. Under these circumstances the power available at the antenna terminals is

$$W = PA$$

$$= \frac{E^2 3\lambda^2}{120\pi \times 8\pi} \text{ watts}$$

where the power density P of the incident field is $E^2/120\pi$ watts/m^2, E being the field strength in volts/m.

If the dipole is not perfectly conducting, the effective area becomes

$$A = \frac{3\lambda^2}{8\pi}\left(1 + \frac{R_L\lambda^2}{120\pi^2 l}\right)^{-1}$$

Without some form of matching, either active or passive, the short dipole will not be matched to its load and the pick-up may be very small indeed.

9.2 The short monopole

The formulae derived for the short dipole will generally apply to a short monopole on an infinite, lossless ground plane. Thus the radiation resistance for a short monopole of height l is $10k^2l^2$ if the antenna is unloaded and $40k^2l^2$ if the antenna is top-loaded to produce uniform current in the vertical element. This condition only applies if the top-loading can be considered non-radiating and the monopole is less than $\lambda/30$ high. For longer elements the current is not quite constant, Fig. 9.1. If these conditions do not apply we have to turn to more general formulae for top-loaded antennas. These abound but examination of them shows that they do not apply over the same range of antenna size.

9.2.1 *The transmission-line antenna*

This is a form of top-loaded antenna in which the top-loading is a rod parallel to the ground plane forming with it a transmission line. Several different forms can be distinguished, two of the more important ones being shown in Fig. 9.2. These can be described as (*a*) the bent monopole and (*b*) the shunt-fed bent monopole.

9.2.1.1 *The bent monopole*

The radiation resistance of the very short monopole is $10k^2h^2$. When $h>\lambda/30$ the current distribution can no longer be considered as linear but as sinusoidal. The

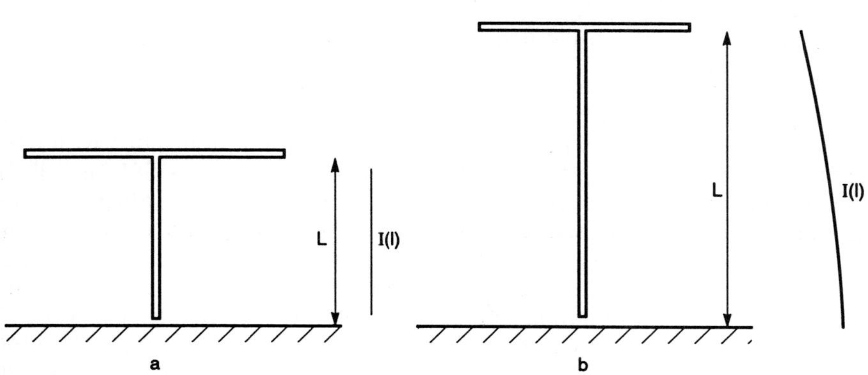

Figure 9.1 *Current distribution on short loaded antennas*

a $L<\lambda/30$
b $L\geqslant\lambda/30$

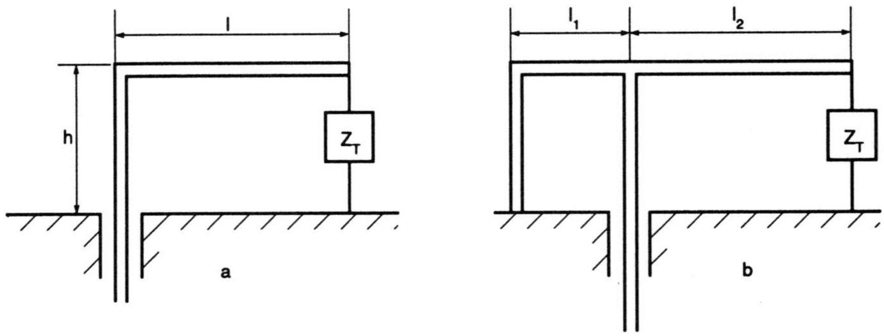

Figure 9.2 *Transmission line antennas L & M*

a Bent monopole
b Shunt-fed bent monopole

radiation resistance is given by a formula due to Laport [7] reproduced by Belrose [1]. According to this,

$$R_v = 0{\cdot}01215\ A^2 \text{ ohms}$$

For a straight vertical antenna with no top-loading, assuming sinusoidal current distribution,

$$A = \frac{180}{\pi}\,\frac{1-\cos G_A}{\sin G_A}$$

Where G_A is the effective electrical height of the antenna.

For a top-loaded antenna as in Fig. 9.2*a*, where Z_T is an open circuit,

$$A = \frac{180}{\pi}\,\frac{\cos(G_A - G_V) - \cos G_A}{\sin G_A}$$

where G_A = electrical height of the whole antenna

G_V = equivalent electrical height of the vertical portion of the antenna

Fig. 9.3 shows the resistance of a radiator, resonated at 30 MHz where $h = 0{\cdot}045\lambda$, compared with that of the same vertical element without top-loading. Fig. 9.4 gives radiation resistances for a radiator of height h with a top element, for h/λ between 0·01 and 0·125. The total length is $\lambda/4$; as will be shown later this is not always the optimum length. A number of other formulae have been tested including those due to Wolff [13], Burton [2], Gouillou [5] and Wanselow and Milligan [12]. All of them give figures which are much too large and these formulae are therefore not recommended. All the equations, including Laport's, assume that radiation from the top-loading is negligible. This is rarely the case. The marker antenna, Fig. 9.5, used in the aircraft Instrument Landing System (ILS) is mounted below an aircraft fuselage to give horizontally polarized radiation downwards. With a dimension h of only 0·01875 λ the measured gain downwards was approximately −21 dB compared with a free space half-wave dipole.

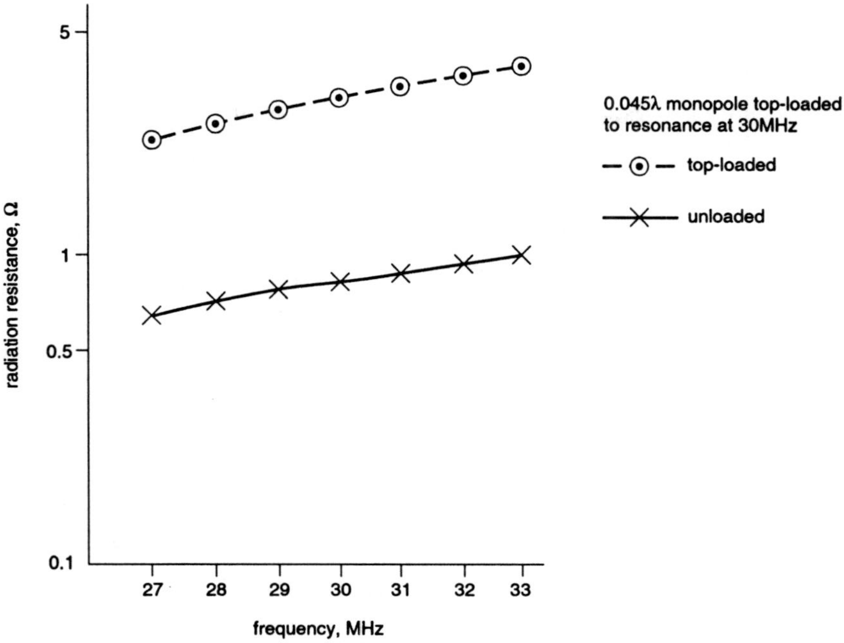

Figure 9.3 *Effect of top-loading on radiation resistance of short monopole*

The feed-point reactance of the short bent monopole can be calculated in one of two ways. In the first of these the vertical portion is considered to have a constant $Z_0^v = 60\ (\ln h/a - 1)$ where a is the radius and h the height.

If the reactance due to top-loading is X, then the feed-point reactance is

$$X_b = Z_0^v \frac{X \cos G_v + jZ_0^v \sin G_v}{Z_0 \cos G_v + jX \sin G_v}$$

where G_V is the electrical length of the vertical portion.

Where the top loading is in the form of a transmission line as in Fig. 9.2*a*,

$$X = -jZ_0^T \cot G_T$$

where Z_0^T is the characteristic impedance and G_T the electrical length of the top section.

At resonance,

$$X \cos G_V = -jZ_0^v \sin G_v$$

hence

$$G_T = \cot^{-1} \frac{Z_0^v}{Z_0^T} \tan G_v$$

It should be noted that $G_V + G_T$ does not equal G_A unless $Z_0^T = Z_0^V$. G_A is obtained by equating the input reactance of the loaded antenna to the reactance which would be seen if the antenna were simply a vertical monopole:

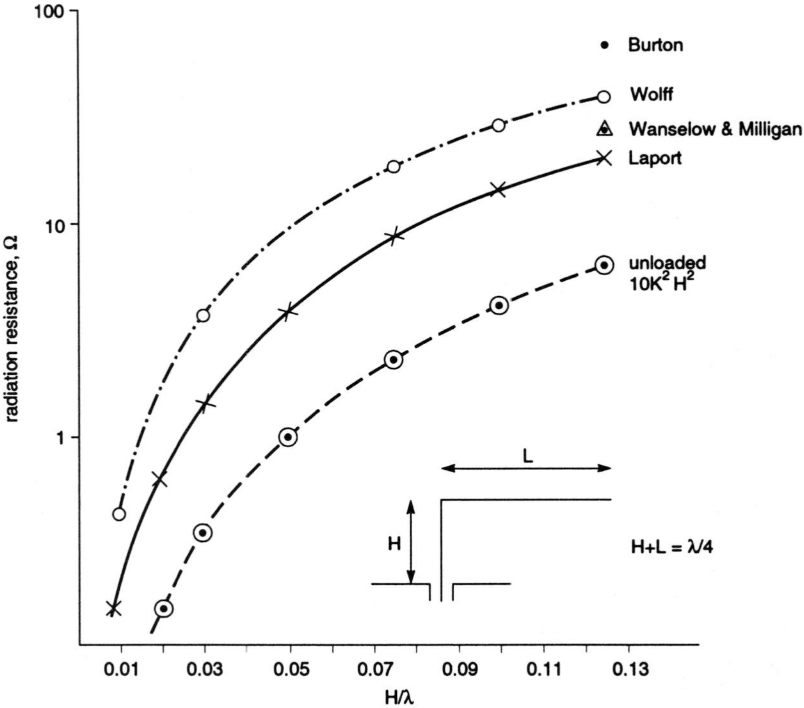

Figure 9.4 *Radiation resistance of top-loaded antenna as a function of h/λ*

$$Z_0^v \frac{X \cos G_v + jZ_0^v \sin G_v}{Z_0 \cos G_v + jX \sin G_v} = -jZ_0^v \cot G_A$$

which reduces to

$$G_A = \cot^{-1} \frac{(Z_0^T/Z_0^v) \cot G_T - \tan G_v}{1 + (Z_0^T/Z_0^v) \tan G_v}$$

and X_b becomes

$$-jZ_0^v \frac{(Z_0^T/Z_0^v) \cot G_T - \tan G_v}{1 + (Z_0^T/Z_0^v) \tan G_v}$$

The alternative method is to calculate the reactance of the vertical portion from the inductance of a straight rod:

$$L = 0{\cdot}02l[\ln 4l/d - 1]\ \mu\text{H}$$

where l = length, mm
d = diameter, mm

X for the top element is calculated as before and the base impedance then becomes

$$X_b = j^{\omega L} - jZ_0^T \cot G_T$$

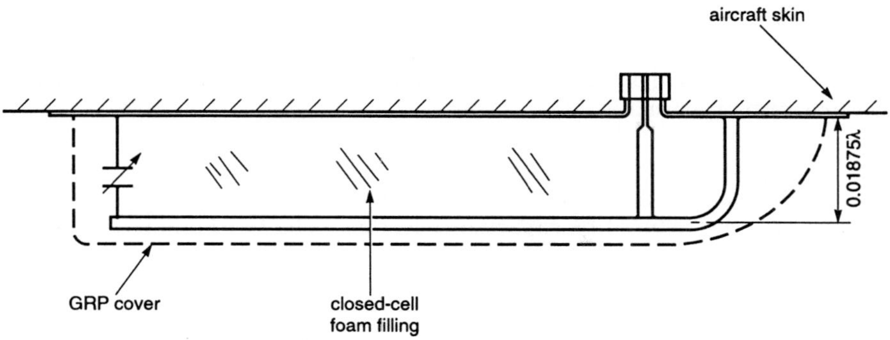

Figure 9.5 *ILS marker antenna*

Both methods were used in calculating the parameters of the antenna of Fig. 9.3. The differences were small: Fig. 9.6 compares the lengths required for resonance calculated over a 3:1 frequency range. In this instance $Z_0^v \simeq 197$ and $Z_0^T = 300$ and it will be seen that the overall length required for resonance is greater than 90°.

For antennas of finite thickness G_V will be slightly greater than the figure given by the physical height. For quite small diameters, $360D/\lambda = 0{\cdot}1$, the increase is about 5% for $G_V = 90°$, increasing to 20° at $360D/\lambda = 20$.

On an infinite ground plane the radiation resistance of these short antennas is still very low compared with 50 ohms. Before adding to the complexity of the

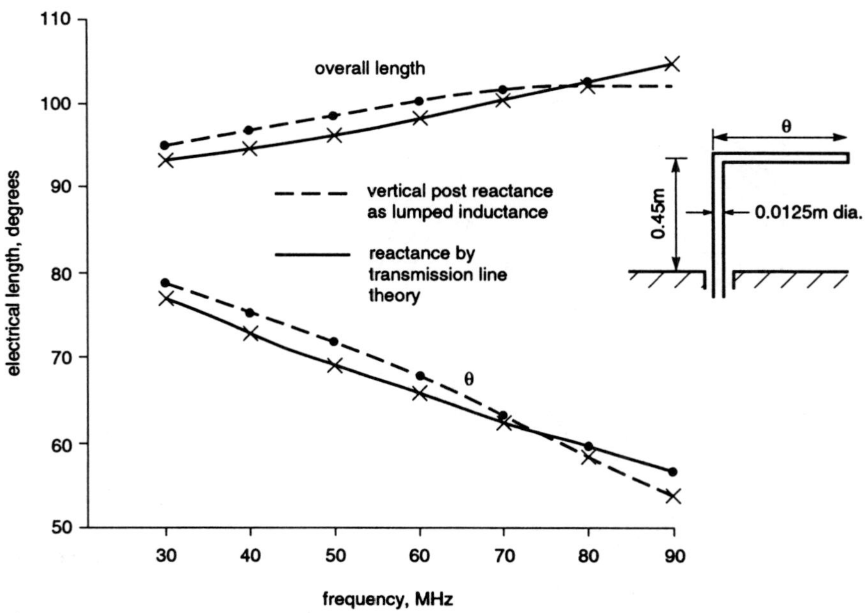

Figure 9.6 *Length of top-loading required for resonance of antenna of Fig. 9.3*

antenna it may be worthwhile to consider the conditions under which it will be used. Fig. 9.7 shows the effect of ground plane size: if the antenna were being used on a small vehicle for example the impedance might well be satisfactory without further matching.

9.2.1.2 *The shunt-fed bent monopole*

Fig. 9.5 shows a typical form of this antenna in which the length of the top section has been reduced by capacitance loading. The top-loading reactance X then becomes

$$X = jZ_0^T \left\{ \frac{Z_0^T \sin G_T - X_c \cos G_T}{Z_0^T \cos G_T + X_c \sin G_T} \right\}$$

If the reactance due to C is converted to an additional length of line θ then

$$X_c = -jZ_0^T \cot \theta$$

and

$$X = -Z_0^T \cot (G_T + \theta)$$

One method of treating the shunt antenna is to consider it as a tee-match monopole, Fig. 9.8, in which the upper portion of the monopole has been bent over. The equivalent circuit is then as shown in Fig. 9.9 where Z_r is the base

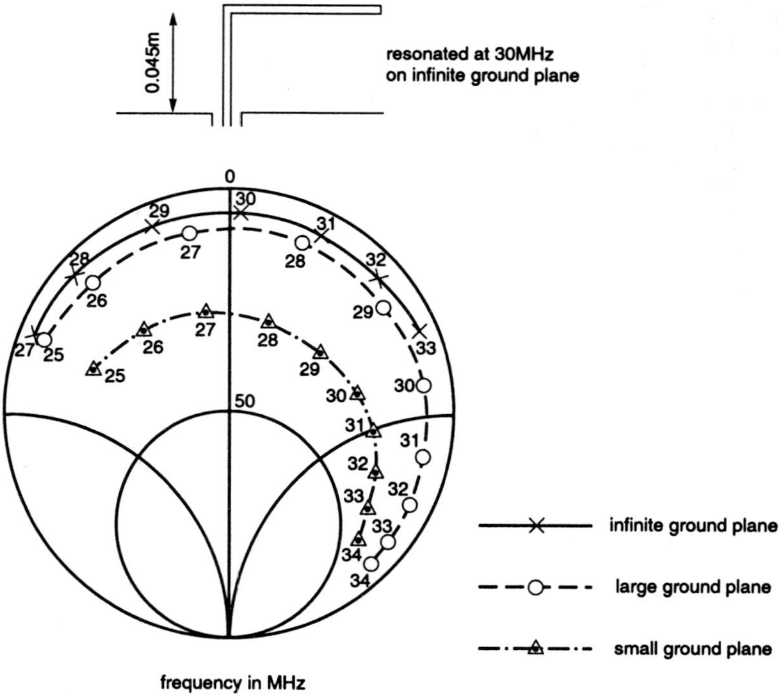

Figure 9.7 *Effect of ground plane size on impedance*

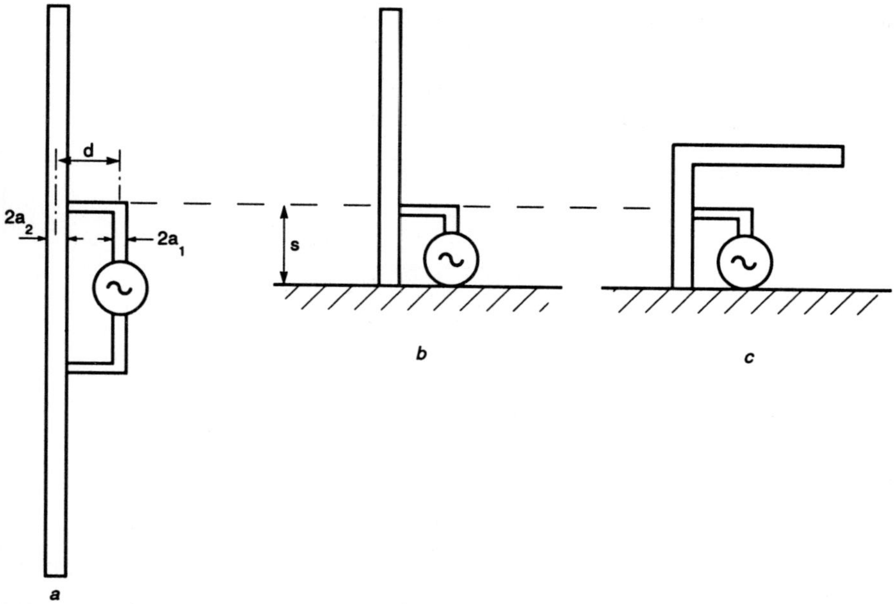

Figure 9.8 *Tee-match monopole*

a Tee-match dipole
b Tee-match monopole
c Tee-match or shunt-fed bent monopole

impedance of the bent monopole, derived as in Section 9.2.1.1 and Z_f is the reactance due to the short-circuited transmission line formed by the parallel elements. Clearly,

$$Z_f = jZ_0 \tan ks$$

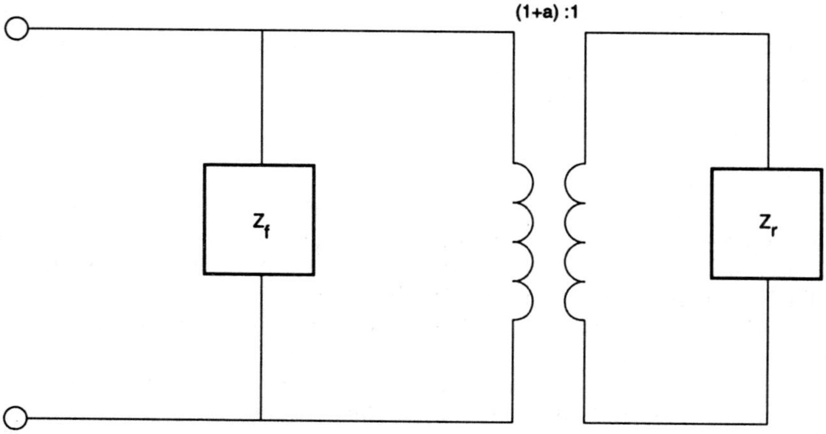

Figure 9.9 *Equivalent circuit of Tee-match monopole*

where Z_0 is the characteristic impedance of this line.

Then

$$Z_{in}=\frac{(1+a)^2 Z_r Z_f}{Z_f+(1+a)^2 Z_r}$$

The 'current division factor' a is given by the equation

$$a=\frac{\cosh^{-1}\dfrac{v^2-u^2+1}{2v}}{\cosh^{-1}\dfrac{v^2+u^2-1}{2uv}}\approx\frac{\ln v}{\ln v-\ln u}$$

$$v=d/a_1 \qquad u=a_2/a_1$$

If, as is often the case, $u=1$ then

$$a=1$$

and

$$Z_{in}=\frac{4Z_r Z_f}{Z_f+4Z_r}$$

It is in fact a comparatively simple process to determine the tapping point experimentally. This may prove easier than calculation if engineering of the antenna introduces physical discontinuities which would make the calculation more complex.

The use of capacitance loading provides easy adjustment if the antenna has to be mounted on finite ground planes differing in size and shape. It also allows the antenna to be tuned over a fairly wide range of frequencies. The instantaneous bandwidth will always be small: the antenna in Fig. 9.5 with a height of 0·01875λ and length of 0·0625λ had a bandwidth of 0·2% between 5 to 1 VWSR points. Care must therefore be taken in choosing a capacitor with an adequate temperature coefficient for the operational use of the antenna.

It is important to ensure a low resistance between the base of the grounded section and the outer of the connector at the feed-point. Even if this is achieved by using a base plate to the antenna with preferably welded joints, it is essential that the base plate itself be well grounded to the ground plane. Failure to follow this rule may result in significant mismatch, particularly with capacitance-loaded antennas.

9.2.1.3 *The hula-hoop antenna*

This is a variant of the shunt-fed bent monopole in which the top element is wound up into a near-circle or polygon, Fig. 9.10. One difference between the hula-hoop and the straight antenna is that there is no upward radiation from the former whereas the contribution from the latter is small but not negligible. There will, therefore, be slight differences in the radiation resistance for the same height and length of top element. Belrose [1] indicates that with careful attention to conductor and ground plane losses an open ended antenna of the height of Fig. 9.5 could have a radiation efficiency of 68%.

A variant of the hula-hoop was designed for mounting on a vehicle, actually a fire-engine, on which there was no convenient ground plane and where the

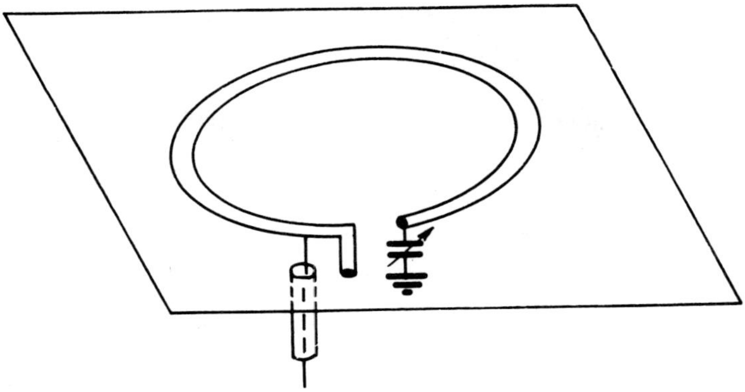

Figure 9.10 *Hula-hoop antenna*

overall height of the vehicle was limited. It took the form of a double hula-hoop, Fig. 9.11, used as a free standing antenna. In theory the feed should be balanced but no problems were found with the arrangement shown so long as the down lead was choked to reduce cable currents. This particular antenna was designed for 83 MHz and had a mean circumference of $0{\cdot}21\lambda$, a height of $0{\cdot}028\lambda$ and an element diameter of $0{\cdot}0028\lambda$.

9.2.2 *Top-loaded folded monopole*

In this variant of the short monopole the top-loading is in the form of a plate, usually but not necessarily circular. A typical antenna is shown in Fig. 9.12. Laport's formula for the top-loaded monopole applies here but there are problems in calculating the reactance due to the top-loading. It is clear, however, that for a short antenna the base radiation resistance will be low. Using a folded monopole has several advantages:

(*a*) The grounded arm provides a physical support for the top plate.
(*b*) A grounded antenna provides lightning protection for the associated radio equipment.
(*c*) By altering the ratio of diameter of the vertical elements different transformation ratios can be achieved.

Experimental work by Seeley [10] demonstrates that the spacing S between the vertical elements has an effect on the resonant frequency at the second resonance, Fig. 9.13. He also showed that the resistance at resonance was practically independent of S for a given ratio Df/Dd. This work was done at frequencies between 200 and 300 MHz. A bandwidth at 2:1 VSWR to 50 ohms ($BW = [(f_{max} - f_{min})/f_{min}] \times 100$) of 13% was achieved with the following dimensions at f_{min}:

$$D = 0{\cdot}1182\lambda$$

$$S = 0{\cdot}0537\lambda$$

$$h = 0{\cdot}09\lambda$$

$$Df/Dd = 1$$

$$Dd/h = 0{\cdot}028$$

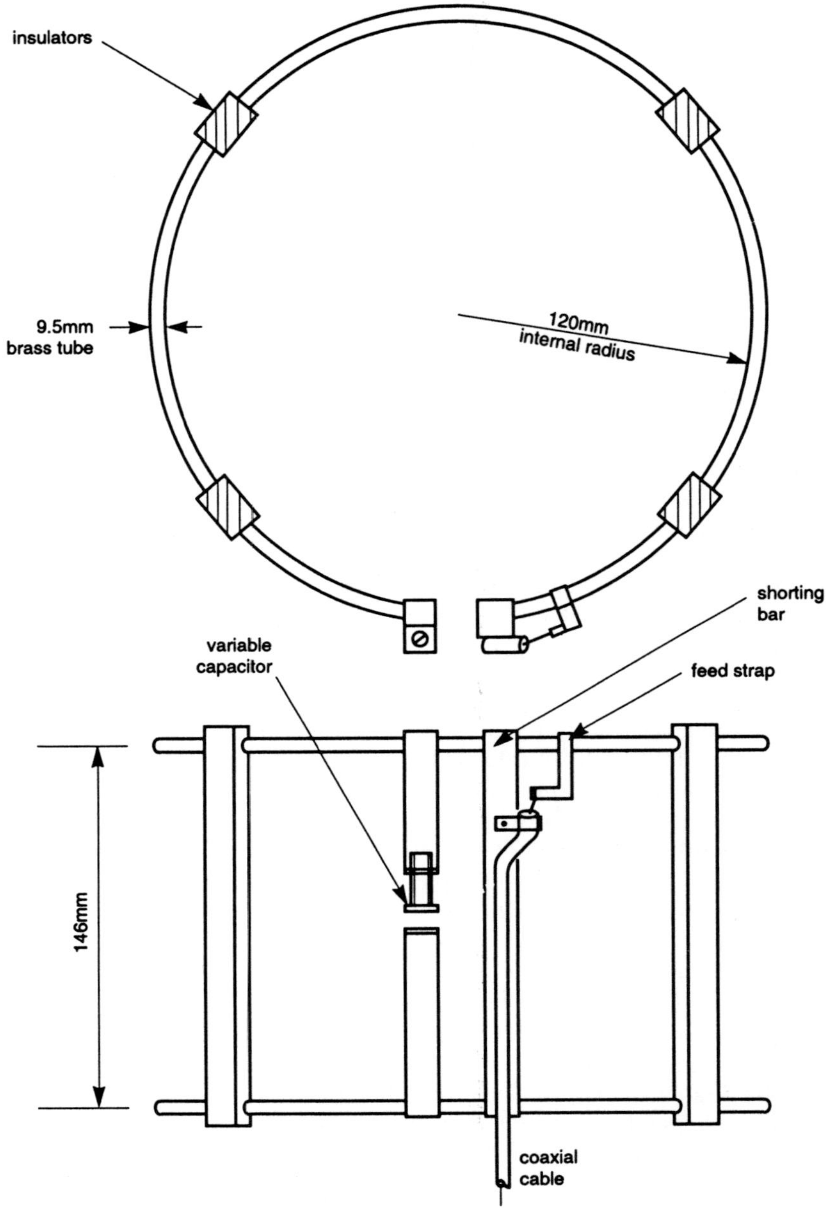

Figure 9.11 *Double hula-hoop*

The azimuth radiation pattern of this type of antenna is essentially circular for 'small' top plates. There may be some pattern distortion if the top plate is large compared with the antenna height and is not central with respect to the folded monopole but there are no guidelines available.

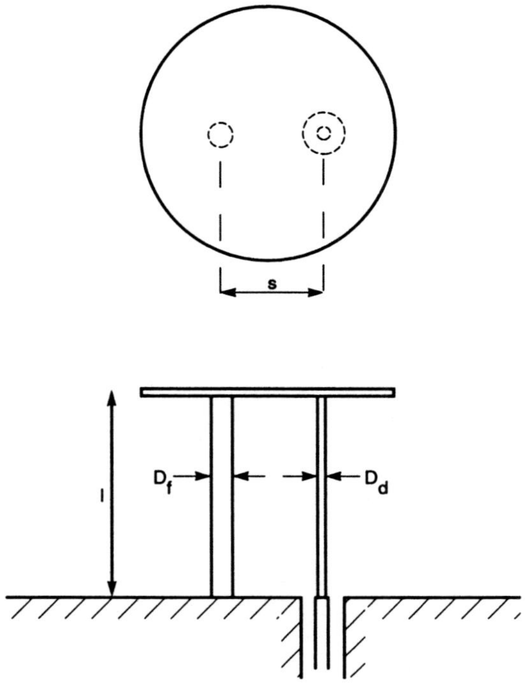

Reference 10

Figure 9.12 *Top-loaded folded monopole*

Another antenna with a 10% BW for 1·5 VSWR between 1·0 and 1·1 GHz had the following dimensions at f_{min}:

$D = 0{\cdot}13\lambda$

$S = 0{\cdot}05\lambda$

$h = 0{\cdot}0733\lambda$

Df/Dd = not given

Dd/h = not given

9.2.3 *Top-loaded sleeve monopole*

In an attempt to provide a robust low profile VHF antenna for armoured fighting vehicles, Newman and Milligan [8] used a disc-loaded sleeve monopole, Fig. 9.14. The sleeve was a 51 mm (2 in) diameter steel tube 610 mm (24 in) high which was surmounted by a steel disc of diameter 406·4 mm (16 in) isolated by a fibre-glass bush. The disc was connected to the inner of a coaxial cable inside the steel tube. A number of fixed matching elements were incorporated both in the antenna and at the end of the coaxial cable and these were followed by a resistive pad which reduced an 8:1 VSWR to 3:1.

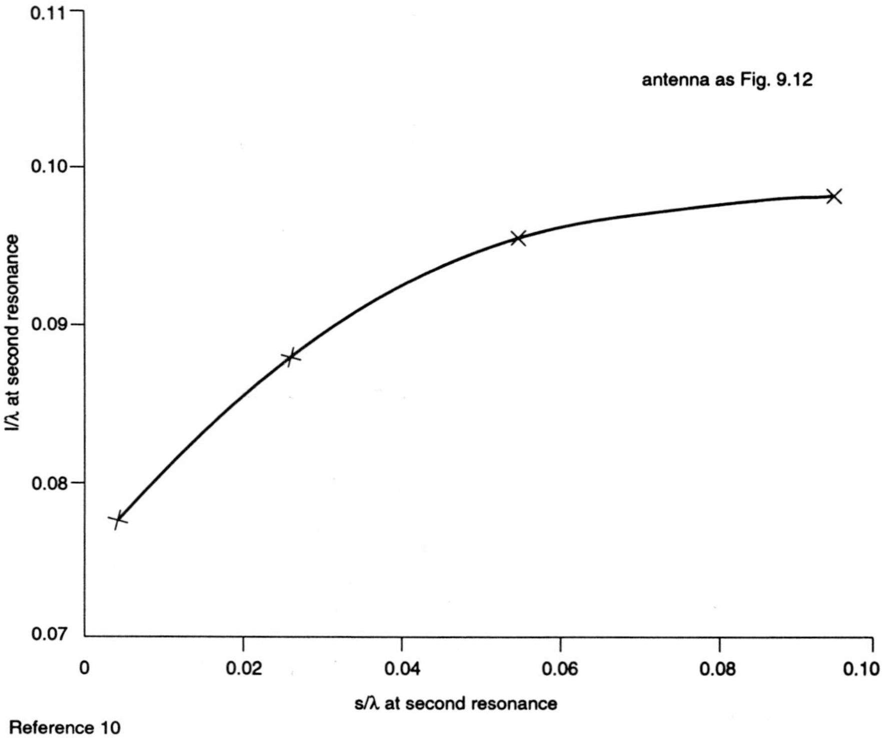

Figure 9.13 *Effect of spacing on resonant length*

With this arrangement it was possible to cover the band 30–88 MHz with a VSWR better than 3:1 to 50 ohms without any tuning but the antenna efficiency proved to be rather low. No figures were given. A sleeve may have been chosen instead of a folded monopole for robustness but it could not provide the step-up which folding can give nor the flexibility in design.

Assuming the top-loading produces resonance, the base radiation resistance would be 5·58 ohms. At the top of the sleeve the resistance would be 6·5 ohms.

9.2.4 *Short monopole with top-loading and inductive loading*

Another very short antenna with a capacitive top and switched series inductance has been described by Cooper [3]. It is shown in Fig. 9.15. Manufactured on a printed circuit board it covers the band 30–100 MHz in a series of narrow bands selected by switching the six series inductances. In this version miniature relays were used for switching: subsequent versions used *p–i–n* diodes and may operate to higher frequencies. With a height of only 0·039λ at 30 MHz this antenna has a low base radiation resistance so a step-up transformer is required. Control of switching may be manual or by a remote logic convertor using frequency information from the radio equipment so the antenna is tuned in the radio-silence mode, i.e. without the need for RF transmission. The power handling of this arrangement is determined by the current that the switches can

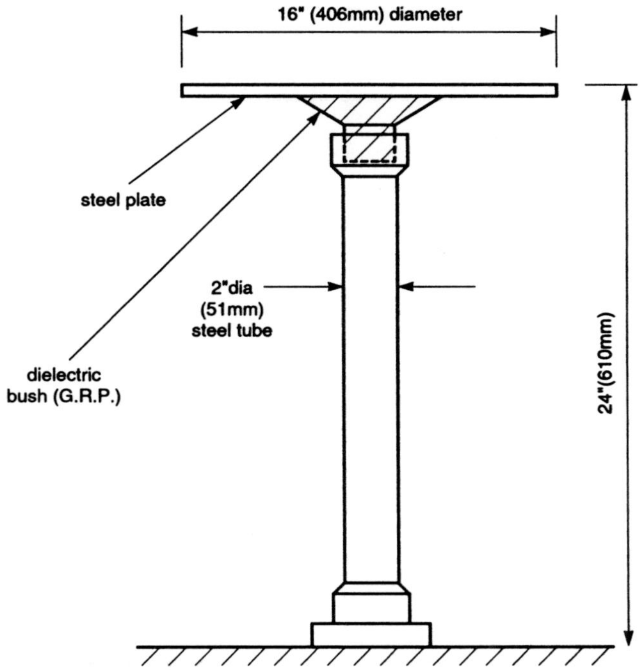

Reference 8

Figure 9.14 *Disc-loaded sleeve monopole*

handle since any closed switch takes the whole antenna current. Nevertheless this type of antenna will handle the moderate powers permitted for mobile communications equipment in this frequency band.

The use of series inductors which are printed probably increases the loss resistance compared with some of the other short monopoles and hence decreases the radiation efficiency. Whether this is important depends on the role for which the antenna is intended; this arrangement appears to be adequate for fairly short range line-of-sight operation. This point needs to be emphasised: to reject an antenna on grounds of low gain without considering the system parameters may be to seek the impossible — a short, broadband low loss antenna. The theory of short antennas clearly shows the problems.

9.3 The small loop

Much of the theoretical work on the radiation patterns of loops assumes constant current. This is true only for very small thin loops. Storer [11] showed that whilst the real component of the current is constant for $kb<0{\cdot}2$ where b is the radius of the loop, the imaginary component is not constant. However, for all practical purposes the radiation pattern in the plane of loops up to $kb=0{\cdot}2$ is circular. The patterns of larger loops are discussed in Chapter 4.

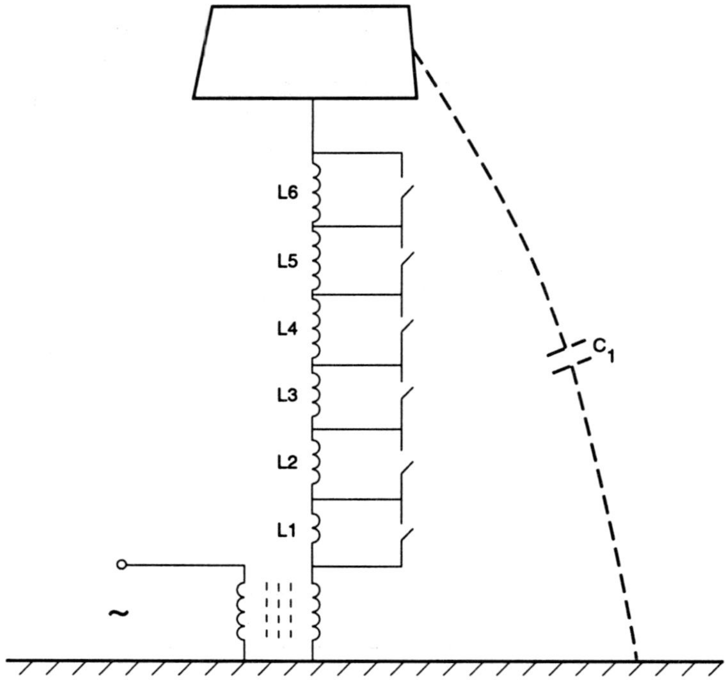

Figure 9.15 *Monopole with capacitance loading and switched series inductance*

The radiation resistance of an infinitely thin small loop is $R_r = 20\pi^2(kb)^4$. For loops of finite thickness this is not exactly true. Storer [11] tabulates resistance for different ratios of b/a where a is the wire radius but his values are very much at variance with the approximate formula. Furthermore, there does not appear to be any trend in the values as b/a is increased so his results need to be treated with some suspicion. There is some evidence from the paper that all the values of R and X are too high by a factor of π. For non-circular small loops $R = 20\ k^4A^2$ where A is the area of the loop.

The inductance of a small loop is given by Schelkunoff and Friis [9] as $L = \mu b \ln b/a$ where μ is the permeability, unity in the case of non-ferrous loops. From this the reactance $X = 300\ kb \ln b/a$. Storer has a slightly different formula which should probably be $X = 120\ \pi\ kb(\ln 8b/a - 2)$. This gives values about 30% higher.

For real antennas the loss resistance is likely to be much higher than the radiation resistance so precise values of the latter are rather academic. This means that the radiation efficiency of a small loop is low but it also means that matching is possible using a two-capacitor system as shown in Fig. 9.16. It will be found by circuit analysis that C_1 needs to be a very high quality capacitor whilst C_2, which is in series with a much higher resistance, is not so critical. Fortunately, high Q variable capacitors are readily available so there is no difficulty in providing an efficient matching system. The setting accuracy

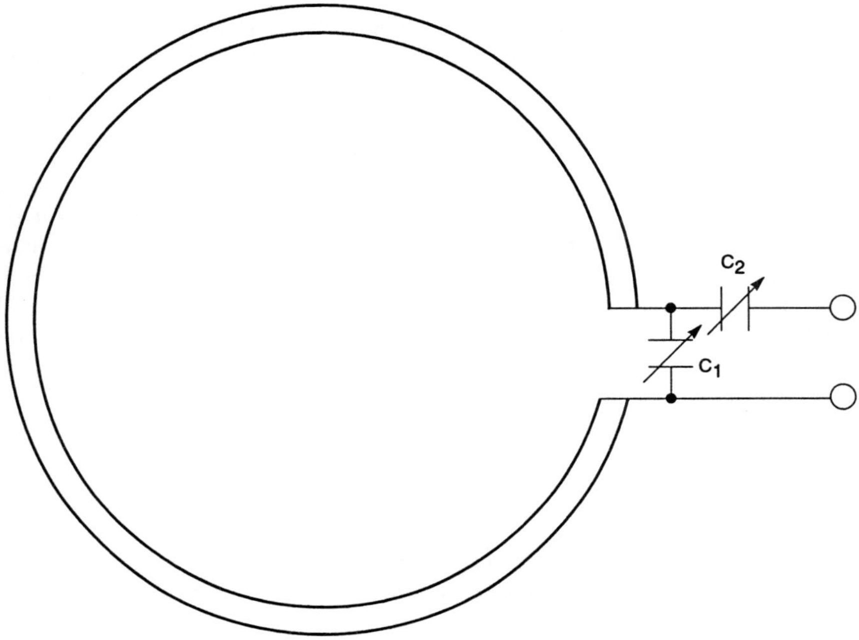

Figure 9.16 *Capacitor matching for small loop*

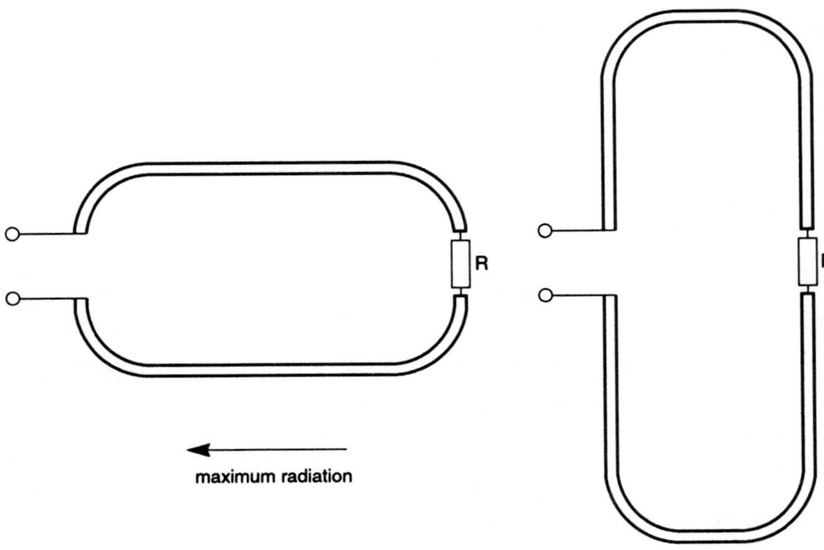

Figure 9.17 *Short loaded loops*

required is very high, particularly for C_1. Typically an accuracy of about 1 part in 1000 will be needed for loops for which $kb \leqslant 0{\cdot}10$, if the ohmic losses have been kept as low as practicable.

9.3.1 *Resistively-loaded loops*

Whilst the small loop which is short-circuited opposite the feed-point has a circular radiation pattern in the plane of the loop this may not be so if the resistance at this point is non-zero. This fact has been used in the development of short, loaded loops and half-loops with directional patterns. Fig. 9.17 shows two extremes of loop shape, but diamond-shaped elements have also been used. For small loops correctly terminated the radiation pattern becomes a cardioid with minimum in the direction of the termination. Jasik [6] gives front-to-back ratios for different shaped loops as their size becomes small:

Diamond: 5·82

Square: 3·0

Circular: 8·31

Elements of this type have a similar radiation pattern in the equatorial plane and can thus be used for homing systems for either vertical or horizontal polarisation by appropriate orientation. It is possible to use a half-loop on a ground plane but this may degrade the performance by having increased response to cross-polarised radiation.

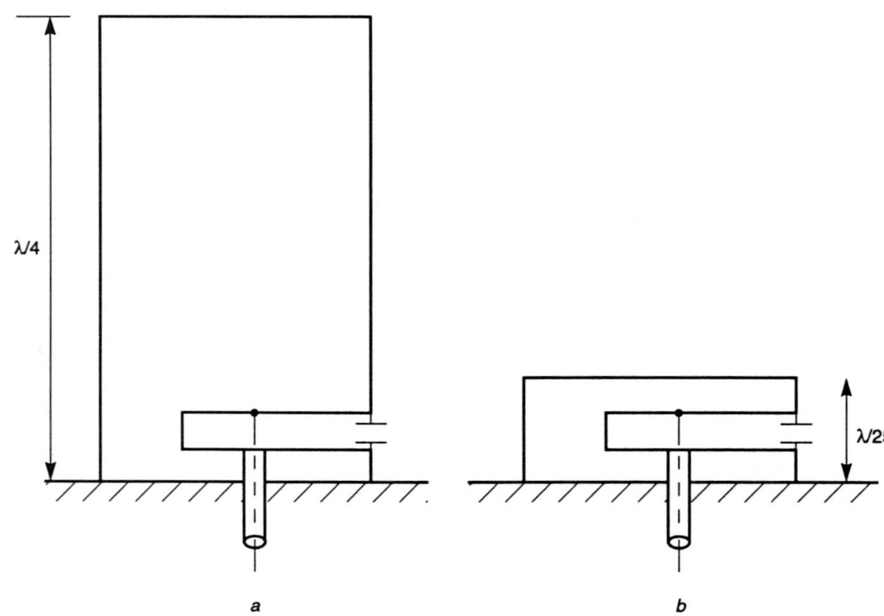

Figure 9.18 *Vestigial notch*

a Notch-fed planar monopole
b Vestigial notch

Because the minimum radiation is in the direction of the termination it is possible to mount two antennas with their terminations close together. There is also some evidence that a metal structure close to and parallel to the 'dead' side of the antenna has little effect on performance. Typically such antennas would have a maximum dimension of about $0{\cdot}1\lambda$ at the lowest frequency but would be usable over a range of 2·5:1 in frequency. The correct value of termination is critical; it depends on the shape of the loop but is likely to be of the order of 300 ohms. Properly terminated, the impedance remains reasonably constant over a wide frequency range but will be of the order of the terminating resistance.

9.4 The short notch

This antenna has been discussed in Chapter 6. It is worth noting, however, that if in a notch-fed monopole such as Fig. 9.18 the metal above the notch is reduced to a thin strip we have the so-called 'vestigial notch' which is effectively a transmission-line antenna.

9.5 References

1 BELROSE, J.S.: 'VLF, LF and MF antennas' *in* RUDGE, A.W. *et al.* (Eds.): '*Handbook of antenna design*', (Peter Peregrinus Ltd., London, 1983) chap. 15

2 BURTON, R.W., and KING, R.W.P.: 'Theoretical considerations and experimental results for the hula-hoop antenna', *Microwave J.*, 1963, **6**, pp. 89–90

3 COOPER, C.E.: 'Airborne low-VHF antennas'. IEE Conf. Publ. 77, 1971, pp. 54–59

4 DUMMER, G.W.A., and BLACKBAND, W.T.: 'Wires and RF cables' (Pitman, London, 1961)

5 GOUILLOU, R.: 'Antenna ultra courte à spirale conique' *in* 'Radio antennas for aircraft and aerospace vehicles'. Technivision Services, Maidenhead, England, Nov. 1967, pp. 251–264

6 JASIK, H.: 'Development of an airborne direction-finding antenna for the 90 to 320 Mc range'. Airborne Instruments Laboratory Inc. Report no. 191–1, May 1950

7 LAPORT, E.A.: 'Radio antenna engineering', (McGraw-Hill Book Co., 1952)

8 NEWMAN, E.M., and MILLIGAN, V.R.: 'A survivable low-profile VHF monopole' *IEEE Trans.*, 1985, **APS–19–5**, pp. 617–620

9 SCHELKUNOFF, S.A., and FRIIS, H.T.: 'Antennas, theory and practice' (John Wiley & Sons, New York, 1952)

10 SEELEY, E.W.: 'Experimental study of the disk-loaded folded monopole' *IEEE Trans.*, 1956, **AP–4**, pp. 27–28

11 STORER, J.E.: 'Impedance of thin wire loops', *Trans. AIEE*, (Communications and Electronics), 1956, **75**, pp. 606–619

12 WANSELOW, R.D., and MILLIGAN, D.W.: 'A compact, low-profile transmission line antenna — tunable over greater than octave bandwidth', *IEEE Trans.*, 1966, **AP–14**, pp. 701–707

13 WOLFF, E.A.: 'Antenna analysis', (John Wiley and Sons, 1966)

Chapter 10

Bodyborne antennas

10.1 Introduction

Initially most bodyborne radio equipment was used by the military, particularly the infantry, and the equipment was cumbersome necessitating the wearer being festooned with a number of radio units from one of which an antenna projected. The advent of transistorised equipment has made it possible to package the equipment in much smaller units allowing more flexibility in its positioning on the body. With the use of higher frequencies than the original 30–80 MHz, personal handsets were introduced. These would not have been practical for significant ranges without the use of elevated base stations; networks of these as in the cellular radio systems permit coverage over a wide area. Without such networks communication ranges are still limited by propagation considerations. For this reason operations in rural areas still rely largely on the lower VHF bands particularly in the military field where mobility may mean that there are no elevated base stations.

Three distinct classes of users can be identified:

(i) The military and some civil and industrial operations where the operator needs the use of both hands so the radio equipment must be carried on the body
(ii) Civil operators using handsets for essentially short range communication, using networks of base stations for area coverage, or operating over very short ranges on isolated sites, e.g. building sites or industrial operations.
(iii) Police and security forces using completely covert equipment.

10.2 Characteristics of the human body

In order to understand the operation of an antenna in close proximity to the human body it is necessary to appreciate the material composition of the body. Belcher [3] gives a table of electromagnetic constants of components of the body compared with water and earth (Table 10.1)

From this it can be seen that the body is essentially a lossy dielectric having properties somewhere between those of sea water and moist earth. The differences between the body components suggest that the effects of the body might vary somewhat from person to person; in fact some tests on a very limited sample tended to confirm this.

Bach Andersen and Balling [1] carried out a simple experiment to assess the impedance and efficiency of the human body as a radiator. A small metal plate

Table 10.1 *Electromagnetic constants of body, water and earth at UHF*

Material	Conductivity (σ) mhos/m	Relative permittivity (ε_r)	Relative permeability
Muscle	1	74	~1
Lung	0·5	35	~1
Fat	0·03–0·1	8	~1
Sea water	5	80	~1
Fresh water	2×10^{-4}	80	~1
Good moist earth	1×10^{-2}	10	~1

300 mm in diameter was mounted 100 mm above a ground plane. The plate was connected to the inner of a coaxial cable whose outer was connected to the ground plane. Admittance measurements were made in the range 30–70 MHz for:

(*a*) thin metal whips 1 mm in diameter
(*b*) a person 1·70 m tall

standing on the circular plate.

Whip lengths of 1 m, 1·5 m and 2 m were used. Fig. 10.1 shows the gain of the human body as a radiator relative to that of the three whips. What this shows clearly is that whilst at the lower end of the range absorption is much greater than radiation this is not true above 60 MHz. In any consideration of bodyborne antennas, therefore, the body has to be considered as a parasitic, if rather lossy, radiator.

Measurements of conductance, Fig. 10.2, made using this experimental arrangement showed that the human body has a near-constant conductance over the frequency range — in other words, there was no significant resonance. For comparison the conductance of the 1·5 m whip is shown; the overall length of the radiator, including the height of the mounting plate, is consistent with the resonance shown. It would have been interesting to compare the conductance of a 300 mm diameter metal cylinder: as a 'fat' monopole this would still have a resonance near 45 MHz but its conductance above 50 MHz would not be very different from that of a person. Also on Fig. 10.2 is a calculated curve for a cylinder of 300 mm diameter and infinite length and having $\varepsilon=80$, $\sigma=$ 0·7 mho/m. The good agreement at the higher frequencies indicates that to regard the body as a cylinder of homogeneous dielectric is a good approximation for this frequency band.

Other studies of absorption confirm a rather broad maximum around 50–80 MHz. They also show a resonance for the arm around 150 MHz and for the head around 375 MHz. It is interesting that the three bands most used for mobile communications and therefore likely to involve bodyborne antennas should be in the bands 30–80 MHz, around 150 MHz and in the 460 MHz band.

Bach Andersen and Hansen [2] derived a computer model in which the human body is approximated by a number of closely-packed dielectric spheres,

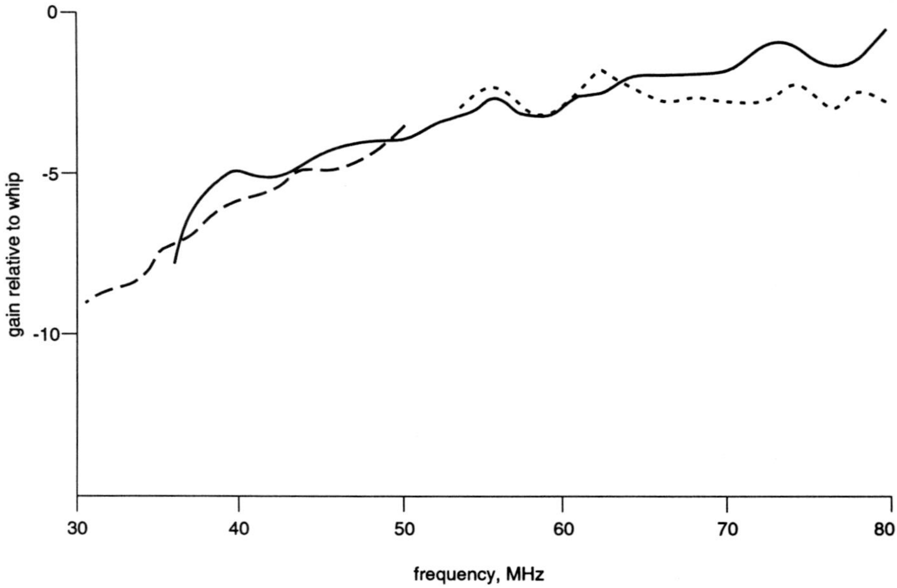

Reference 1

Figure 10.1 *Gain of the base-fed human body compared with a metal radiator*

- - - 2 m whip
—— 1·5 m whip
· · · · 1 m whip

100 mm in diameter. Using this model reasonable correlation was obtained with experiment in horizontal radiation patterns of a small antenna in front of the body. Computation of front to back ratio as a function of 'body' length showed that below $\lambda/2$ the 'body' acted as a director, above this as a reflector, thus paralleling the effect of a metal cylinder.

10.3 Man and antenna

We can distinguish three separate arrangements of radio equipment on the body:

(i) Pack-set in which equipment is on the wearer's back and antenna may project above the wearer's head. Primarily used in the 30–80 MHz band.
(ii) Handset in which the radio equipment and antenna are held more or less in front of the operator's face, i.e. higher than in the preceding cases.
(iii) Pocket set in which the radio equipment is kept in the user's pocket or on his belt. May have a separate microphone and earpiece the cable to which will act as part of the antenna either intentionally or by accident.

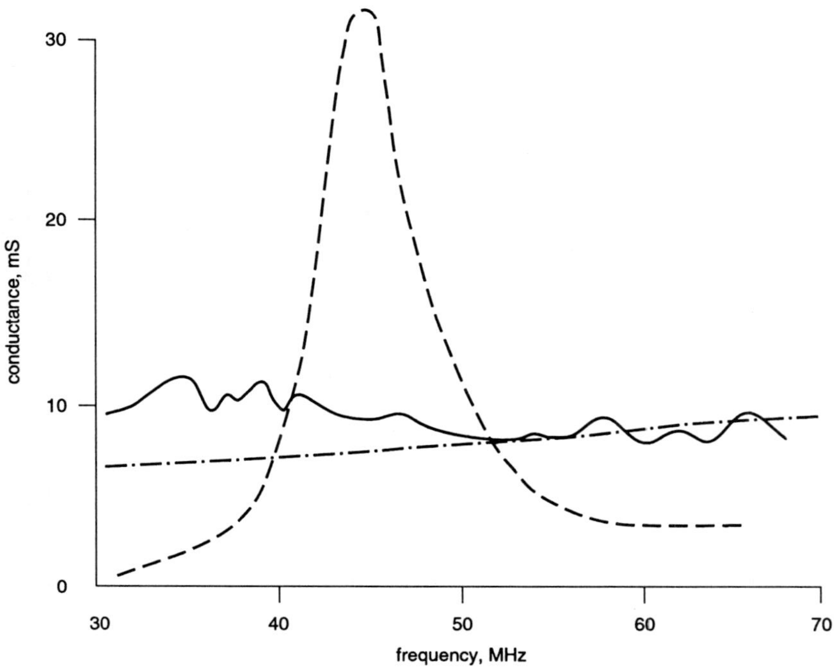

Reference 1

Figure 10.2 *Conductance of the human body when end-fed*

—— 1·7 m tall man (measured)
– – – 1·5 m whip 2mm diameter (measured)
–·–·– infinite dielectric cylinder 300 mm diameter $\varepsilon = 80$ $\sigma = 0{\cdot}7$ S/m

10.3.1 *Pack-set antennas*

A most important study of this arrangement was carried out by Krupka [5]. His representation of the electrical situation is shown in Fig. 10.3. The antenna and pack-set together can be considered as an asymmetric sleeve dipole. The pack-set coupling to the body, Z_{st}, will largely be capacitive, the other, Z_{at}, is a strong radiative coupling.

Measuring the impedance of such a system requires special care to prevent the measuring cable becoming part of the radiating system. A high impedance is needed between the outer of the coaxial cable and the pack-set to eliminate currents on the outside of the latter travelling down the outer of the cable. This impedance needs to be as close as possible to the base of the pack-set; it may be provided by a coaxial sleeve tuned to quarter-wave resonance or by ferrite beads or any other convenient means. Further chokes are needed along the cable to reduce radiated pick-up. The pack-set itself is simulated by an empty case of the same size. From such measurements with a 10 ft (3·05 m) whip antenna on a pack-set, Rashid [7] deduced that in the range 30–55 MHz the apparent length of the radiating system was increased by 1·94 ft (0·59 m) if the

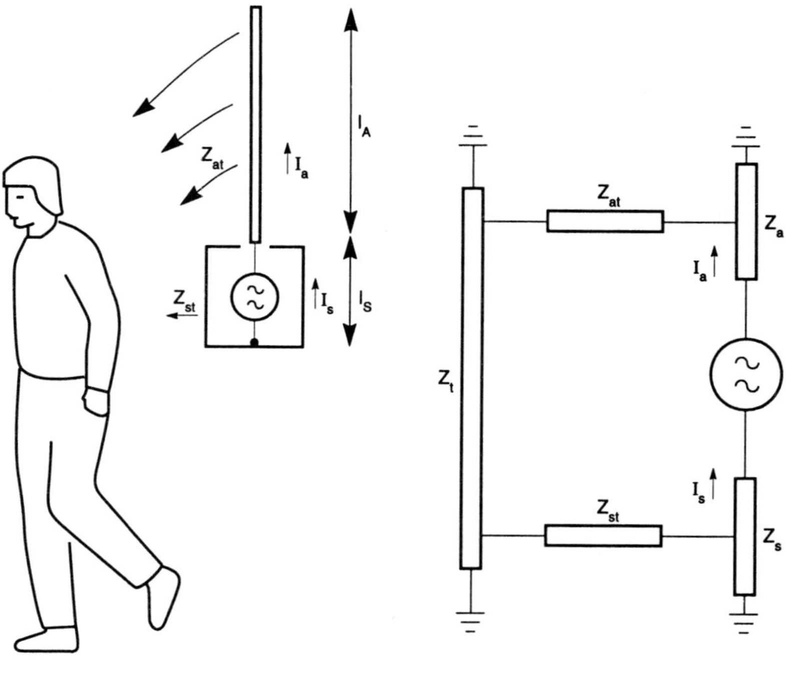

Figure 10.3 *Electrical representation of pack-set and operator*

operator was standing and by 2·5 ft (0·76 m) if the operator was kneeling. It is not obvious why the kneeling man apparently has a larger effect but presumably this is due to a change in the whole body impedance. The dimensions of the pack-set were not given.

Krupka measured the impedance and radiation patterns of portable (i.e. pack-set) radios carried at the front and the side of the body. Antennas were matched in the side position using a two-element system in the dummy equipment box to VSWR$\leqslant$1·6. The change of position from side to front was found to cause significant change in impedance. Of the four antenna lengths used, 0·1λ, 0·15λ, 0·27λ and 0·5λ, the 0·27λ one was least affected by position on the body. Gain measurements were made by comparison with a ground plane monopole whose base was at the same height as the base of the pack-set antenna. Results for the peak gain, Fig. 10.4, show that the side position is generally better below 80 MHz but worse above.

It is difficult to summarise the effect of the human body on the radiation patterns. In general the body acts as a director below 60 MHz and as a strong reflector above 150 MHz. In the intermediate range the effects appear to depend on antenna length and are rather indeterminate. Fig. 10.5 shows an azimuth pattern for a hip mounted equipment in which the antenna was built into the carrying harness and extended below the equipment. The frequency was about 80 MHz. It was noted that there were differences in signal level if the

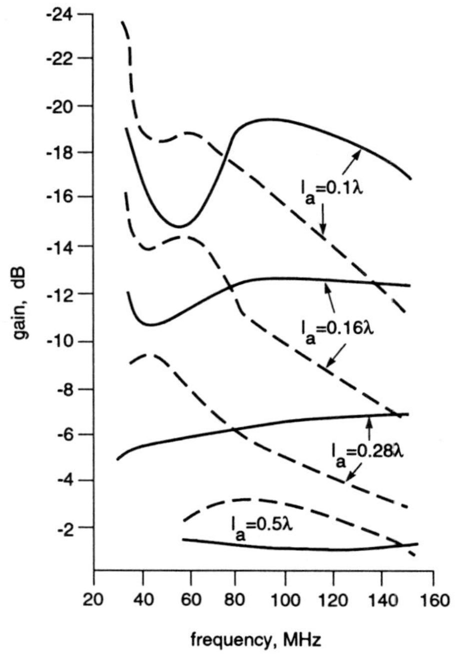

Reference 5

Figure 10.4 *Peak gain of pack-set antenna as a function of antenna length, frequency and position on the human body*

—— side position
– – – front position

operator's hand holding the microphone crossed the harness. These effects were more significant at 150 MHz — not surprising in view of the arm resonance around this frequency — and operators were advised accordingly.

So far we have only considered base-fed antennas and seen that their impedance is very much modified by the proximity of the human body. Mink [6] explored alternative feed arrangements and showed that there were less sensitive systems. In order to measure impedances, a resonant-bridge impedance meter complete with battery-driven crystal controlled signal source was housed in a dummy box of manpack size. The bridge controls could be operated remotely by non-conducting cords. With this arrangement measurements could be made at 20 ft (6·1 m) distance, sufficiently far from the antenna to cause no detectable perturbation of impedance.

The three systems considered are shown in Fig. 10.6: (*a*) is the standard base-fed whip in which the maximum current is at the antenna base and coupling to the body is high, (*b*) is the centre-fed whip isolated from the pack-set by a high impedance circuit. This arrangement is commonly used on vehicles to minimise vehicle currents; a typical example is the US Army AS-1729/VRC antenna. Technically this arrangement is ideal as the current at the pack-set is a minimum and the point of maximum current has been raised above the body.

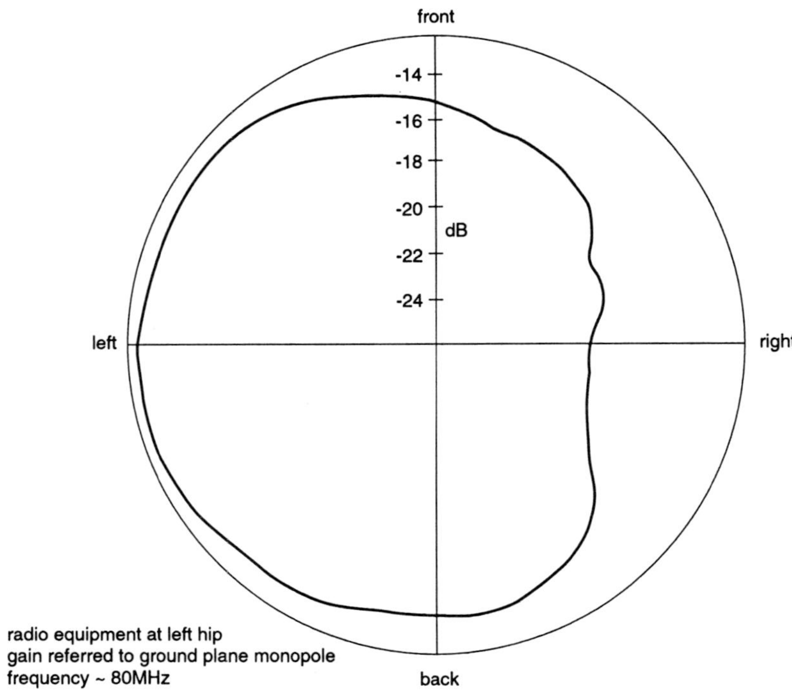

Figure 10.5 *Azimuth radiation pattern of hip mounted VHF radio with antenna in harness*

However, this is a complex arrangement: the isolating circuit has to have a high *Q* at each frequency and a tuning unit is therefore necessary, (*c*) represents a compromise with the antenna fed part-way up its height. This reduces the base current and hence the coupling effect of the body.

The best height was determined experimentally using as the antenna a 48 in (1·22 m) coaxial cable from which the outer could be progressively stripped. The outer was bonded to the pack-set at its lower end. Measurements were made with the pack-set on and off the body and with different configurations of the handset which were found to be significant. From these experiments a sleeve of 20 in (510 mm) was seen to be optimum and to yield an impedance curve which only required simple band-switching to provide an adequate match over the band 30–80 MHz.

10.3.2 *Handset antennas*

Whilst the position of the pack-set radio is substantially constant with respect to the body this is not the case with the handset. We are considering here a system in which the whole equipment and antenna have to be held close to the face in order to communicate, the separation being determined by equipment sensitivity and ambient noise.

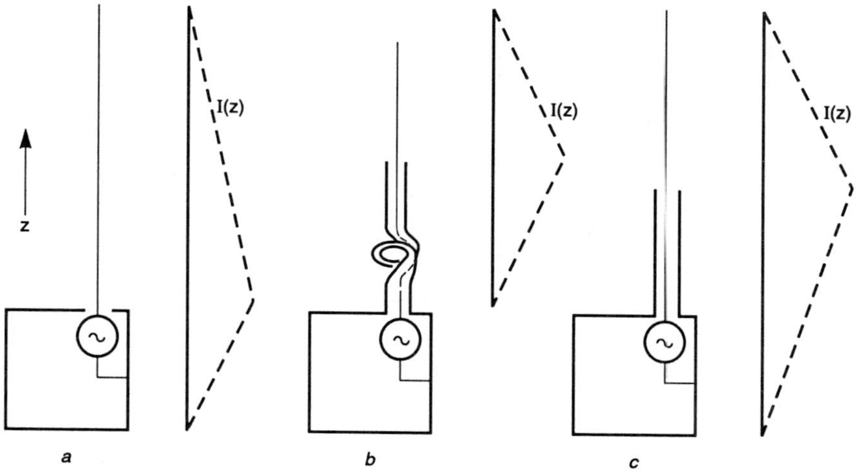

Reference 6

Figure 10.6 *Alternative feed systems for VHF manpack whip antenna*

a Base fed
b Isolated centre fed
c Sleeve fed

The effects of proximity to the body and hand contact are not the same. Fig. 10.7 shows the effects on the impedance of a UHF handset antenna suspended at various distances in front of the chest or hand-held in the same position. The chest is not the best place for a UHF antenna. Fig. 10.8 shows an azimuth pattern for a UHF (460 MHz) set, chest mounted. This should be compared with Fig. 10.9 where the set was held at the normal speaking position in front of the head. In the course of assessing a number of UHF handsets for the British Home Office it was shown conclusively that much higher gains were achieved if the antenna was at the top of the handset and the microphone near the bottom rather than the reverse. In other words, the higher up the body the antenna is, the lower the loss. Proximity to the body is very critical and quite small increases in separation can have a measurable effect. This is demonstrated by Fig. 10.10 in which the same equipment is worn in a shirt pocket and in the top pocket of a jacket.

Numerous antenna types have been investigated for handsets. For antennas wholly external to the equipment flexible helical antennas operating in the side-fire mode are often used. Several experiments claim that short squat helices are to be preferred to long thin ones. If the antenna is to be enclosed in a plastic cover at the top of the equipment some form of top-loaded monopole, probably folded and with the top-loading coiled up, is a possible solution. In any case the

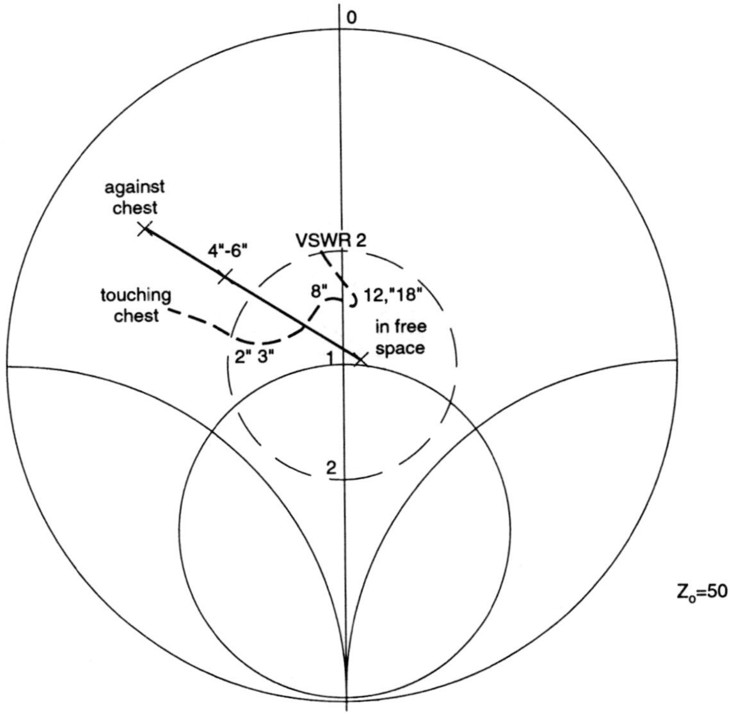

Reference 3

Figure 10.7 *Effects of proximity and hand contact on impedance of a UHF handset antenna*

—— handset suspended in front of chest
– – – hand held in similar position

aim should be to use the antenna with the lowest Q that the available space will allow and to ensure that the equipment is used as high up the body and as far from it as is practicable.

10.3.3 *Concealed antennas*

For security reasons no examples can be given of such antennas. It will be obvious from what has been said in the previous sections that the closer the proximity to the body, the more absorption there will be. Also, the higher the antenna the higher its gain. Because these systems have to be worn by people of different size and in different clothing it is not practical to use electrically short antennas which require tuning. The aim must be to use low-Q systems which will tolerate different surroundings without too large VSWR changes. By implication, radio transmitters used with such antennas must be able to tolerate mismatch: a VSWR of 5:1 would be a reasonable target. It may be desirable to incorporate some attenuation between transmitter and antenna. It will be seen from Chapter 15 that 3 dB attenuation will reduce any VSWR to no more than 3:1; knowing the likely range of VSWR to be found in service a suitable amount of attenuation can be chosen.

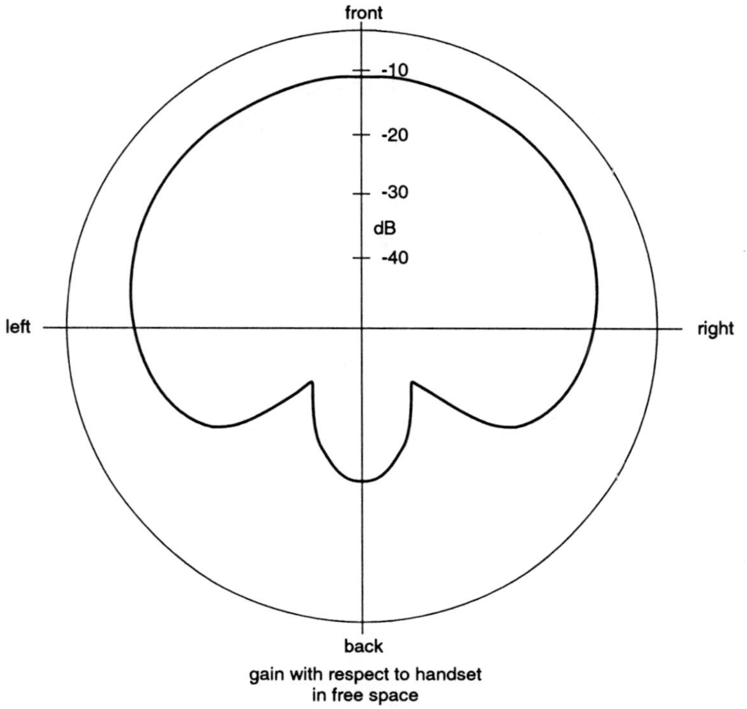

Figure 10.8 *Azimuth pattern of chest mounted UHF handset antenna*

10.4 Antennas for radio tracking animals

Radio transmitters for attaching to animals have to satisfy a number of criteria:

(i) The equipment must have sufficient range to make radio tracking worthwhile.
(ii) The equipment and antenna must not endanger the animal so equipped or handicap its activities.
(iii) If possible the antenna coverage in azimuth should be as near uniform as possible.
(iv) The active life of the transmitter should be compatible with the planned study period. Frequent re-trapping to replace batteries is traumatic for the animal.

For most mammals some type of collar around the neck appears to be the most suitable arrangement. For the animals with small necks, a loop antenna in the collar is practicable. Since the coupling to the body will be minimal this arrangement should give reasonable efficiency. The azimuth patterns appear to be approximately figure-of-eight which would be expected from a vertical loop in free space. The reactance will be inductive for a perimeter of less than a half-wavelength. A perimeter of between $0{\cdot}35\lambda$ and $0{\cdot}40\lambda$ would require only series compensation which could easily be achieved with a capacitor. This may well be impractical since tracking frequencies are allocated in defined bands. It

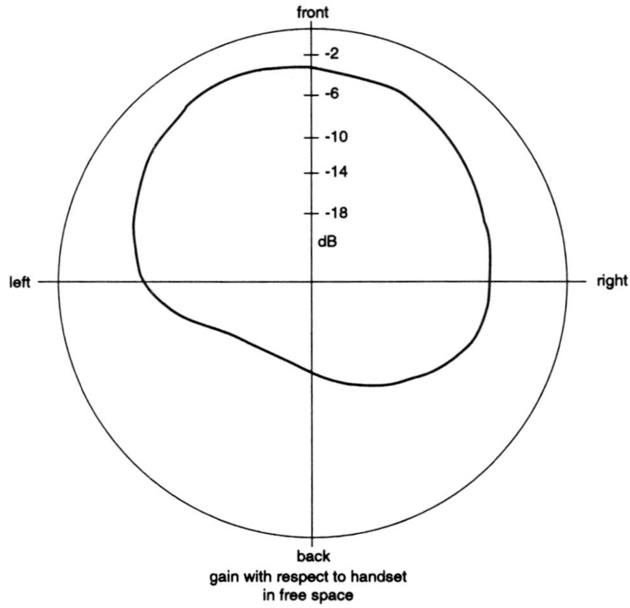

Figure 10.9 *Azimuth pattern of hand held UHF handset antenna in front of face*

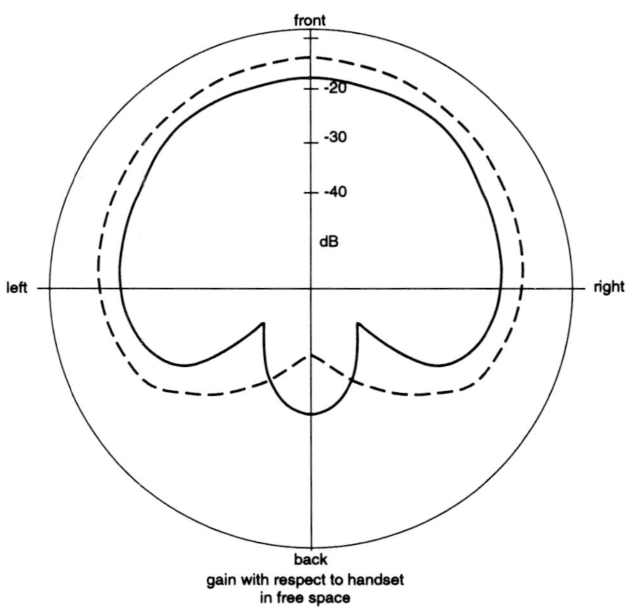

Figure 10.10 *Azimuth patterns of chest mounted UHF antenna showing effect of clothing*

—— set in shirt pocket
– – – set in top jacket pocket

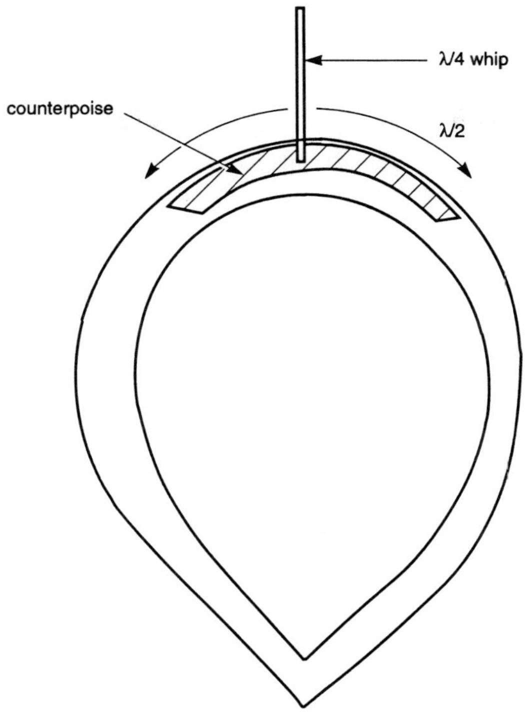

Figure 10.11 *Whip antenna with counterpoise for radio-tracking*

should be noted that on animals whose necks are more vertical than horizontal, e.g. giraffe, the radiation from a neck loop will be horizontally polarised not vertically as with most animals. If the neck is too large for a half-wave loop, e.g. elephant, it may be better to use a vertical quarter-wave wire. If possible a half-wave 'earth-bar' counterpoise should be incorporated in the equipment collar as in Fig. 10.11. Whilst a single wire would be adequate a better arrangement would be a number of fine wires, or even a loose mesh, attached to the collar.

Radio tracking is occasionally used with some species of large birds, the transmitter and antenna usually being attached to the wing near the shoulder. The density of a bird's wing is so low that it can have little effect electrically. The body may have more effect in the case of a swan or goose but its resonance is likely to be in the range 100–150 MHz for the swan, 250 MHz for the goose. It is probable that the transmitter and antenna should be considered as an isolated, vertical, asymmetric dipole.

10.5 Simulation of the human body

Carrying out experimental work on bodyborne antennas is expensive if a real person is used to hang the antennas on, as well as being extremely tedious for the subject. For initial experiments a reasonable model is a plastic tube filled with sea water. The tube should be about 300 mm diameter and 1·8 m tall.

More precise simulations, known as phantom models, can be constructed. Tissue-equivalent mixtures have been developed for muscle tissue, brain, fat and bone. For example see Chou and Chen [4] describing muscle-tissue equivalents for the range 50–2450 MHz. It should be noted that some workers have pointed out that these mixtures degrade within a few days due to the onset of bacterial growth. Improved materials have been developed by the University of Ottawa; a life of at least a year is claimed (unreferenced).

10.6 References

1 BACH ANDERSEN, J., and BALLING, P.: 'Admittance and radiation efficiency of the human body in the resonance region', *Proc. IEEE*, July 1972, pp. 900–901

2 BACH ANDERSEN, J., and HANSEN, F.: 'Antennas for VHF/UHF personal radio: a theoretical and experimental study of characteristics and performance', *IEEE Trans.*, 1977, **VT–26**, pp. 349–357

3 BELCHER, D.K.: 'Human proximity effects on small antennas'. Proceedings of Carnahan Convention on Crime Countermeasures, May 1976, pp. 171–178

4 CHOU, C.K., CHEN, G.W., *et al.*: 'Formulas for preparing phantom muscle tissue at various radio frequencies'. 5th Annual Meeting of the Bioelectromagnetics Society, 1983, Boulder, Colorado

5 KRUPKA, Z.: 'The effects of the human body on radiation properties of small-sized communication systems', *IEEE Trans.*, 1968, **AP–16**, pp. 154–163

6 MINK, J.W.: 'Experimental investigation of manpack whip antenna characteristics and proximity effects'. Proc. ECOM-ARC Workshop on electrically small antennas, Oct. 1976, pp. 177–181

7 RASHID, A.: 'A representation of cylindrical antennas for manpack installation', *IEEE Trans.*, 1967, **AP–15**, pp. 699–700

Chapter 11
Direction finding antennas

Two classes of direction finders can be distinguished. These are:

(i) Terrestrial or airborne types requiring omniazimuth but limited elevation coverage
(ii) Systems for tracking space vehicles and radio stars where essentially hemispherical coverage is required.

11.1 Azimuth systems

We can distinguish here between antennas mounted on the ground and those on some form of vehicle. In the latter case the effects of the finite ground plane will be considered. In both cases the great majority of systems use vertical polarisation, since there are few horizontally polarised transmitting systems apart from broadcast television and radio stations.

11.1.1 *Fixed DF systems*

11.1.1.1 *Rotatable H-Adcock array*

The directional part consists of two vertical elements, usually centre-fed half-wave dipoles, separated by a horizontal distance $2D$ and with their outputs coupled together in antiphase. Fig. 11.1 gives the arrangement and shows a centre element which is used to resolve ambiguities. At broadside to the outer pair, assuming that they are in free space and are identical elements, their combined signal will always be zero. The sharpness of the null increases with spacing up to $2D = 0{\cdot}75\lambda$ above which secondary nulls start to appear, Fig. 11.2. Wider spacing would be possible if the element patterns were non-circular, preferably cardioid.

The pattern of the pair of elements is given by

$$E(\theta) = 2Eh_e \cos\,(kD \cos\,\theta + \pi/2)\cos\,\omega t$$

where $E \cos\,\omega t$ is the incident field at the centre of the array, h_e is the effective height of each element and θ is the azimuth angle from the plane of the array. It can be seen that $E(90°)$ is always zero.

The outer elements are rotated to find the null directions; it should be noted that null directions are always sharper than peak ones so give more accurate bearings. This improvement in accuracy is at the expense of range: in practice there will be some uncertainty since the signal will disappear into the receiver noise. However, even on a manual system the accuracy can be restored by 'bracketing'. Fig. 11.2 shows two opposing nulls. By combining this output with

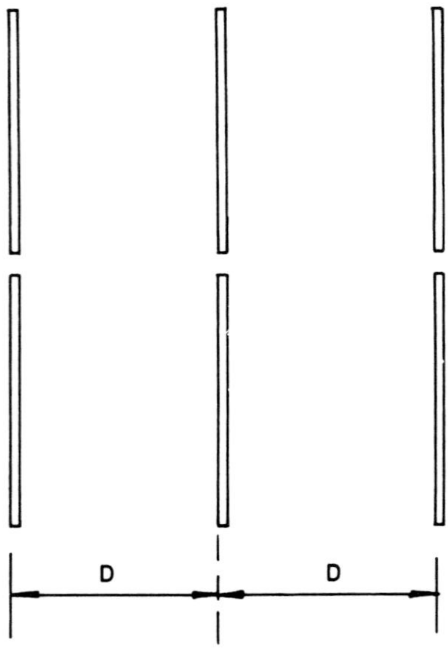

Figure 11.1 *Arrangements of elements for rotating H-Adcock array*

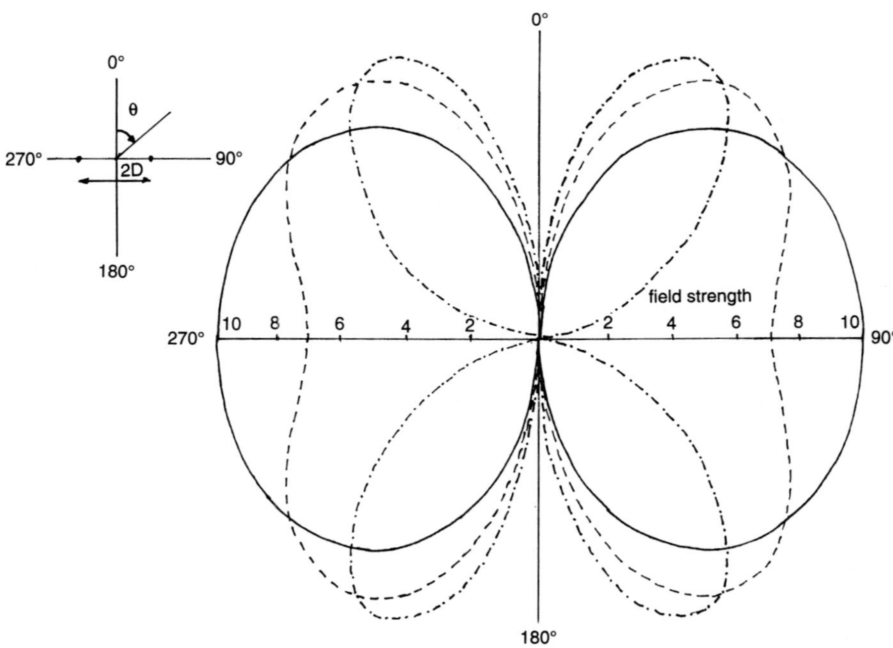

Figure 11.2 *Equatorial plane patterns of a pair of equal parallel dipoles coupled in anti-phase*

—— D = 0·25λ
– – – D = 0·375λ
–·–· D = 0·5λ

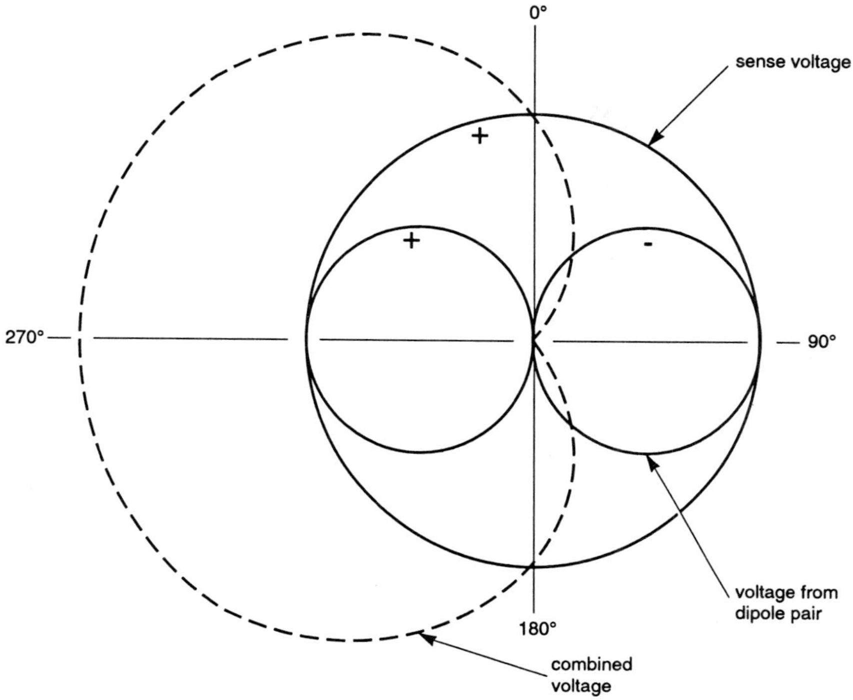

Figure 11.3 *Combination of Adcock pair with sense antenna*

that of the central omnidirectional antenna a cardioid pattern, Fig. 11.3, can be obtained. A 90° phase shift has to be introduced into one of the signals and the two amplitudes must be equal to give a sharp null to the cardioid. By noting whether the signal increases for a rotation in one direction or decreases, the ambiguity can be resolved. Griffith and Rosinski [4] have discussed the factors controlling the performance of a VHF system in more detail than can be given here. A further important analysis is given by Hopkins and Horner [5].

This type of antenna needs to be raised above the ground as high as possible, (*a*) to increase range and (*b*) to reduce the unbalance in capacitance to ground between the upper and lower arms of the dipoles. This unbalance will enhance the pick-up of cross-polarised radiation. The maximum height allowable is determined by the range of elevation angles to be covered. If h is the height of the antenna system above ground the first null occurs at $\sin \phi = \lambda/4h$.

11.1.1.2 *Fixed Adcock array*

Instead of rotating a directional pair it is possible to obtain bearings with two fixed orthogonal pairs of antennas as shown in Fig. 11.4. By comparing phase and amplitude of two outputs as well as that of the centre sense antenna, the bearing of the transmitter can be determined. As with the rotating system, good balance and equality of impedance are important. Broadband dipoles are usually preferred for this reason.

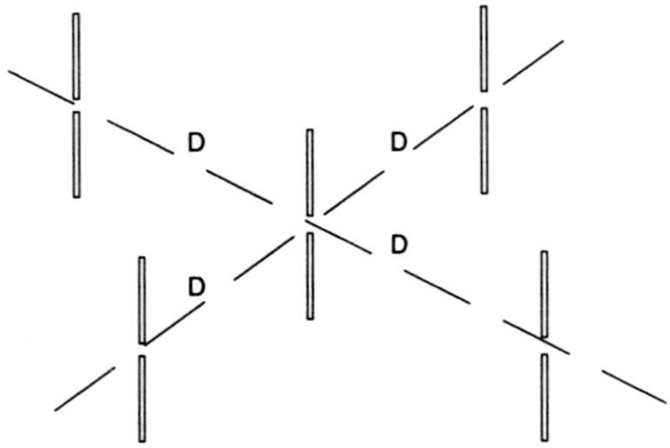

Figure 11.4 *Fixed Adcock array*

One of the main disadvantages of these very compact antenna systems is that they require a good reflection-free site to eliminate bearing errors. This problem arises from the short base line of the system which means that all the antenna elements are equally affected.

11.1.1.3 *Commutated antenna direction finders* (*CADF*)
The use of the commutated antenna system was pioneered by Earp and Cooper Jones [3]. Fig. 11.5 shows a typical arrangement of 18 dipoles equally spaced

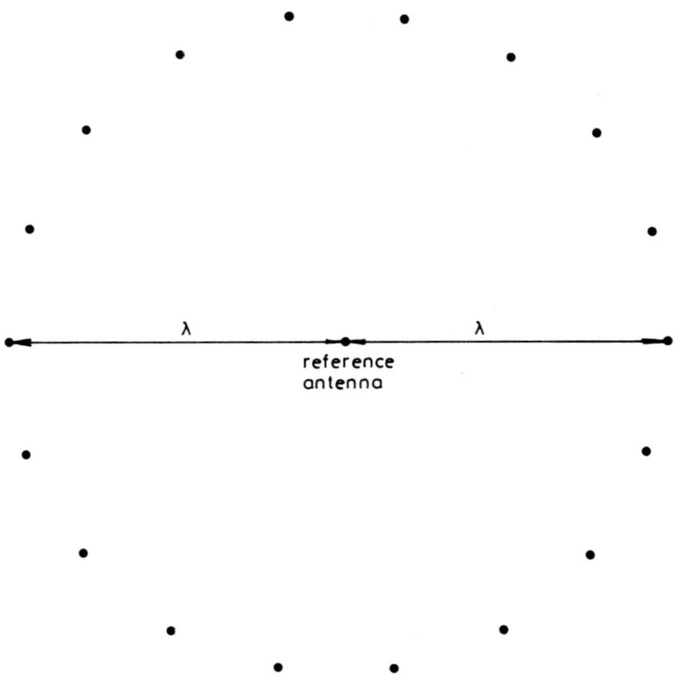

Figure 11.5 *Commutated DF antenna system*
Elements are $\lambda/2$ vertical dipoles

around a 2λ diameter circle, the antenna-to-antenna spacing being approximately $\lambda/3$. Sampling signals around the circumference at constant angular velocity gives a Doppler shift varying sinusoidally with angular position in azimuth and a phase shift equal to the azimuth. The bearing can then be resolved using normal phase measurement techniques. A reduction in RMS site error by a factor of 7 compared with a conventional Adcock system has been achieved with the system described. It is usual to terminate idle elements or open-circuit them to minimise coupling effects.

Although dipoles have been mainly used at VHF, arrays have been built with monopoles on a common ground plane at UHF (225–400 MHz). This can introduce some phase shift errors because, as each element is asymmetrically disposed on the ground plane its phase centre shifts with azimuth angle because of diffraction from the edge of the ground plane. It is possible that the phase centre movement could be reduced by mounting the antennas on an annular ground plane or by providing a skirt around the edge but the author is unaware that either of these have been tried. The ground plane does help to reduce high-angle reflections from the ground if it extends sufficiently; see Chapter 3 for the effects of finite ground planes on monopole patterns.

11.1.1.4 *Doppler DF antennas*

A similar result can be obtained by rotating a single element around the sense antenna. The antenna would usually be a dipole on a horizontal arm with a dummy antenna diametrically opposite to provide aerodynamic and mechanical balance. The author worked for some time adjacent to a system of this type on life test which was rather like being close to a helicopter blade test rig as the antenna was rotated at about 30 Hz.

11.1.1.5 *Horizontally polarised systems*

Though these are rarely required in the VHF and UHF bands, some possible antennas are suggested below.

(*a*) *Horizontal crossed dipoles*

There is no advantage in using dipoles longer than $\lambda/2$ as the null is narrower for this length of antenna than for longer elements. One problem is in obtaining and positioning a suitable sense antenna. The only possibilities appear to be a small loop, a slotted cylinder or a turnstile raised above the crossed dipoles. Interaction is likely to be a problem. Perhaps the best scheme would use four quarter-wave radial elements around an axially slotted cylinder (Figs. 5.14–5.16). Since the impedance of the slotted cylinder is so different from that of the dipole and the reactance changes with frequency in the opposing direction, obtaining good amplitude balance over any significant bandwidth may be difficult.

(*b*) *Slotted cylinders*

An array of two slotted cylinders for a rotatable system or four for a fixed system with another central slotted cylinder as the sense antenna appears to make more sense than (*a*).

(*c*) *Small loops*

Small horizontal loops perhaps terminated or open circuit are another possibility but probably not as good as (*b*).

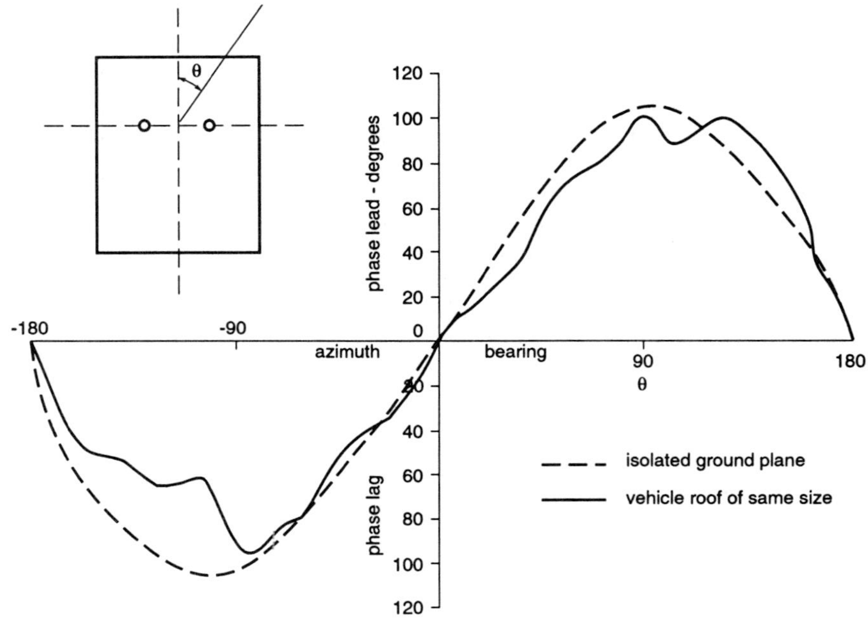

Figure 11.6 *Effect of vehicle on phase shift of a vertical polarised DF system*

11.1.2 *Mobile DF systems*

For land-vehicle based systems the best arrangement is undoubtedly a conventional system as in Section 11.1.1 mounted as high as possible above the vehicle on a mast which can either be retracted or pivoted when the vehicle is on the move. Attempts to use monopole type systems on a finite ground plane, i.e. the vehicle roof, have not proved successful because of re-radiation from the remainder of the vehicle. Fig. 11.6 shows the effect on the phase shift between a pair of monopoles on a car roof compared with the results on a finite but regular ground plane.

On aircraft it is only possible to use antennas of low height. This inevitably means that the horizon gain of horizontally polarised antennas will be reduced because the angle of peak gain will be away from the conducting plane. A scheme using the terminated loops described in Chapter 4 was developed by Jasik [6]. This was a rotating system using a diamond-shaped loop for horizontal polarisation and a half diamond for vertical both mounted on the same rotating platform. Fig. 11.7 shows the arrangement. Some measurements were made of the performance of the vertical system on a scale model aircraft but not of the horizontal system. From measurements on a finite ground plane the horizontal antenna gain appeared to be about -9 dB referred to the vertical at an elevation of 10° to the ground plane.

The problems of siting on an aircraft are highlighted by Kunachowicz [8]. He contrasts the results for a pair of monopoles on a circular cylinder with those on an aircraft fuselage. The bearing errors in the latter case were approximately

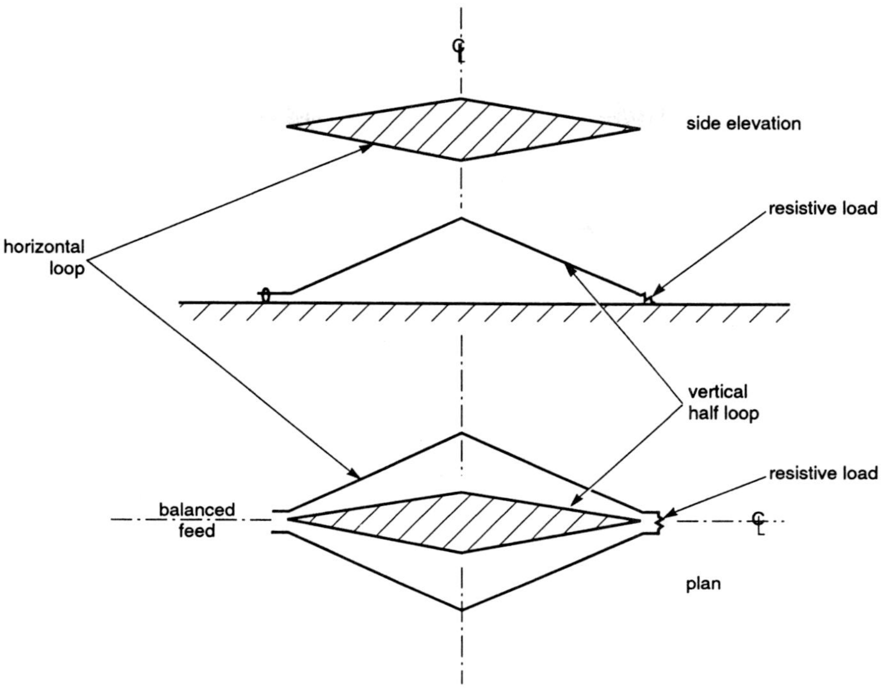

Figure 11.7 *Airborne rotating DF system using terminated loops*

five times those in the former. It has to be said, however, that this was a difficult installation in which the antenna system was on the same side of the fuselage as the wings, always a worst case.

11.1.3 *Circular arrays*

The use of circular arrays for direction finding goes back many years and, in a sense, the four-element fixed Adcock array is a circular array if a rather sparse one. The type of array to be described below dispenses with the central reference antenna and uses two or more excitation modes to produce a unique null.

A simple arrangement using four vertical dipoles is described by Davies and Rizk [1]. The four antennas are arranged at 90° spacing around a circle of small diameter. In the original work, the circle was of radius 0·16 λ but subsequently experiments were done at 0·04λ; these are described in the reference. In the simplest arrangement just two modes are used:

(*a*) *Zeroth order*: all elements combined in phase and equal amplitude.
(*b*) *First order*: elements combined in equal amplitude but with a progressive 90° phase shift between them.

The first order then has one cycle of phase rotation around the array. When this output is combined with the zeroth order output, the two outputs having been balanced in amplitude by appropriate attenuation, a null will be produced in

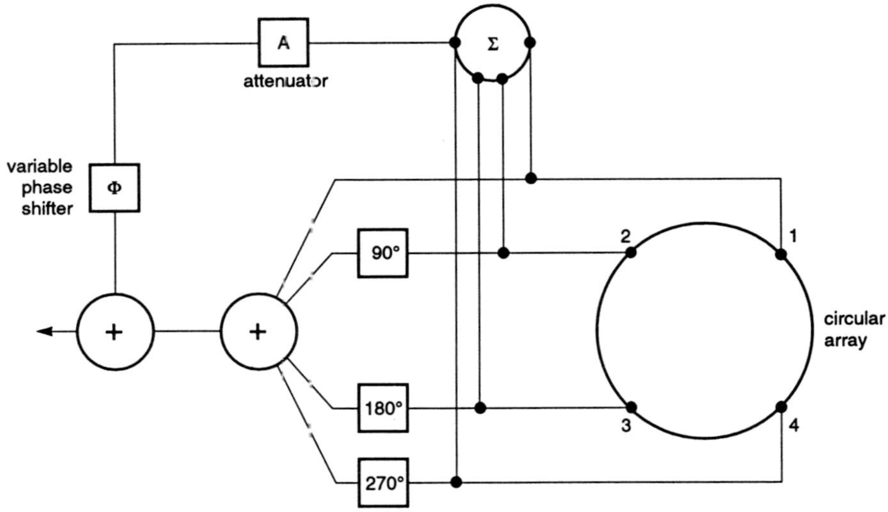

Figure 11.8 *Feed system for a four-element circular array using modes 0 and 1*

one direction. By inserting a variable phase shift in one output the null can be steered around 360°. Clearly if the phase shift is calibrated against null direction, a DF system can be obtained.

Over a wide frequency band the gain of the first order mode will vary considerably if omnidirectional elements are used. Davies [2] shows that using directional elements with patterns of the form $(1 + \cos\phi)$ considerably improves the uniformity of mode gain. Cyclic phase errors arise because the array has a finite number of elements. There is, therefore, a limit to the spacing between adjacent elements: it should be less than $\lambda/2$. The use of directional elements does not improve this so the limitation is clearly fundamental.

In view of the phase errors shown in Fig. 11.6 it is interesting to note that an array of four monopoles each at the corners of a square equal in side to the width of a vehicle roof i.e. two monopoles along each side of the roof, gave a good working system over the range 30–76 MHz. The spacing was about $0{\cdot}15\lambda$ at 30 MHz. Undoubtedly the non-circular patterns of each of the elements improved the performance. The feed arrangement for a four-element system using two modes is shown in Fig. 11.8.

11.2 Tracking in two planes

The two main methods are:

(*a*) Orthogonal arrays of omnidirectional elements whose phase is varied continuously to sweep a beam over the search area
(*b*) Circular arrays of comparatively narrow beam antennas combined to give a conical scan.

The first method includes an antenna system associated originally with radio astronomy but subsequently used for satellite tracking — the interferometer.

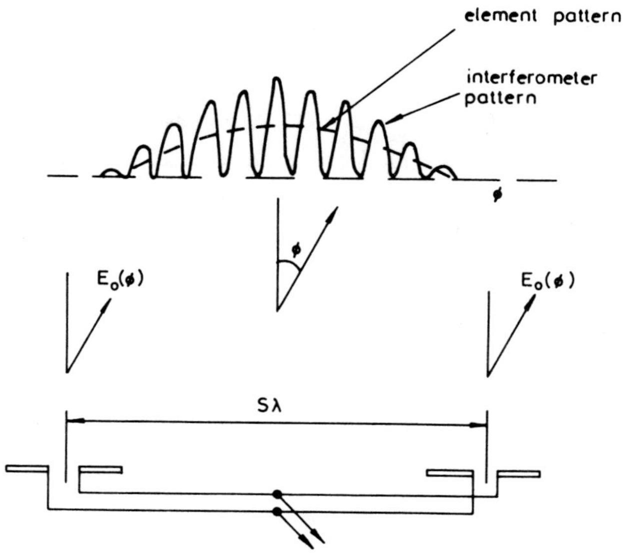

Figure 11.9 *Basic two-element interferometer*

11.2.1 *Interferometers*

11.2.1.1 *The basic two-element system*

Fig. 11.9 shows a basic system of two widely separated identical antennas. The radiation pattern in the vertical plane through the two antennas consists of a number of narrow lobes which are the product of the individual antenna patterns and the grating pattern given by two isotropic sources at the same spacing. In the transverse plane the patterns are that of a series of fan beams of amplitude decreasing away from the plane of symmetry (Fig. 11.10).

If the individual antennas are combined in phase then the far field pattern becomes

$$E(\phi)_i = E_0(\phi)[\exp(j\psi/2) + \exp(-j\psi/2)]$$

where $E_0(\phi)$ is the normalised pattern of an individual element and $\psi = 2\pi S_\lambda \sin\phi$.

Provided the individual patterns have peaks at $\phi = 0$ then the central lobe of the combined system will also have a peak at $\phi = 0$. If, however, the two arrays are combined in anti-phase.

$$E(\phi)_0 = E_0(\phi)[\exp(j\psi/2) - \exp(-j\psi/2)]$$

which has a null on axis.

By altering the phase between the two arrays the beam can be swept. By altering from the in-phase to the out-of-phase arrangement, a source located in the main beam can be more precisely located using the null which, as we have seen earlier, is much sharper than the maximum. This method was first proposed by Ryle [10].

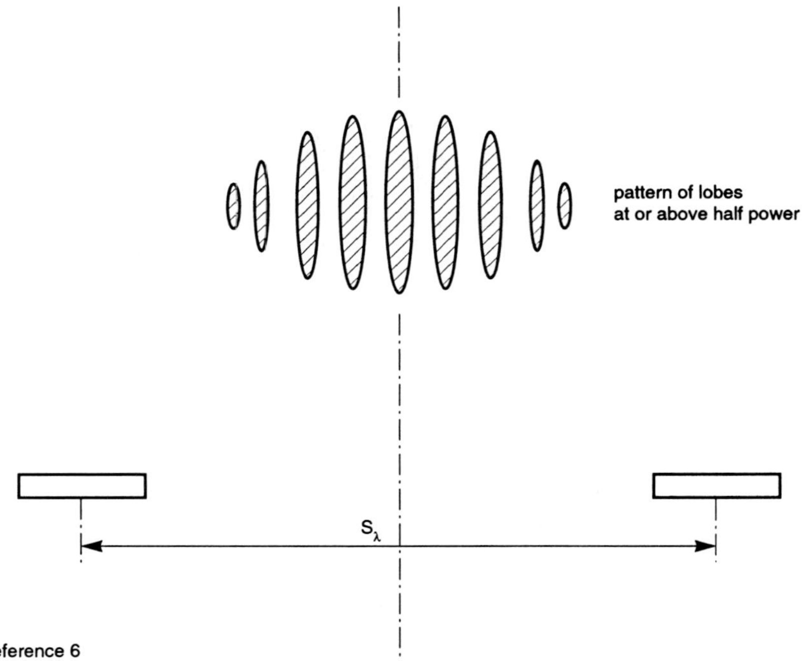

Figure 11.10 *Patterns of interferometer in a plane parallel to the ground*

For angles near the central lobe, the lobe width between first nulls (LWFN) is given by

$$\text{LWFN} = \frac{57{\cdot}3}{S_\lambda} \text{ degrees}$$

This is half the width of that obtained from a uniform array of the same length S_λ.

11.2.1.2 *Multi-element systems*
If additional pairs of elements are added as in Fig. 11.11 the amplitude of some of the lobes can be reduced by appropriate choice of spacing. This gives better discrimination where two sources are close together.

11.2.1.3 *Mills Cross*
Fig. 11.12 shows an interferometer in which the two arrays are unequal but all polarised the same way. A special case in which $S_\lambda = 0$ is the Mills Cross (Mills and Little [9]). A number of radio-telescope antennas based on this arrangement are described by Kraus [7]. The basic system is shown in Fig. 11.13.

11.2.1.4 *General comments*
Clearly any convenient antenna types can be used. Helices and short backfires have certain advantages in giving moderate gain from a single feed. Dipoles with or without parabolic reflectors have also been used, sometimes in the form

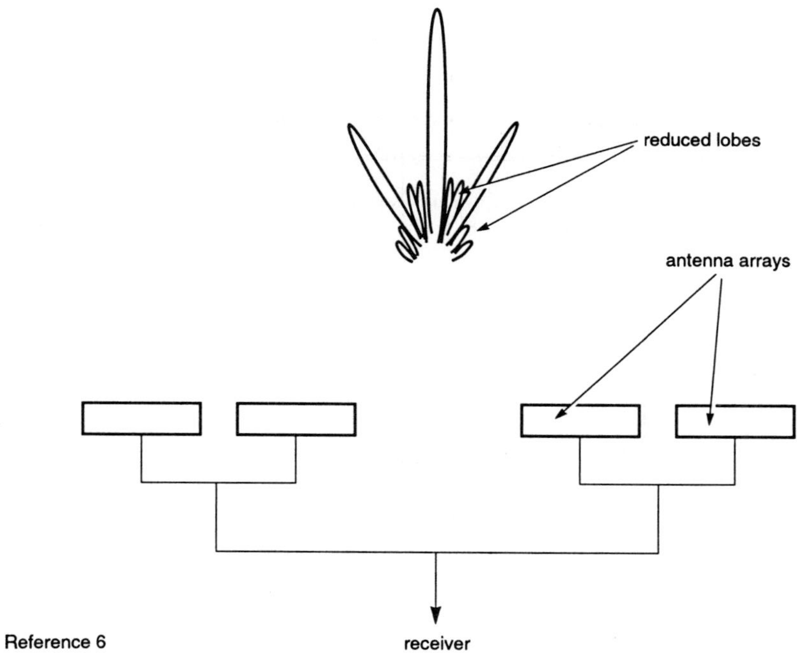

Figure 11.11 *Multi-element interferometer*

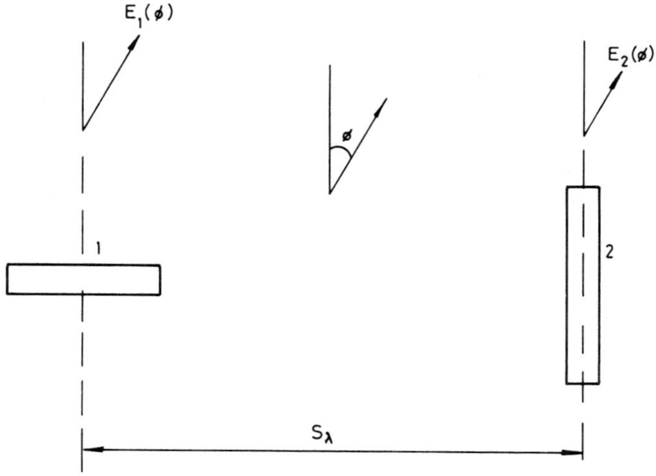

Figure 11.12 *Asymmetric interferometer*

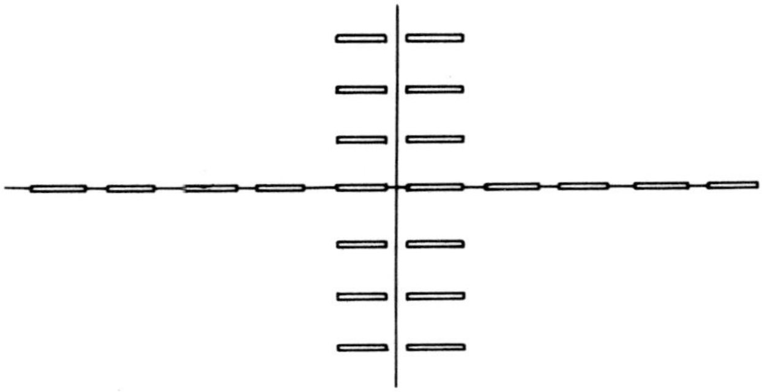

Figure 11.13 *Mills Cross*

of a parabolic cylinder. A scheme explored by the author used slotted cylinders of rectangular cross-section each of which provided an aperture of 2λ and required only a single feed. Whatever the antennas used care must be taken to ensure phase stability of antennas and feed system under all weather conditions.

11.3 References

1 DAVIES, D.E.N., and RIZK, M.S.A.S.: 'A small radius circular array antenna with 360° null steering capability'. IEE Conf Publ. 169, Nov. 1978, pp. 60–64

2 DAVIES, D.E.N.: 'Circular arrays', in 'Handbook of antenna design' (Peter Peregrinus Ltd., 1983) chap. 12

3 EARP, C.W., and COOPER-JONES, D.L.: 'The practical evolution of the commutated aerial direction-finding system', *Proc. IEE*, 1958, **105** Pt B Suppl. 9, pp. 317–325

4 GRIFFITHS, R.M., and ROSINSKI, W.: 'The extension of wireless direction-finding techniques to very high-frequencies for naval use', *J. IEE*, 1947, **94** Pt IIIA, pp. 727–740

5 HOPKINS, H.G., and HORNER, F.: 'Rotating H-type Adcock direction finders for metre and decimetre wavelengths', *Proc. IEE*, 1951, **98** Pt IV (Monograph 11)

6 JASIK, H.: 'Development of an airborne direction finding antenna for the 90 to 320 Mc range'. Airborne Instruments Lab Inc. Report No. 191–1, May 1950

7 KRAUS, J.D.: 'Radio astronomy' (McGraw Hill Book Co., New York, 1967) chap. 6

8 KUNACHOWICZ, K.J.: 'Model testing of airborne VHF direction finding antenna system'. IEE Conf. Publ. 128, June 1975, p. 223

9 MILLS, B.Y., and LITTLE, A.G.: 'A high resolution aerial system of a new type', *Australian J. Phys.*, 1953, **6**, pp. 272–278

10 RYLE, M.: 'A new radio interferometer and its application to the observation of weak radio stars', *Proc. Roy. Soc.*, 1952, **211A**, pp. 351–375

Chapter 12

Mobile antennas

General

Many systems for communication and navigation involving vehicles of any kind use the VHF and UHF bands because propagation characteristics permit the use of simple, low-gain, omnidirectional antennas. In general the siting constraints of such antennas are less severe than with more complex systems and there is more scope to suit the antenna to the vehicle.

Vehicle in this context means any form of transportation and three broad classes can be distinguished:

(i) *Land mobile*
Wheeled and tracked road and off-road vehicles
Railway trains
(ii) *Marine mobile*
Ships
Hovercraft
(iii) *Aerospace mobile*
Aircraft and helicopters
Rockets
Space vehicles and satellites

Personal radio antennas have been considered separately in Chapter 10.

In all instances the performance of the antenna, both as regards radiation pattern and impedance, will be modified to some extent by the vehicle on which it is mounted. The antenna itself may be simple but the combination of vehicle and antenna can be an electrically complex structure. A feel for the behaviour of an antenna on a vehicle can be obtained by considering its performance on appropriate regularly-shaped conductors such as flat sheets, cylinders or cubes. To get more precise information on radiation patterns recourse must be made either to measurements on, for example, scale models or to numerical methods. These are dealt with in Chapters 15 and 14, respectively, but some results will be shown in this chapter.

12.1 Antenna siting

Choosing the right antenna and the right place to put it on a particular vehicle can be very straightforward or may involve a number of attempts. This is particularly true when the vehicle has to carry a number of antennas for different functions where each antenna has to perform satisfactorily without affecting the others or being affected by them. The worst example is probably that of a small military aircraft or helicopter which might require 30 separate antennas several of which will compete for the same positions. What is possibly even more difficult is to add antennas for new equipment to a vehicle which already has a good complement of antennas all carefully positioned to give satisfactory performance. To meet this challenge may well involve some lateral thinking and the use of antennas not previously associated with this type of vehicle in positions not normally considered.

Apart from the normal electrical requirements of VSWR, gain and radiation pattern coverage over a defined frequency range there is a wide range of vehicular constraints some of which are listed below.

12.1.1 *Vehicular constraints*

These can be divided into three classes:

(*a*) Physical
(*b*) Mechanical
(*c*) Environmental

(*a*) *Physical*: These impose limitations on the height and size of antennas for any of the following reasons.

Ground clearance, e.g. underneath aircraft
Overhead clearance: road vehicles, trains, helicopters
Hazard to vehicle operators and passengers
Interference with other vehicle functions
Launch-vehicle constraints: spacecraft, missiles
Concealment of vehicle
Disguise: plain vehicles for law enforcement

(*b*) *Mechanical*

Aerodynamic drag
Wind loading
Ice loading
Wave force: ships, submarines
Damage by cleaning equipment: cars, trains, public transport vehicles
Weight: aerospace vehicles

(*c*) *Environmental*

Vibration
Shock, e.g. spacecraft or missile launch, gunfire
High temperature
Low temperature
Effect of ice on electrical performance
Driving rain
Humidity
Low atmospheric pressure causing flashover and corona
Contamination by fuels and lubricants
Corrosion by salt, exhaust gases, etc.

The effect of these constraints may not be immediately obvious as in some instances they arise from the possibility of consequential damage to the vehicle.

An example of this would be the build-up of ice on a blade antenna on an aircraft causing either blade failure or ice shedding; in either case fragments entering an engine intake could cause catastrophic failure. Equally serious might be broken antennas falling to ground from an aircraft in flight.

12.1.2 *Siting procedure*

Even though it may not always be done formally the procedure will follow these general lines:

(i) Definition of:
Shape of vehicle including moving parts
Radio installations required (the 'radio fit')
Operational constraints

(ii) Formulation of required antenna coverage from radio system and operational requirements. Coverage will be in terms of gain within specified azimuth and elevation angles.

(iii) Study of possible antennas to satisfy (ii)

(iv) Construction of antenna installation diagram showing preferred arrangement of antennas

(v) Submission to vehicle constructor or user for approval

(vi) Modification of (iv) until agreement is reached with user. Since the process may begin before a vehicle is fully defined, especially in the case of an aircraft, subsequent changes may require several repetitions.

In one particular study in which the author was involved, the undercarriage of a naval helicopter was changed six times each of which required a reappraisal of the antenna siting. On a new military aircraft the whole process could take 5–10 years.

(vii) Performance prediction either by numerical methods or scale modelling or, if performance is not critical, by reference to similar installations on similar vehicles. Good judgment is needed here to decide whether previous installations are sufficiently similar.

(viii) Engineering of new antennas if no suitable types exist.

(ix) Full-scale trials with engineered antennas.

It should be noted that (viii) includes qualification testing if required for the particular use. It may be necessary to extend the qualifications of otherwise suitable existing designs by further testing (see Chapter 15).

12.2 Land vehicles

12.2.1 *Cars and vans*

The electrical representation of an antenna on a wheeled vehicle on the ground is shown in Fig. 12.1, where the impedances to ground include wheel-bearings and the effect of tyres. These are rather indeterminate but Webster [39] indicates that in the 20–70 MHz band currents through these paths are important. Some workers have claimed that on cars the effects can be seen in the 150 MHz band but there is little published information to support this. For

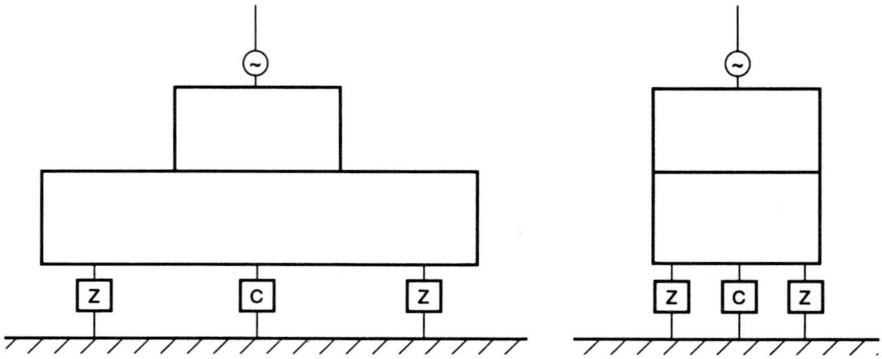

Figure 12.1 *Electrical model of antenna on wheeled vehicle*

a medium-sized saloon car the capacitance to ground is of the order of 250 pF so this may be as important as the impedances through the wheels. The effect of ground constants becomes of little practical importance above 100 MHz; below this frequency the effective height of the antenna will vary depending on the depth of the water table and the ground conductivity. It seems probable that the performance of an antenna on a vehicle roof will not be dissimilar to that of a monopole on a thick cylinder as discussed in Chapter 3.

For ground-to-ground communication only a small range of angle above the horizon is required. Davidson and Turney [10] showed that, in a heavily built-up area, propagation into streets from an elevated base station could involve angles of up to 20° above the horizon. Simple $\lambda/4$ VHF antennas on cars will have broad elevation patterns, Fig. 12.2, so this is no problem. In the higher UHF bands, attempting to overcome the increased propagation loss by using vertically stacked antennas with narrow elevation patterns may prove ineffective.

In purely urban areas the strongest signal may not necessarily come by the most direct path from the ground station. This is particularly true where a street is continuously lined by tall buildings: here the strongest signal may be along the street. With a number of signals arriving at the vehicle severe fading can occur and there would be some advantage in using an adaptive antenna if continuous communication with high data rates is required. Results suggest that very good omnidirectional coverage from the vehicle antenna is unnecessary in city communications. This is not, however, true in open terrain: trials with good and poor horizontal coverage antennas on a vehicle clearly show the need for good omnidirectional coverage for rural use.

For ground communication the antenna needs to be as high as possible to gain the advantage of the height factor in the propagation equation. For cars in particular the centre-roof position gives the best omnidirectional coverage as Fig. 12.3 indicates. It will be noted that a forward position gives enhanced rearward coverage and vice-versa. The reason is clear from Fig. 12.2: the off-centre position tilts the elevation pattern. It can be demonstrated that the elevation pattern is a function of the whole length of the car and not just of the roof. Apertures such as the front and rear screens and side windows cause

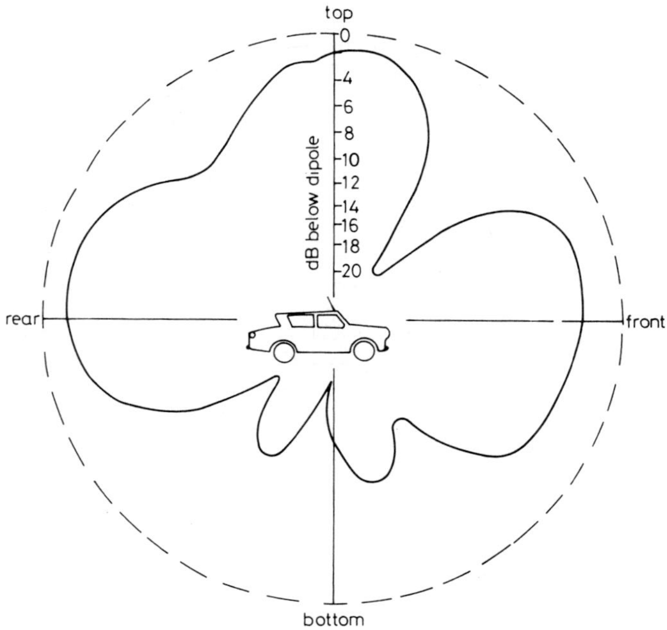

Figure 12.2 *Elevation pattern of roof-mounted whip measured on scale model in free space*

strong current concentration in the conducting members joining the roof to the remainder of the body. Above about 100 MHz it is possible to mount an antenna inside the normal saloon car and to obtain reasonable coverage with little loss of gain although in the UHF band there may be some pattern break-up because of the electrically-large separation of the roof pillars. Measurements on delivery vans with fewer and smaller windows show less variation in pattern with roof position.

The impedance variation for the same antenna in different parts of a vehicle may be considerable, certainly more than the makers of many car antennas anticipate, to judge from their installation instructions. Fig. 12.4 shows just how much variation may occur; the same antenna was moved to different positions as shown.

Apart from this variation there may be appreciable change of impedance with time because of corrosion. This is a particular problem where the antenna is installed on a curved surface and arises from problems in ensuring adequate grounding of the feed cable outer to the body of the vehicle. Experience suggests that the best method is to ensure clean metal-to-metal contact in the initial installation and to protect the joint with petroleum jelly replaced at intervals. Without these precautions corrosion may lead to significant loss of efficiency.

Mounting on the car wings produces more distortion as Fig. 12.5 shows and a mean loss of about 3 dB compared with the roof positions. Some use has been made of wing mirrors adapted to serve also as antennas. Essentially these are

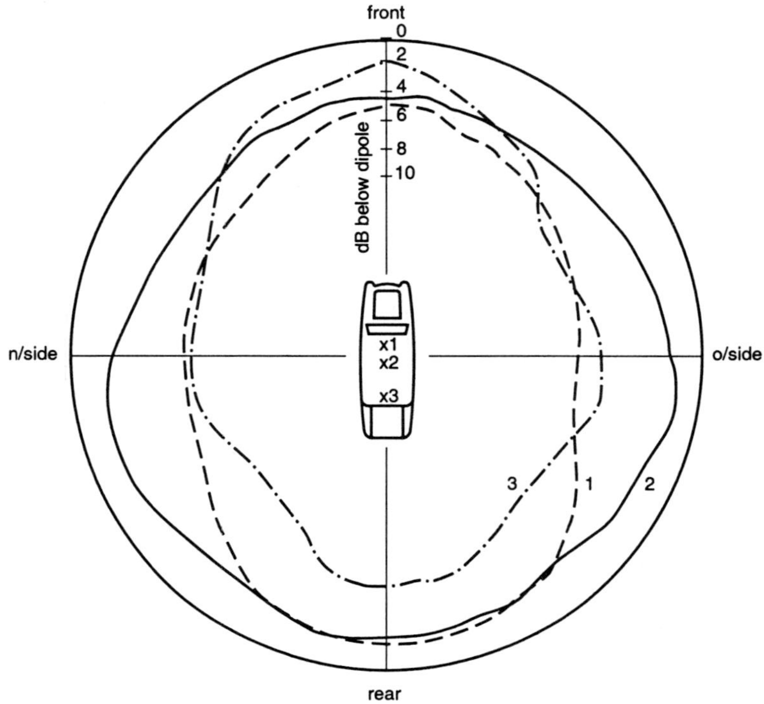

Figure 12.3 *Azimuth patterns of a VHF monopole at three positions on car roof*

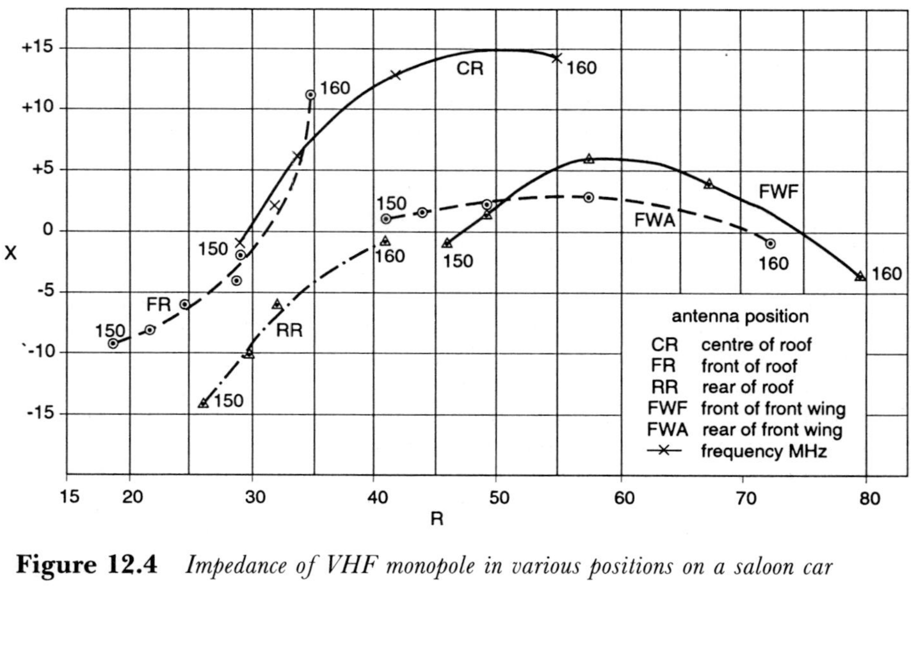

Figure 12.4 *Impedance of VHF monopole in various positions on a saloon car*

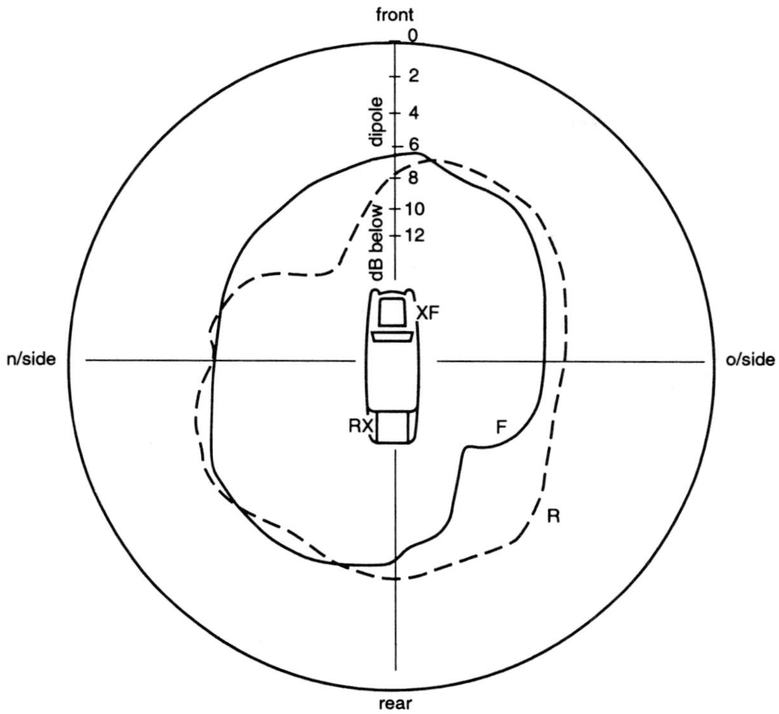

Figure 12.5 *Azimuth patterns of a VHF monopole on front and rear wings of car*

short monopoles insulated from the car body and incorporating a preset matching section in their bases. They tend to be rather less efficient than a $\lambda/4$ monopole since some inductive components are usual in the matching network.

Inconspicuous antennas either of thin wire or of conducting film have been fitted to front and rear windscreens. These tend to be rather unidirectional in the VHF and UHF bands since they involve some radiation into the vehicle and out of other apertures.

Although their low cost makes whip antennas attractive for the civil user they are vulnerable to damage in several ways:

Vandalism
Low overhead clearances: bridges, garages
Automatic car washes.

On public service vehicles it is usual to fit some form of low height antenna. This may be a type of transmission line antenna (Section 9.2.1) or a pancake annular slot (Section 5.5).

Door mirrors have been tried as antennas but are not very satisfactory. They are usually polarised horizontally rather than vertically and have irregular radiation patterns.

12.2.2 *Other wheeled civil vehicles*

Vehicles which do not have continuous metal upper surfaces present special problems. Examples include fire appliances, flat-bed lorries, earth-moving equipment, tractors and articulated vehicles. In some instances some form of coaxial dipole may be the best solution. As they tend to be taller than cars and vans, the height gain available can compensate for irregular radiation patterns. The double hula-hoop of Fig. 9.11 was designed for just such a situation.

A trend towards the use of non-conducting materials can have two effects on vehicle-mounted antennas. In the first place it can affect radiation patterns and possibly impedance, in the second place it may reduce the shielding between engine ignition systems and electronics and the radio system. Interference by the ignition system on the radio and interference by the radio transmission on the car electronics are both possible and should be investigated in any new installation.

12.2.2.1 *Motor cycles*

Motor cycles are very unsatisfactory vehicles on which to fit antennas. The main problems are:

(i) Small electrical size
(ii) Tubular structure
(iii) Effect of rider
(iv) Low height

Even if transmission is not attempted in motion, the presence of the rider can affect the antenna performance in reception. If transmission is restricted to times when the motorcycle is stationary — which will improve speech quality by removing wind noise and reducing ignition noise — the rider will probably still be on the machine so there may be no antenna improvement.

The best position, both from the performance aspect and operationally, is at the rear of the machine but this undoubtedly gives a distorted radiation pattern. The tubular structure means strong currents in individual members; there would be some advantage in ensuring that the return path to the radio equipment is as broad as possible, perhaps by the addition of a grid of wires or a mesh on which the antenna base is fixed.

The low height means that at best the performance will be comparable with that of an antenna on a car wing.

There should be some advantage in using a coaxial dipole, perhaps with a shortened lower section as in Fig. 12.6. This would raise the feed-point and ought to reduce the effect of the vehicle as a ground plane. This may well have been tried but the author has no evidence of such an attempt.

12.2.3 *Military vehicles*

Whilst civil systems can use an elevated base station with relay facilities for car-to-car operation this is not always possible for tactical communications systems used by the military. For this reason, ground-to-ground communications have traditionally been in the range 25–76 MHz although this range is being edged up towards 100 MHz. The lower propagation losses favour the use of the lower VHF bands although against this must be set lower antenna heights in terms of wavelength.

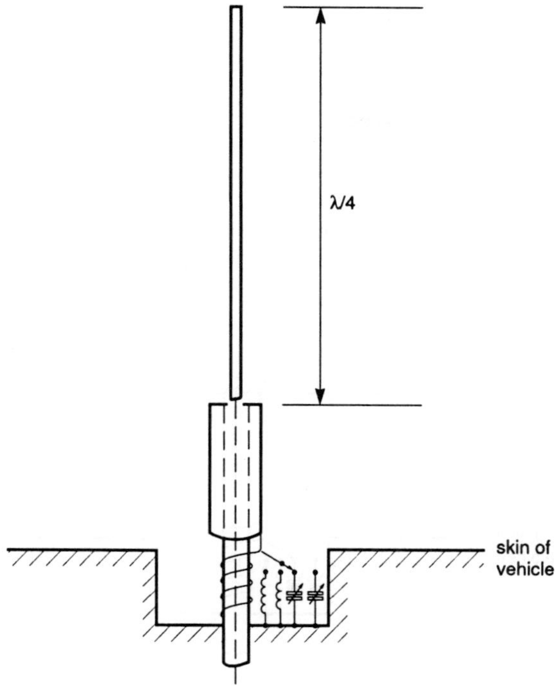

Reference 7

Figure 12.6 *Loaded coaxial dipole on land vehicle*

The main requirements for antenna systems for military vehicles are:

Efficiency
Good omnidirectional coverage
Broadband coverage
Robustness
Adaptability to a wide range of vehicles
Inconspicuousness
Ease of replacement

These requirements are not always compatible and to some extent logistics have favoured the use of $\lambda/4$ whip antennas or coaxial dipoles loaded to reduce height. Considerable work has been done in the USA with the latter type; it is claimed that the impedance and radiation pattern are less sensitive to vehicle shape than are base-fed antennas. This work has been done by or for the US Army Electronics Command at Fort Monmouth, New Jersey, or the US Navy NOSC at San Diego. A typical example of the loaded coaxial dipole is shown in Fig. 12.6. Among the many reports on this subject is Brueckmann [7]. Some of the other work is reported in Proc. ECOM–ARO Workshop [42].

These antennas do not have broadband impedance performance and required complex tuning units or possibly a number of switched pre-tuned sub-bands. An automatically tuned shortened version of the standard US type AS 1729 is reported by Andes [4]. These whips and dipoles suffer from two main disadvantages:

Vulnerability
Visibility

Tests show that a whip antenna of no more than 6 mm diameter on a tank can be seen at distances of up to 8 km against the horizon when the rest of the vehicle is hidden. The straight vertical line of the antenna attracts attention as being out of place in natural surroundings. Although these antennas usually have flexible bases or are themselves flexible to minimise impact damage the number of annual replacements even in peace time must run into hundreds of thousands, if not millions world-wide. This is why there has been a search for a robust low-profile antenna for many years. One such candidate is shown in Fig. 9.14.

The introduction of fast frequency-hopping radios has simply compounded the problems of the military antenna designer. Systems using switched-band or continuous tuning are incompatible with fast hopping rates. There appear to be a number of different approaches:

(*a*) To achieve broadband operation by resistive loading. The penalty here is loss of range because of reduced antenna efficiency.
(*b*) In a switched-tuned system to replace any mechanical switches by diode types with fast response times and to use coded frequency information to determine when to switch. A possible problem is power handling limitation. Microprocessor control is described by Demmel and Stark [11].
(*c*) In a continuously tuned system, to substitute as many fixed tuned elements as required to cover the whole frequency band together with fast switching. The system requires to be calibrated at sufficient frequencies for each antenna on each vehicle and the frequency information stored. This sounds formidable but could be done when the equipment is first installed.

It is clear from the above that these problems really arise in the transmit mode because of the tighter VSWR requirements. The more these can be relaxed the easier the task will be for the antenna designer.

12.2.4 *The vehicle as an antenna*

A number of attempts have been made to use the vehicle itself as a radiating system. Some of these have been for camouflage, others to remove an antenna from an operationally restricting position. The slot antenna appears an attractive candidate in that slots naturally occur in a number of places on vehicles. Shaffer and Ikrath [35] claimed some success with door slot antennas but the author considers this to be over-optimism. The position of most natural slots is unlikely to give good radiation patterns or anything but narrow impedance bandwidth. The author has tried shunt-feeding vertical members on motor cars but the radiation patterns are generally poor and the structural work required to obtain a good VSWR is out of proportion to the benefits obtained.

The use of a pancake type annular slot (Fig. 5.21) will probably result in a more effective radiator although it does not meet the requirement of being totally hidden.

12.2.5 *Railway vehicles*

The main constraint on antennas for railway vehicles is overhead clearance. This usually means that for the VHF band some form of top-loaded antenna is necessary. This could be one of the transmission line antennas described in Chapter 9 or the pancake annular slot (Fig. 5.21). Corrosion may be a problem particularly where steam locomotives are still in use and antennas may need to be sufficiently robust to withstand automatic washing systems. In most instances the top of locomotive, carriage or van will have an adequate metal surface to act as a ground plane. It may be necessary to check that good RF continuity exists if the metal surface consists of a number of sections. Naturally-occurring slots could ruin what would normally be good all-round coverage.

In some cases line-side communication may be by leaky feeders. In this instance a small transmission line antenna on either side of the train will be required. It should be possible to provide a flush-mounted antenna, for example by recessing the antenna of Fig. 9.5 in a shallow cavity.

12.3 Marine antennas

Apart from specialised military requirements, the VHF and UHF bands are used mainly for communications — in the 160 MHz band for general ship-to-shore and from 225 to 400 MHz for Western navies. The main problem is avoiding the shadowing caused by the superstructure. Kubina [22] illustrates the problem with computed radiation patterns of a UHF antenna near a tubular mast. Unfortunately the dimensions are not given in this reference. The minimum due to the mast is clearly shown as is the ripple due to re-radiation from it. It is not surprising to find that this pattern is very similar to that of a vertical antenna on an aircraft fuselage forward of the tail fin.

To minimise these effects antennas will be mounted out on yard-arms as far from the mast as possible. On vessels where the mast-head is not required for a higher priority system such as a DF antenna, the mast-head will be the best position. Because of the lack of a counterpoise, VHF and UHF antennas will normally be coaxial dipoles. Law [24] gives details of many systems used by the US Navy. Kitchen [20] describes the design of a UHF bi-cone (Fig. 2.18) which, with variants in construction, forms the basis of many naval installations.

The idea of a circular array around a cylinder is not exactly new but the main problem has been to achieve a broadband solution. Wyatt [41] has described arrays of four broadband dipoles and a more complex array is described by

Rahim *et al.* [30]. This consists of 16 broadband dipoles in front of individual reflectors. By use of a Butler matrix feed network this array could also provide a single rotatable null.

12.4 Aircraft antennas

This section covers both fixed- and rotary-wing aircraft.

12.4.1 *Antenna coverage*

The coverage requirements can be divided into three groups for simplicity:

(i) *Ground-to-air and air-to-air*

Equipment functions	*Coverage*
Communications	Omnidirectional in azimuth and
Navigation	near-azimuth angles
Identification and air traffic control	Linear polarisation
Telemetry and guidance	
EW	

(ii) *Satellite-to-air*

Equipment functions	*Coverage*
Communications	Upper hemisphere
Navigation	Circular polarisation

(iii) *Position fixing*

Equipment functions	*Coverage*
ILS marker and glideslope	Directional
Sonobuoy on-top indicator	Linear polarisation
Radio altimeter	
Homing	

At first sight it might appear that ground to air operation would require lower hemisphere coverage. Simple link budget calculations will show that the range at an elevation angle of 30° from the ground (30° depression from the air in level flight) is so much less than the horizon range that the path loss has reduced by far more than the reduction in any real radiation pattern at VHF and UHF. Table 12.1 gives typical figures for an aircraft height of 100 000 ft.

In calculating coverage with reference to the aircraft axes, the range of aircraft manoeuvre has to be considered. Two classes of aircraft are recognised:

Table 12.1 *Reduction in free space path loss referred to horizon range for a height of 100 000 ft*

Depression angle	Reduction in path loss
10°	11 dB
20°	16·6 dB
30°	20 dB

(*a*) Highly manoeuvrable with coverage up to ±65° elevation with respect to aircraft horizontal
(*b*) Less manoeuvrable with coverage up to ±30°

Commonsense has to be applied because the maximum manoeuvre angles may apply for so short a time that they may be neglected for some radio systems.

The electrical requirements for many aircraft antenna systems are not defined and a specification has to be derived for each system for each aircraft type or at least each class of aircraft. This specification has to be based on the operational scenario and the aircraft performance. One must not lose sight of the need for aircraft to communicate while on the ground; this might well dictate an antenna on the upper rather than the lower surface of the aircraft because of height gain and shielding by undercarriage. These factors would not be relevant in flight although height gain could be important in hedge-hopping military helicopters.

12.4.2 *Siting constraints*

12.4.2.1 *Physical*

The height of an antenna which can be fitted to various part of the airframe may be limited by:

- Ground clearance, including flat tyres and fully compressed oleo legs and allowance for uneven ground for some aircraft and most helicopters.

Other areas of airframe may be restricted by any of the following:

Undercarriage
External stores
Access panels
Fuel tanks
Ground handling equipment
Pitot static tubes

It may also be impractical to mount antennas in other areas because of the need to maintain airframe structural integrity or because any antenna would significantly affect the aircraft performance or would be within the cone angle of jet efflux.

12.4.2.2 *Mechanical*

Any aircraft antenna will be subject to a wide range of external factors which will govern the type of antenna used and will influence its detailed design. These factors can be grouped under the following headings:

(*a*) *Aerodynamic*
Drag
Side load
(*b*) *Climatic*
Wide ambient temperature range
Humidity
Driving rain
Ice
Pressure differential

(*c*) *Dynamic*
Vibration
Shock and acceleration
(*d*) *Chemical*
Hydraulic fluids
Fuels and oils
Salt

(*a*) *Aerodynamic loads*
The wind loads on an external antenna can be resolved into two orthogonal components:

- Drag, along the line of flight
- Side load, normal to the line of flight.

Both are proportional to:

- Air density
- Local air velocity squared
- Frontal area
- A factor dependent on the cross-section or degree of streamlining of the antenna.

Increasing the streamlining to reduce the drag inevitably increases the side load when the local air flow is not parallel to the axis of symmetry of the antenna. For example, the side load and drag of the antenna of Fig. 3.16*b* are shown in Fig. 12.7. Since the stiffness in bending of a hollow aerofoil is proportional to the cube of its thickness there are practical limits to the degree of streamlining possible. If the chord (front to back dimension) is too large the azimuth radiation may cease to be omnidirectional. Fortunately antenna drag is perhaps not quite as critical to aircraft speed as it once was but it still has an effect on range and payload and hence the economics of operating.

(*b*) *Climatic factors*
The temperature range depends on the following factors which will vary for each aircraft type according to its role:

- Ground low temperature may be as low as −60°C in polar regions
- Ground high temperatures including solar heating up to +85°C
- Low temperature at altitude, down to −55°C above 30 000 ft
- Kinetic heating due to aircraft speed, up to $0{\cdot}9\,(V/100)^2$ where V is speed in miles per hour.

Thus the range might vary from −40°C to +50°C for an aircraft used in temperate regions to as much as −60°C to +130°C for supersonic aircraft such as Concorde or high performance fighters. Such wide temperature ranges call for high standards of material technology in antenna design.

(*c*) *Dynamic loads*
Vibration fatigue used to be one of the most serious problems encountered by the aircraft antenna designer. Whilst it certainly cannot be overlooked, there is now a much better understanding of it and failures in service are comparatively rare. Vibration frequencies up to 2000 Hz may be encountered in jet aircraft but their transmissibility through the airframe may be very variable. Frequencies around 10 Hz may be associated with the fundamental resonance of helicopter rotor systems.

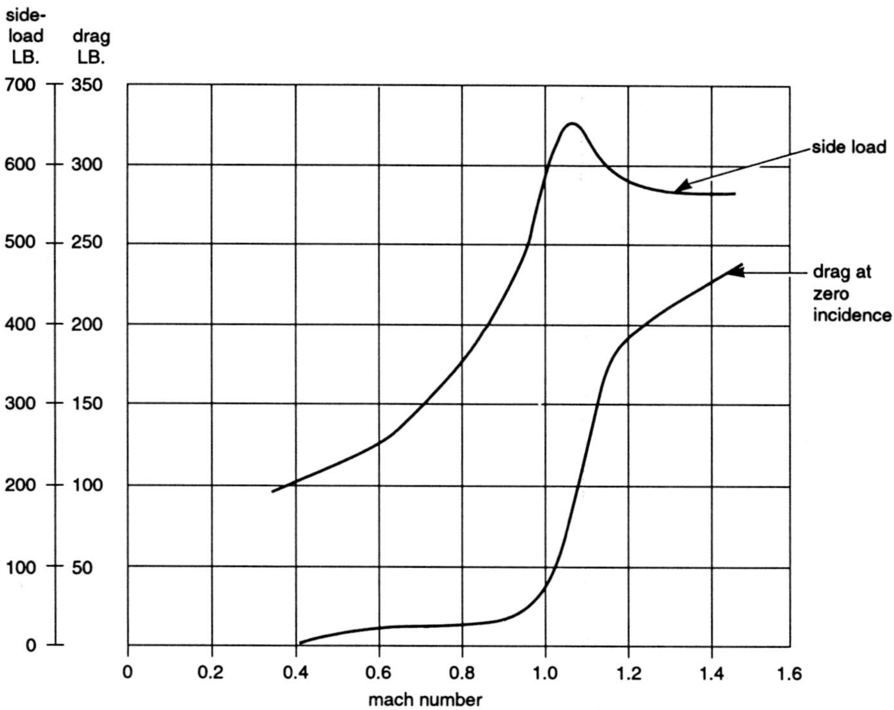

Figure 12.7 *Drag and side load of VHF bent sleeve antenna for transonic flight*

Acceleration and shock loads are, in general, not serious problems, the loads being small compared with the aerodynamic and vibration forces. Acoustic noise is rarely a problem since areas of high noise are seldom suitable for antennas from siting considerations.

Ice accretion can be important. Ice tends to build up on the front edge of a blade in flight; it alters the antenna's natural resonances and may cause serious torsional vibration. Two possible hazards may follow:

- The antenna may break and cause airframe damage or fall to earth
- Ice shed from the antenna may enter engine intakes and cause catastrophic damage.

(*d*) *Chemical factors*

Some of the many fluids used on an aircraft have a deleterious effect on a wide range of insulating and sealing materials and thus limit the choice which is already much circumscribed by other environmental factors. Non-flammable hydraulic fluids in particular attack many rubbers and paints. These exotic fluids are not the only ones which have to be considered: the bilge of a pressurised airliner may contain a nauseous mixture of water, sweat, grease, nicotine — to name just a few constituents.

The underside of any aircraft is covered with oils and grease and in certain areas there may be conductive rubber from the tyres, as well as mud and, on occasions, salt from icy runways.

12.4.3 *Effects of the airframe*

12.4.3.1 *Omnidirectional antennas*

12.4.3.1.1 *Vertical polarisation*

Most antennas for this function will be monopoles, including electrically short ones, mounted on the fuselage. Provided that the antenna is not close to the end of the fuselage, the pitch plane or fore-and-aft elevation pattern will approximate to that of a monopole on a flat ground plane (see Figs. 3.4–3.8) if the antenna is on the underside of the fuselage. On the upper surface, diffraction caused by the tail fin will result in one or more minima rearward but the forward signal will be comparatively smooth providing there are no obstructions. Fig. 12.8 shows the yaw plane (azimuth plane) pattern of a 1000 MHz monopole on the upper surface of a helicopter. The characteristic lobing due to the rotor pylon in the forward direction and the tail pylon in the rearward direction are clearly seen. Any conductive obstacle parallel to the plane of polarisation will perturb the pattern. As Fig. 12.9 shows, both the chord of the obstacle and its distance from the antenna determine the effect. It is to be noted that a VHF blade antenna close to a 1 GHz antenna will have the same effect on the latter as the tail-fin will have on a VHF antenna at an electrically

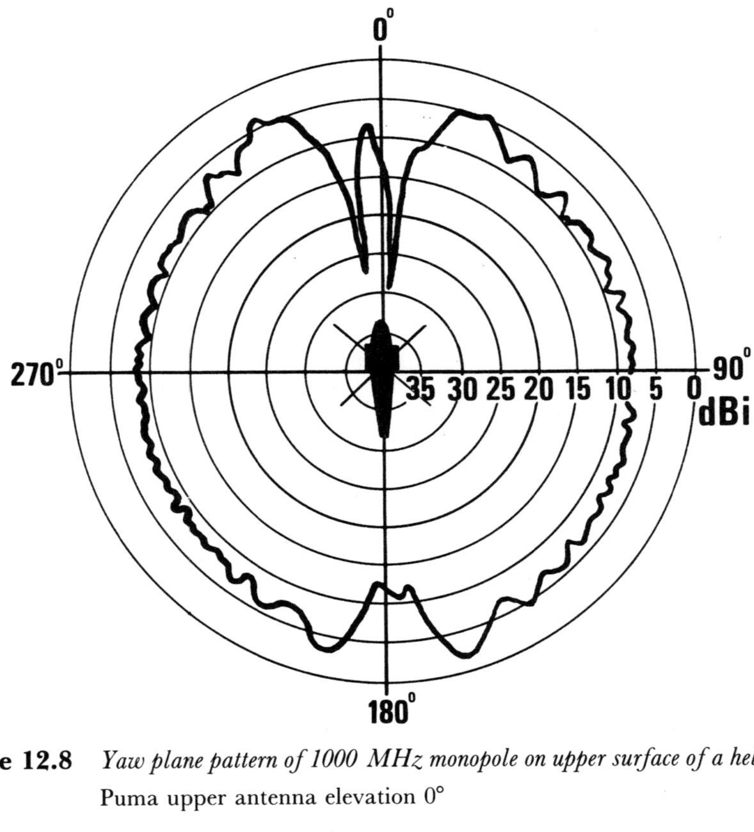

Figure 12.8 *Yaw plane pattern of 1000 MHz monopole on upper surface of a helicopter*
Puma upper antenna elevation 0°

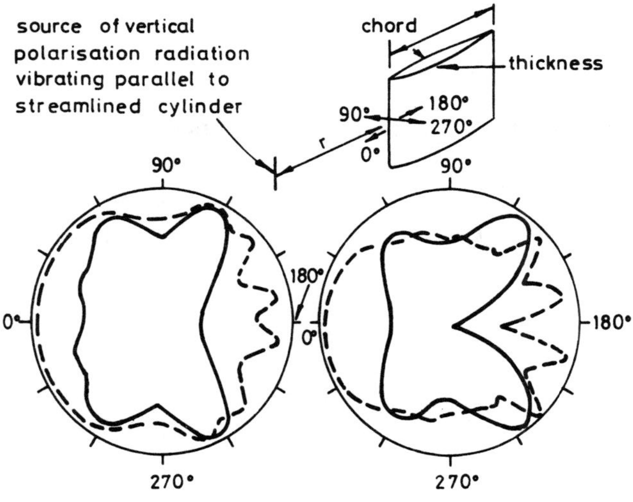

Figure 12.9 *Effect of a streamlined obstacle on radiation pattern of dipole parallel to its own axis*

a Horizontal polar diagram with streamlined cylinder of 0·1λ thickness 1·0λ chord
b Horizontal polar diagram with streamlined cylinder of 1λ thickness 10·0λ chord
--- r = 10λ
—— r = λ

comparable distance. Fig. 12.10 shows the devastating effect on the yaw plane pattern of a 1 GHz monopole when two UHF homing antennas were placed within half a wavelength of it.

In the roll plane the pattern will be similar to that of a monopole on a cylinder unless the antenna is in line with the wings. In this case, depending on the height of the antenna base above the wing surface, there will be both reflection and diffraction effects. Examples of these effects are given by Burnside *et al.* [9]. It is sensible not to mount an antenna on the middle third of the fuselage. Offsetting the antenna will tilt the pattern as in Fig. 12.11.

If the fuselage varies considerably in cross-section then the current flow on the fuselage forward and aft of the antenna may differ considerably if the fuselage is small in terms of wavelength. Such a situation occurs on a small helicopter with antennas in the band 30–70 MHz. Fig. 12.12 shows the remarkable changes recorded at three not very different frequencies for a $\lambda/4$ monopole mounted under the fuselage near the junction of the main body and the tail cone. The overall length of the fuselage was 9·24 m, almost exactly one wavelength at 32 MHz.

Close to the extremities of the fuselage, pattern break-up may occur sometimes producing a minimum in the azimuth plane in the direction of the metal edge. The notch in the pitch plane pattern is shown in Fig. 12.13 for a UHF monopole just behind a completely plastic canopy.

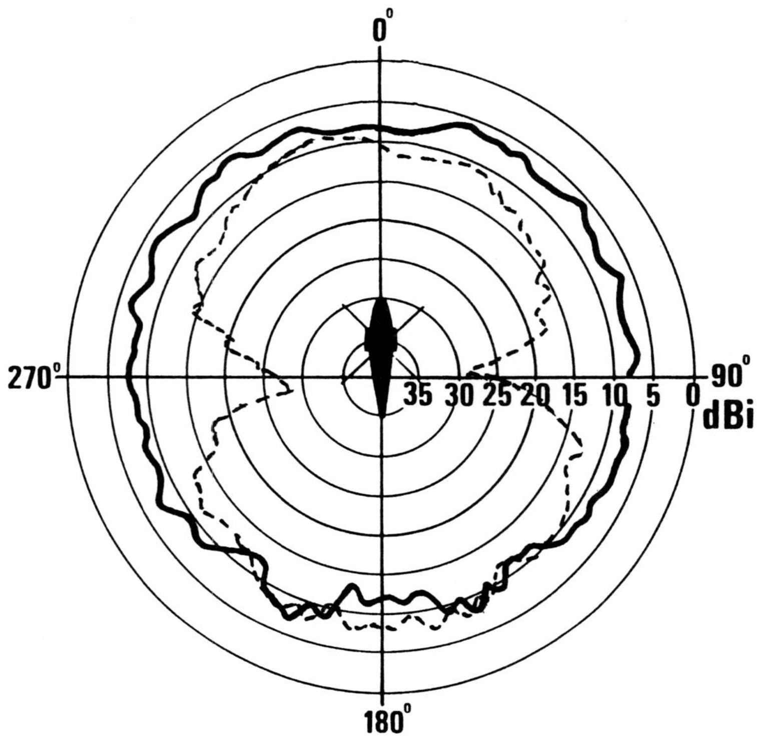

Figure 12.10 *Effect of a pair of UHF blades on 1000 MHz monopole*

—— aircraft clean
– – – with pair of UHF blades

For identification systems requiring cover both above and below the horizontal, rather more care is necessary in siting the antennas. Fig. 12.14 shows three arrangements of two 1 GHz antennas and it is clear that the conventional arrangement of one antenna (2) above and (1) below the fuselage at mid-length actually leaves gaps close to the horizontal in both forward and rearward directions. The combination of antennas 3 and 4 is better and will often be the most suitable for installation. However, two separate antennas 5 and 6 each with yaw plane coverage of about 240° give the best possible arrangement. Antenna 5 could be a horizontal slot with cavity or a triple notch as Fig. 6.16, which would also be suitable for antenna 6. This position has also been paired with position 3 when there was no suitable position 4 for structural reasons. In one instance the tailwheel on a helicopter completely ruined position 4 and an add-on triple notch was a successful alternative.

Lack of suitable space on the fuselage sometimes necessitates the use of the tail-fin cap for antennas. Structurally this is often a convenient position if the antenna can be completely covered by a dielectric cap. There are, however, a number of problems in siting and in impedance. The effect of parasitic notches on radiation patterns was demonstrated by Figs. 6.14 and 6.15. It is not always

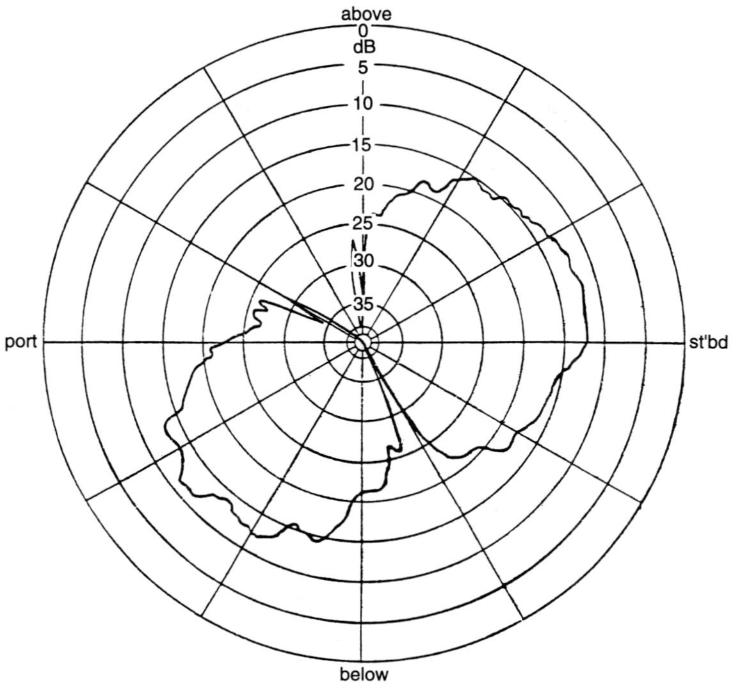

Figure 12.11 *Roll plane pattern for antenna with 30° offset from centre line*

appreciated that movements of rudder and of the horizontal tail plane if high on the fin can also affect the VSWR. A paper by Mahoney 1971 [26], a one-time colleague of the author, demonstrates effects of both moving surfaces and also provides an excellent summary on fincap antennas. Mahoney mentions the problem of current flow down the leading and trailing edges of tail-fins with cap antennas. This problem was analysed by Granger and Bolljahn [15] who demonstrated the existence of a travelling wave down the fin which was then reflected by the top of the fuselage. Fig. 12.15 shows a pitch plane pattern of a VHF antenna on the fin of a small fighter. Since the fin is not very large electrically (about 0·6λ), reflection effects are not as serious as in Grainer's examples.

In one class of aircraft the fincap is an ideal position. These are the aircraft whose horizontal tail surfaces are at the top of the fin or close to it. An antenna mounted above the tail planes is effectively on a separate ground plane and the problems of tailfin are no longer important. Excellent all-round coverage has been obtained from VHF and UHF antennas in such positions.

12.4.3.1.2 *Horizontal polarisation*

Only two systems are in common use. These are the VHF omnirange (VOR) and the localiser function of the Instrument Landing System (ILS). Both operate in the range 108–118 MHz, VOR providing en-route guidance and ILS

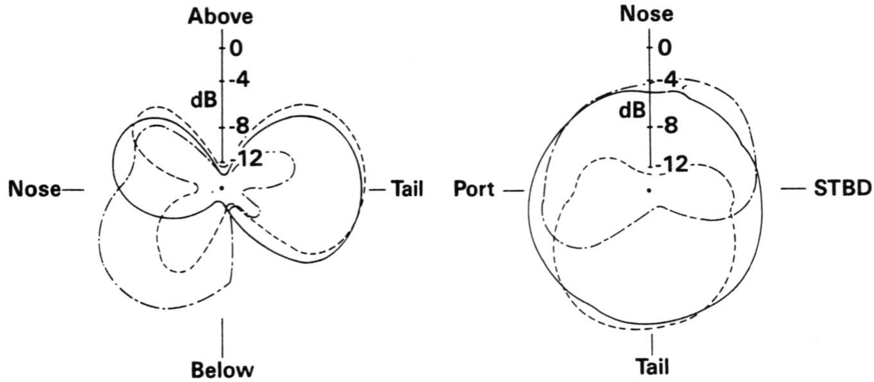

Figure 12.12 *Yaw and pitch plane patterns of VHF monopole under small helicopter*

—— 26 MHz
- - - 32 MHz
–·–· 42 MHz
Vertical polarisation

localiser terminal guidance in azimuth. Whilst VOR requires full omnidirectional coverage, ILS localiser requires primary forward cover plus some rearward cover for overflying the ground system and some cover to the side for acquiring the signal from the ground. It is, therefore, common for the same antenna to be used for both functions especially when the receiver is a common one. Special

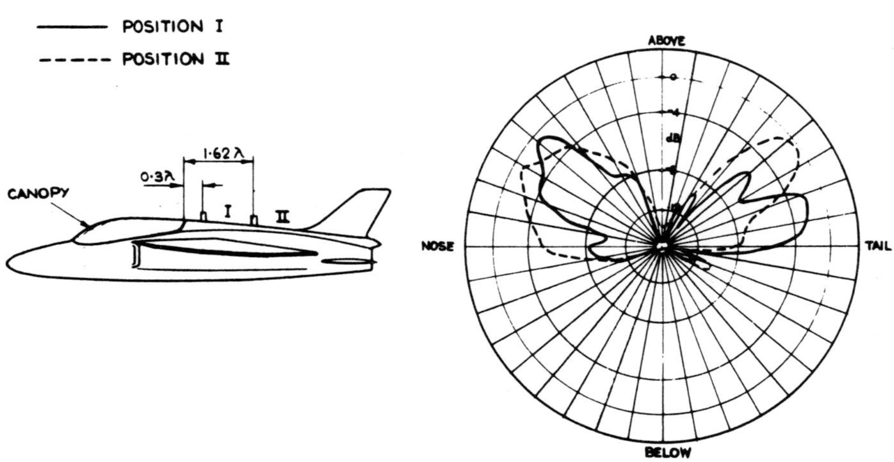

Figure 12.13 *Effect of proximity of metal edge on pitch plane pattern of UHF monopole*

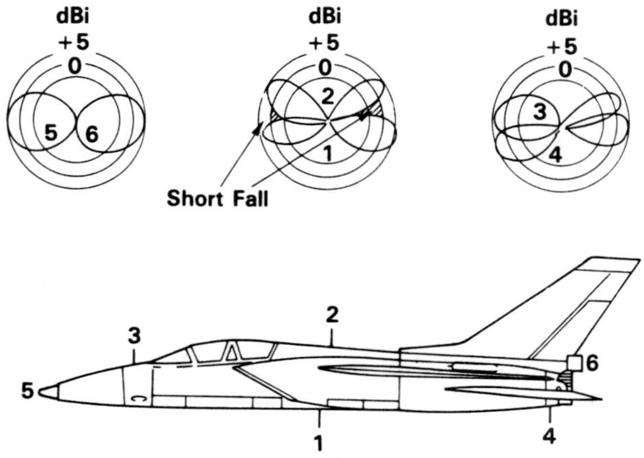

Figure 12.14 *Combined coverage of two antennas at 1 GHz*
Courtesy RAeS

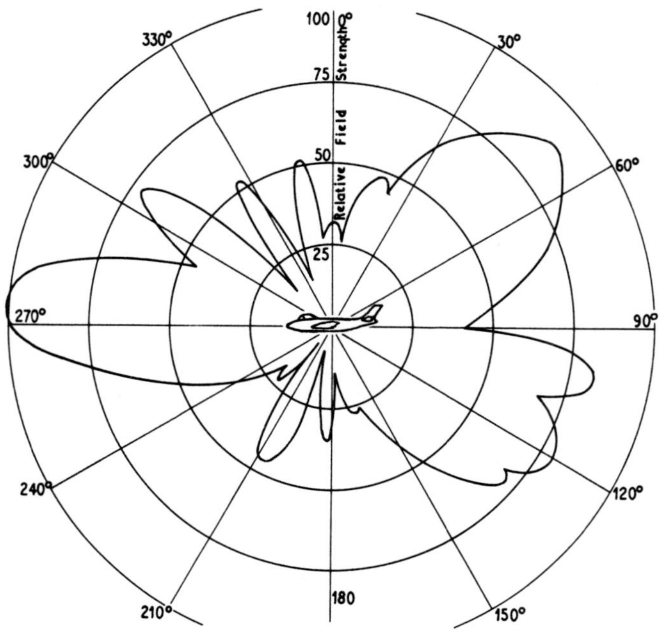

Figure 12.15 *Pitch plane pattern of fincap VHF antenna*

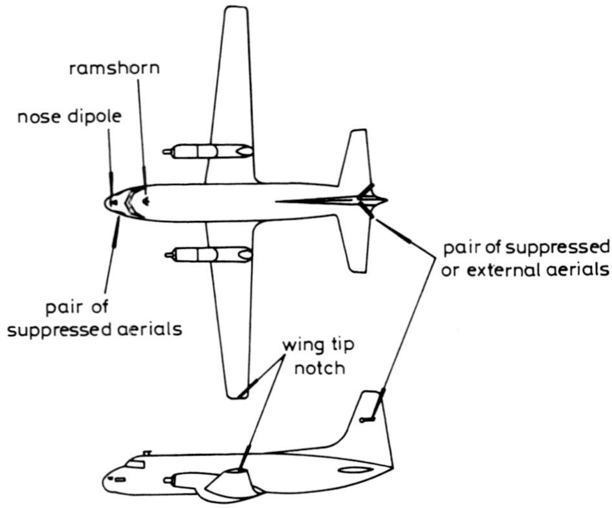

Figure 12.16 *Positions for airborne localiser antennas*

requirements for localiser arise when automatic landing is required as the integrity of the antenna and of the guidance information has to be of a higher order than for normal operations.

The general requirements are for a low level of cross-polarisation and for the apparent phase centre of the antenna to remain substantially constant whatever the aircraft attitude. If the phase centre moves then the apparent position of the aircraft with respect to the guidance system will also move, a phenomenon known as 'course push'. The whole process is discussed by Jones [18]. The antennas he studied are shown in Fig. 12.16 taken from the definitive paper on ILS antennas by Burberry [8]. What emerges clearly from these studies is that asymmetric antennas such as the wing tip notch suffer severely from cross-polarisation when the aircraft rolls. The general conclusions are:

(i) The antenna should be symmetrical about the fuselage centre line.
(ii) The two elements of the antenna, whether physically separated or combined in some form of dipole, must be balanced and fed in antiphase.

These precautions minimise cross-polar radiation from the airframe particularly along the axis of symmetry, the line of flight. They do not of themselves produce the system with minimum phase shift. This is achieved by putting the antenna where the reflections due to the airframe are a minimum in the forward direction. This is in the nose of the aircraft as shown in Fig. 12.17. Fig. 12.18 compares the yaw plane pattern of this antenna with a fin-mounted system. Although the pattern of the latter has some interference lobes in the forward sector, its general signal level is higher than that of the ramshorn also shown in Fig. 12.16 as Fig. 12.19 demonstrates. In fact the ramshorn pattern in Fig. 12.19 was for an antenna under the fuselage which would give better rearward cover than the position above the fuselage. Fig. 12.20 shows why the horizontal gain of the ramshorn is so low — the peak signal is directed upwards. Whilst for most applications a fin mounted system is convenient as well as giving a high signal

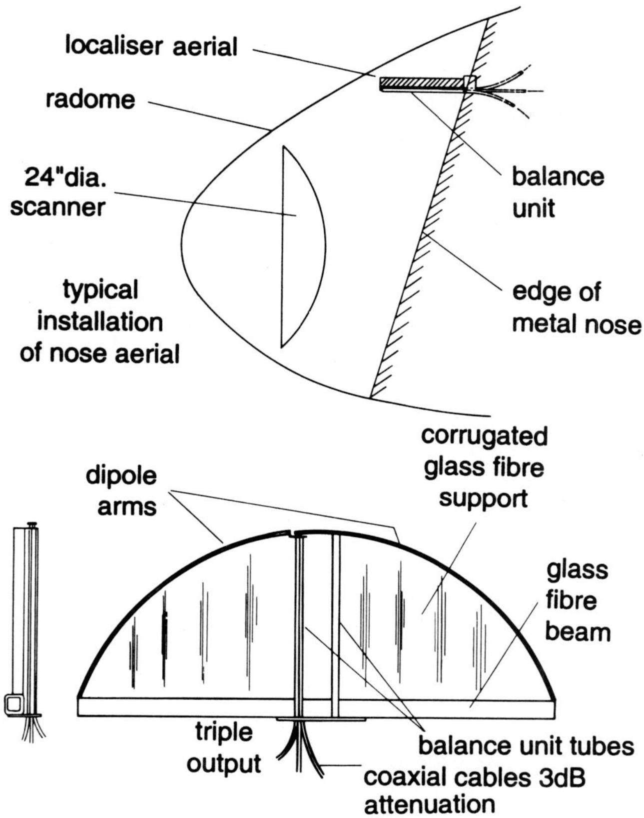

Figure 12.17 *Nose mounted localiser aerial for automatic landing*

level, there can be problems if the aircraft has a high tailplane. Fig. 12.21 shows that considerable care is needed to achieve adequate signal in the horizontal plane because of reflections from the fuselage and the tailplane.

Two of the antennas shown in Fig. 12.16 have physically separated elements — those on the sides of the fin and the forward fuselage. A variety of elements have been used:

Raked $\lambda/4$ monopoles
Bent sleeve antennas as Fig. 3.16
Terminated half-loops
Notch-fed plates
Flush-mounted (suppressed) transmission line antennas

Slots are too large for this frequency band but vertical pannier slots as in Fig. 5.12 could be used at higher frequencies. These separate elements on the

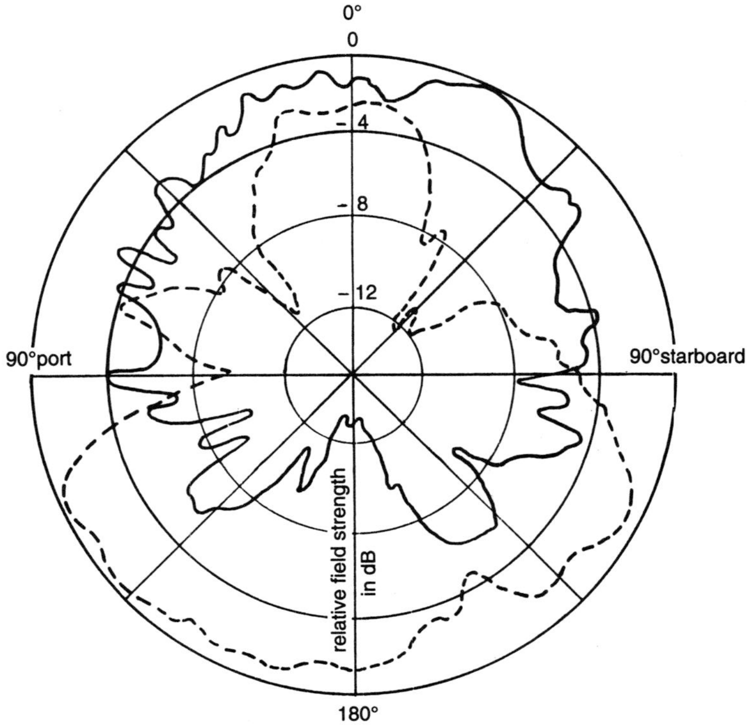

Figure 12.18 *Yaw plane patterns of nose-mounted localiser and fin-mounted pair of monopoles*

--- fin-mounted antenna
—— nose dipole

opposite sides of a conducting surface need to be fed in antiphase. Several different methods of achieving this are possible:

(*a*) Coaxial cables differing in length by $\lambda/2$ electrically connected at a tee-junction
(*b*) Equal length coaxial cables connected to a balun which may be $\lambda/4$ or a shortened loaded type
(*c*) Rat-race

These are discussed in more detail in Chapter 13. It should be noted that in (*a*) the antenna impedances, transferred through the coaxial lines, are added in parallel at the tee junction whilst in (*b*) the impedances are in series at the balun. The translated impedances therefore need to be higher than Z_0 in (*a*) and lower than Z_0 in (*b*) to optimise the VSWR. Scheme (*a*) is really only suited to comparatively narrow bands and is the case here. In a similar arrangement of two bent sleeve monopoles, one above and one below a helicopter tail boom, coupled by a $\lambda/4$ balun the band 100–156 MHz was achieved within a 2·5:1 VSWR to 50 ohms.

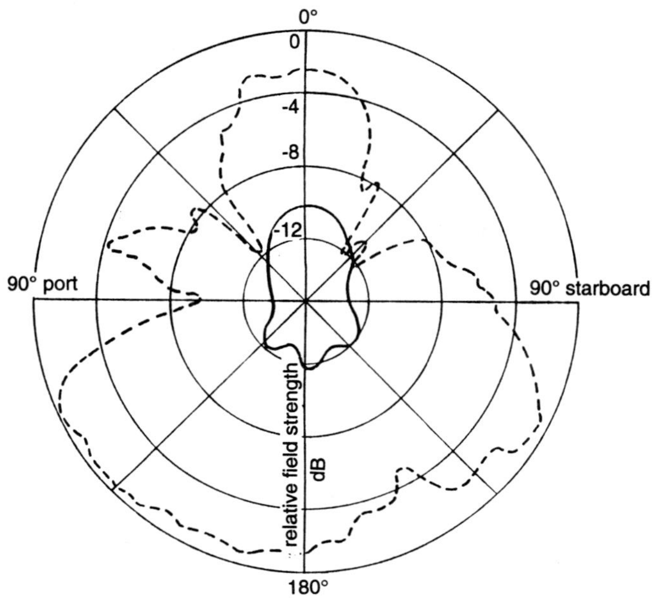

Figure 12.19 *Yaw plane patterns of fin-mounted antenna and ramshorn antenna*

--- fin-mounted antenna
—— ramshorn

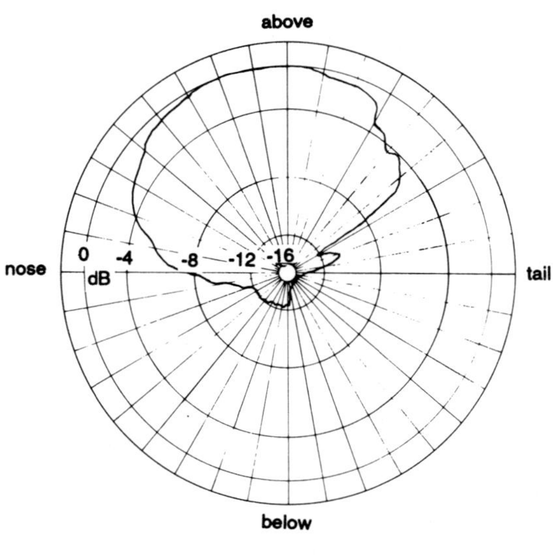

Figure 12.20 *Pitch plane pattern of ramshorn antenna*

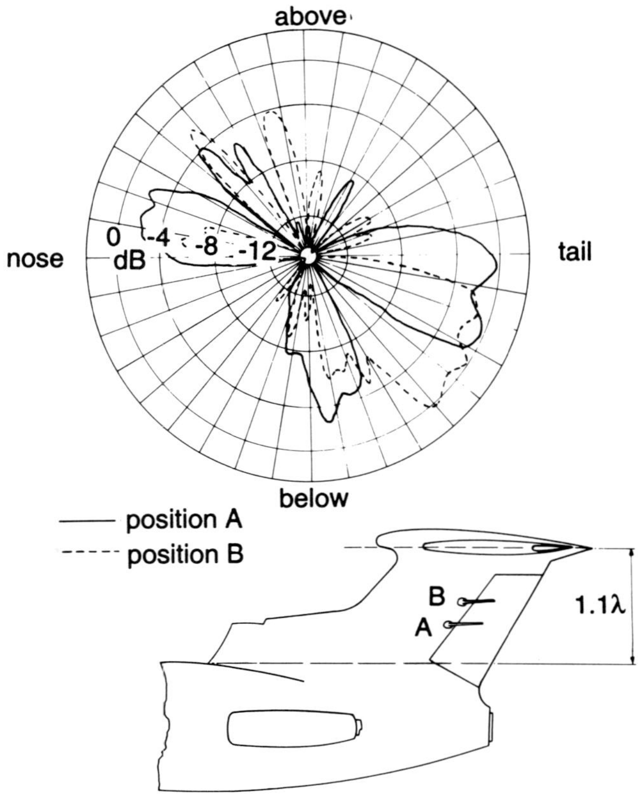

Figure 12.21 *Pitch plane patterns of fin-mounted localiser antennas on fin with high tail plane*

If the separate elements are too far apart interference lobes will occur. Fig. 12.22 shows a pattern for two antennas separated 0·9 m (0·34λ at mid-band). Wider spacings will bring the minima closer to the centre line and will deepen them.

12.4.3.2 *Satellite-to-air communication and navigation*

For operation in normal horizontal flight, aircraft to satellite systems require almost hemispherical coverage, from 10° above the horizon to 90° upwards. This implies that if operation is required during aircraft manoeuvres rather wider coverage with respect to the aircraft is required. Circular polarisation is necessary, some communication systems requiring one hand for transmission and one for reception. Navigation systems which are generally receive-only at the aircraft require only one hand of polarisation.

Anywhere on top of the aircraft will give good upward coverage but, at low angles with respect to the fuselage, systems will degrade to mainly vertical polarisation. It might be argued that mounting at the top of the tail-fin would be an advantage but this is not necessarily true. Any reflection from wings and

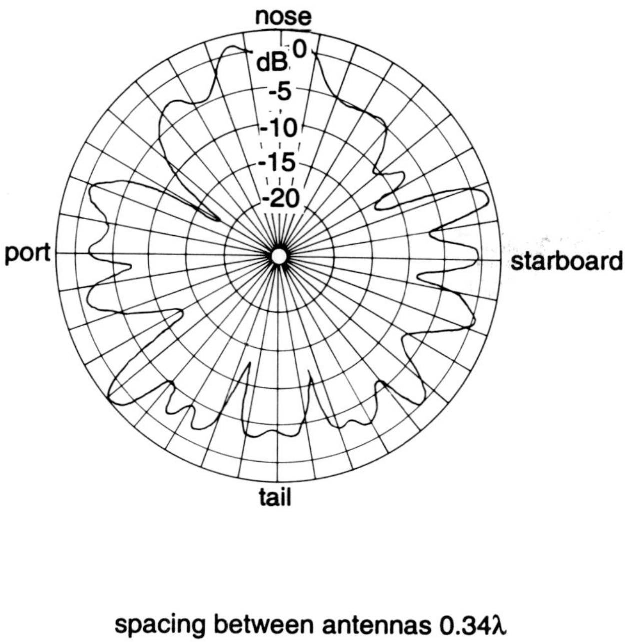

Figure 12.22 *Yaw plane pattern of pair of antennas on side of aircraft nose*

fuselage would cause lobing with consequent reversal of the hand of polarisation and this is very undesirable. Unless downward cover can be restricted, as for example on an aircraft with high tailplanes, the top of the tail-fin does not appear to be a satisfactory position for this function. From the top of the fuselage there will inevitably be some degradation rearward because of the tail-fin.

12.4.3.2.1 *Antenna types*
A number of antenna elements have been used. They include the following:

(*a*) Canted turnstile
(*b*) Crossed slots with backing cavity
(*c*) Crossed dipoles
(*d*) Slot–dipole combinations
(*e*) Patch antennas

Whilst (*a*) is electrically satisfactory it is rather a tall antenna and is not particularly robust so needs a radome. This tends to make it unduly large even at 1560 MHz. One arrangement which minimised the size of the radome was to set the arms of the dipoles at 45° to the line of flight. It was then possible to print two adjacent elements on each of two boards which were themselves canted like the roof of a house (Fig. 12.23). This only required a long but narrow radome which was aerodynamically more acceptable. Another application is described by Werstiuk *et al.* [40].

Generally antennas of lower profile have been preferred. For frequencies up to about 400 MHz comparatively modest gains have been required so single elements have been adequate. A simple arrangement of crossed slots is described by King and Wong [19] and shown in Fig. 12.24. In an alternative feed arrangement the two slots are tapped at the 50 Ω feed-point and the two cables are connected in quadrature. This tends to have a narrower bandwidth. Using a ring of folded dumb-bell slots as in Fig. 12.25 gives 8 dB gain on axis and a bandwidth of 7%. Outside the main beam, however, the hand of polarisation reverses so this has limited application. Herskind [16] describes a circular slot which could be fed by two orthogonal striplines: with appropriate phasing this would give circular polarisation and Herskind claimed that the ellipticity ratio off beam was better than for linear crossed slots. Results of model measurements on a crossed slot installation are given by Bland and Clarke [5].

Crossed dipoles do not appear to have been considered as single antennas because of their comparatively narrow beam width. They have, however, been used in steerable arrays disposed at various angles to provide the required coverage with a higher gain than single elements could give. Different schemes are described by Fraser and Williams [13], Dorier [12] and Ancona and Froidure [3].

Various slot–dipole combinations have been designed. One of these was tested by Bland and Clarke [5]; this consisted of a slewed dipole lying in the plane of a broad slot. The neatest arrangement is that due to Sidford [36] (Fig. 12.26). In this design a folded dipole, coaxially fed, lies in the plane of the slot to which it is coupled by one or two capacitive tabs. The whole slot/dipole assembly can be printed on one double-sided board. The cavity is of the pocket type (Fig. 5.10) and is therefore only $\lambda/12$ deep. This makes it acceptable for mounting on the outside of the aircraft. The antenna has been used in arrays of three elements mounted at 45° to the vertical on either side of an aircraft fuselage. For further details see Sidford [37].

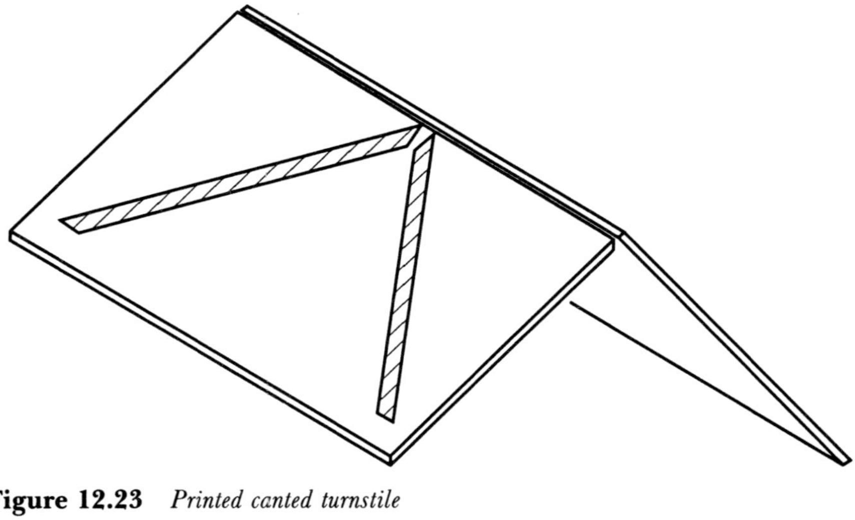

Figure 12.23 *Printed canted turnstile*

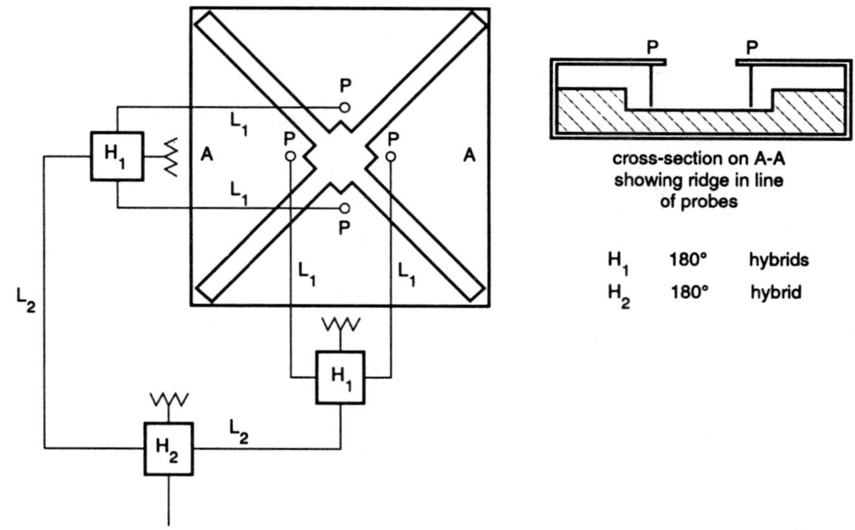

Reference 19

Figure 12.24 *Crossed slots with ridged cavity*

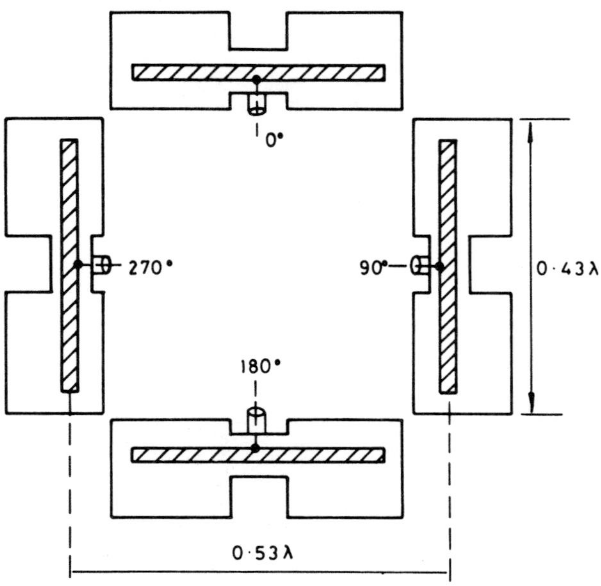

Figure 12.25 *Ring of four folded dumb-bell slots*

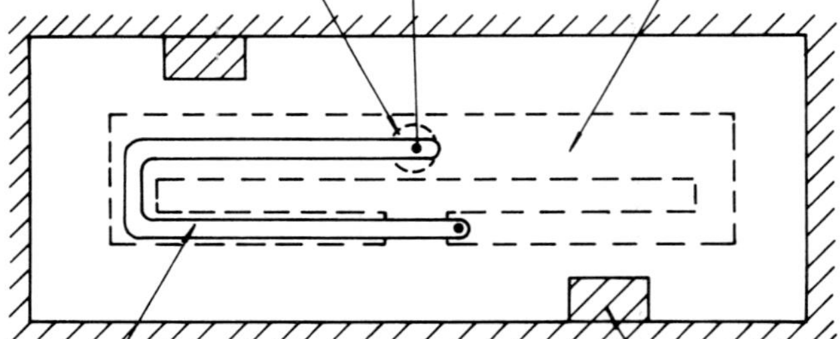

Figure 12.26 *Slot-dipole antenna*

Several types of patch antenna have been suggested. The immediately obvious solution to give circular polarisation is a square patch fed on adjacent sides, the two feeds differing in phase by 90°, Sanford and Klein [32]. Sanford and Munson [33] showed that a single point feed could be used if the patch was not quite square, the feed-point being at a corner or at some point along the diagonal. These three arrangements are shown in Fig. 12.27.

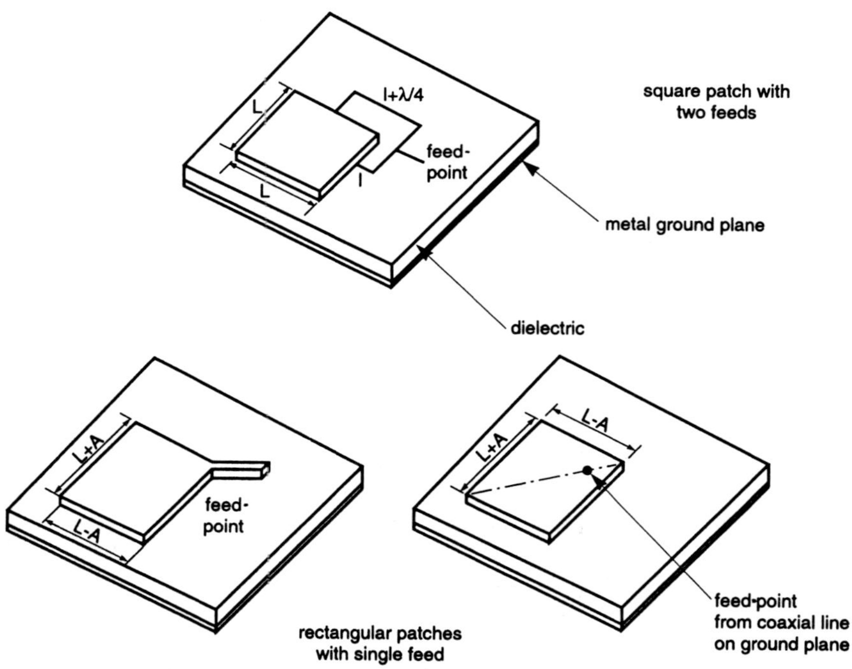

Figure 12.27 *Square and rectangular patches*

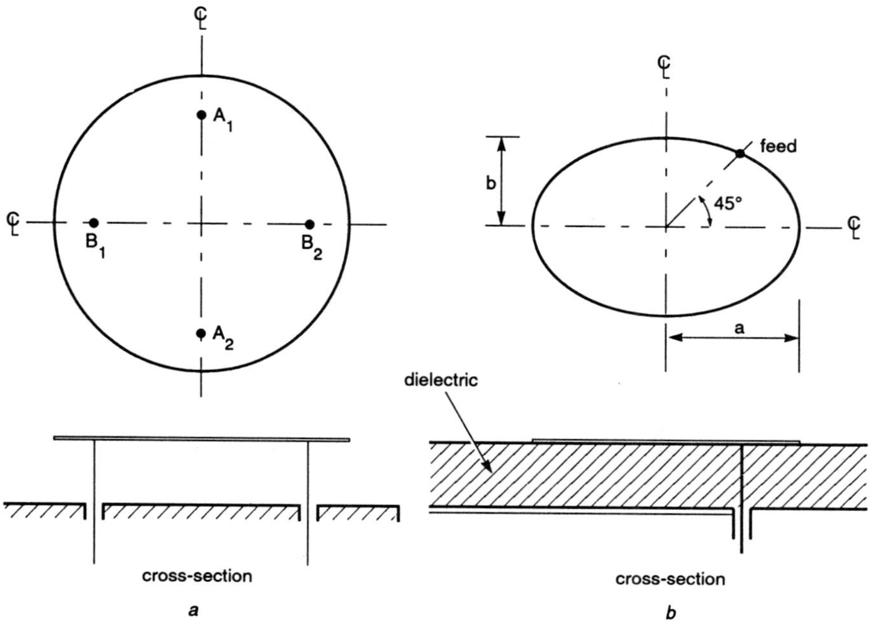

Figure 12.28 *Circular and elliptical patches*

a Circular patch
b Elliptical patch

A number of non-rectilinear patches have been developed. Brain and Mark [6] used a four point feed with 90° progression between feed-points. It is interesting to compare this with Herskind's circular slot. Long and Shen [25] showed that with a slightly elliptical patch ($b/a \approx 0{\cdot}96$) good circular polarisation could be achieved. The bandwidth is, however, small — only 1·5% for 6 dB axial ratio — compared with 12% for Brain and Mark's design. The two arrangements are shown in Fig. 12.28.

12.4.3.3 *Position fixing antennas*

These have very specific requirements so only a few systems are mentioned which indicate specific airframe effect.

(*a*) *ILS marker: 75 MHz*

This requires radiation vertically downward to receive the signal from ground beacons. In theory the sensitivity to the airborne receiver should be adjusted so that the signal received at a given altitude exceeds a set level for a fixed period at a given speed. This is rarely possible because of the different power output and antenna gain of ground stations and the gain and pattern shape of the airborne antenna. Often the latter, which may be a half-wave dipole $0{\cdot}0375\lambda$ below the fuselage or a transmission line antenna with 10 dB lower gain, is merely placed at a convenient point under the fuselage. As Fig. 12.29 demonstrates, even on a fuselage 4λ long the position still has a significant effect on the pitch plane pattern.

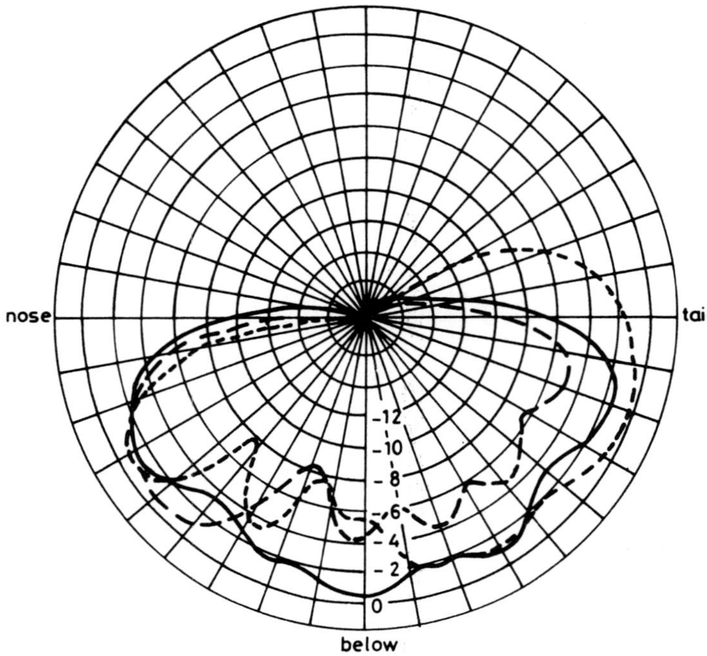

Figure 12.29 *Effect of position on pitch plane pattern of fuselage-mounted ILS marker antenna*

Fuselage length $\lambda 4$, antenna $\lambda/2$ parallel to fuselage, spacing $0{\cdot}0375\lambda$
Antenna positions:
– – – $1{\cdot}15\lambda$ from nose
—— $1{\cdot}91\lambda$ from nose
----- $2{\cdot}67\lambda$ from nose

(*b*) *ILS glideslope: 328–336 MHz*
This requires good forward coverage and horizontal polarisation. The nose of the aircraft is the obvious position for an antenna but its performance can be ruined by the proximity of a weather radar dish (Fig. 12.30). It may therefore be necessary to mount the antenna under the nose or to shield it in some other way from re-radiation by the scanner.

(*c*) *Azimuth homing*
Two types of homing are in general use. They are:

Phase comparison
Amplitude comparison

For phase comparison a pair of identical elements are mounted symmetrically on either side of the fuselage upper or lower centre line. They are linked by a fixed phase shift which can be switched to either antenna. This generates a pair of mirror image slewed patterns. Kunachowicz [23] gives results for a VHF pair on an aircraft model. It must be said, however, that to place a homing pair on the upper surface of an aircraft with a high wing betrays a lack of familiarity with antenna siting.

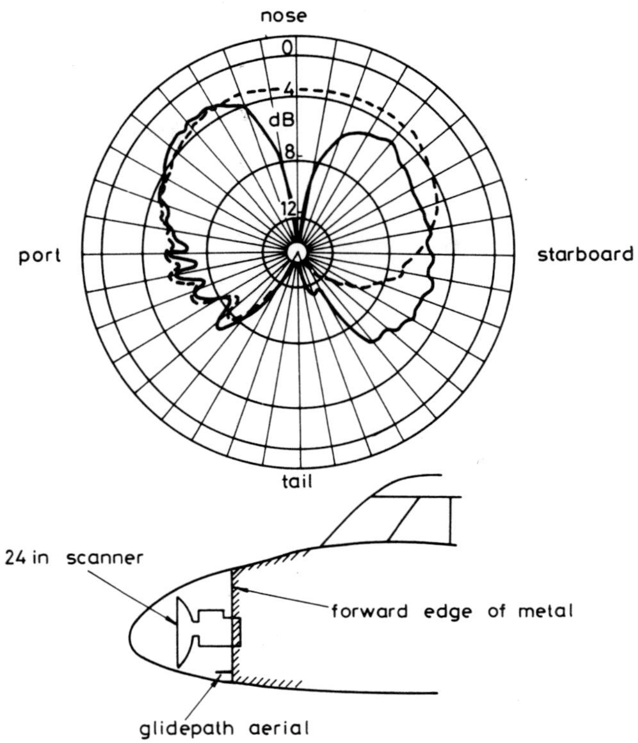

Figure 12.30 *Effect of weather radar dish on ILS glideslope antenna*
—— scanner in central position
– – – scanner removed

Because there is often a lack of good sites for both homing and communications antennas, the author evolved a scheme by which a pair of UHF homing antennas could also be used for omnidirectional communication. The scheme requires a pair of broadband blade antennas, the particular type not being important. For the communications mode the two antennas are coupled together and fed through a fixed matching section to the radio equipment. For homing the antennas are connected to a phasing unit and thence to the homing system. The arrangement is shown schematically in Fig. 12.31.

In amplitude comparison systems the two antennas are still symmetrically mounted but are independent, each producing an asymmetric pattern. This may be created by proximity to the side of the fuselage or may be inherent in the antenna itself — a terminated loop or a short Yagi for example. The main problems are in ensuring that the airframe is electrically symmetrical; doors and removable panels can create some odd effects. Fig. 12.32 gives one example taken from a system using flush-mounted horizontally-polarised wing tip antennas at 176 MHz. The nacelle on the starboard wing skews the equi-signal line and also flattens the left/right ratios on the port side.

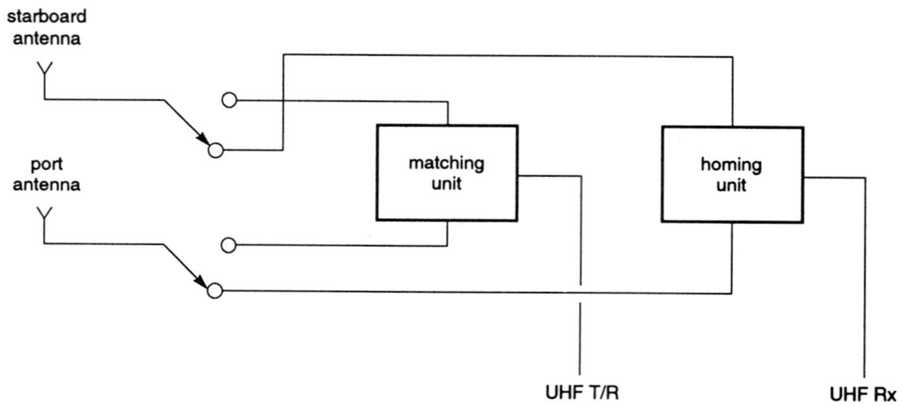

Figure 12.31 *Use of pair of UHF antennas for homing and communications*

12.4.3.4 *Some general airframe effects*
In Chapter 6 the effect of parasitic notches was described. It is important to appreciate just how widespread such discontinuities are. Fig. 12.33 shows typical ones. Many are not as obvious as these: it is not uncommon to find that wings and tail units are not connected to the fuselage at their leading and trailing edges, the gaps often being hidden with non-conductive fairings.

In aircraft, particularly helicopters, with large non-metallic areas on a metal framework there may be strong currents in the comparatively narrow metal members just as in the roof pillars of a motor car. These may have a serious effect particularly on low-VHF homing systems.

12.4.4 *Antennas for missiles, rockets and targets*

VHF and UHF antennas are rarely used operationally on missiles but are used for telemetry and control during test firings. Similar systems are used on target drones and pilotless aircraft. Unless launch constraints require the use of flush-mounted or at least low-profile antennas simple monopoles and notch-fed antennas are practicable. The coverage requirements are usually omnidirectional in the roll plane and good side and rearward coverage in the yaw plane since most tracking systems will be to the side or behind the missile. A very good analysis of the problems and discussion of suitable antennas is given by Mahoney [27]. One antenna not mentioned in this paper is the spike projecting forward from the nose. Clearly this forms with the vehicle body an asymmetric dipole and can have no axial coverage. Fig. 12.34 shows the yaw plane pattern of such an antenna on a 4λ body.

If the vehicle is considered as a finite cylinder certain general conclusions can be drawn:

(i) A monopole normal to the cylinder will give some axial radiation but have a null along its own axis.
(ii) A diametrically-opposed pair of monopoles fed in antiphase will give good broad fore and aft coverage but will have roll-plane minima close to the axis of the pair.
(iii) Two opposed monopoles fed in phase will have an axial null.

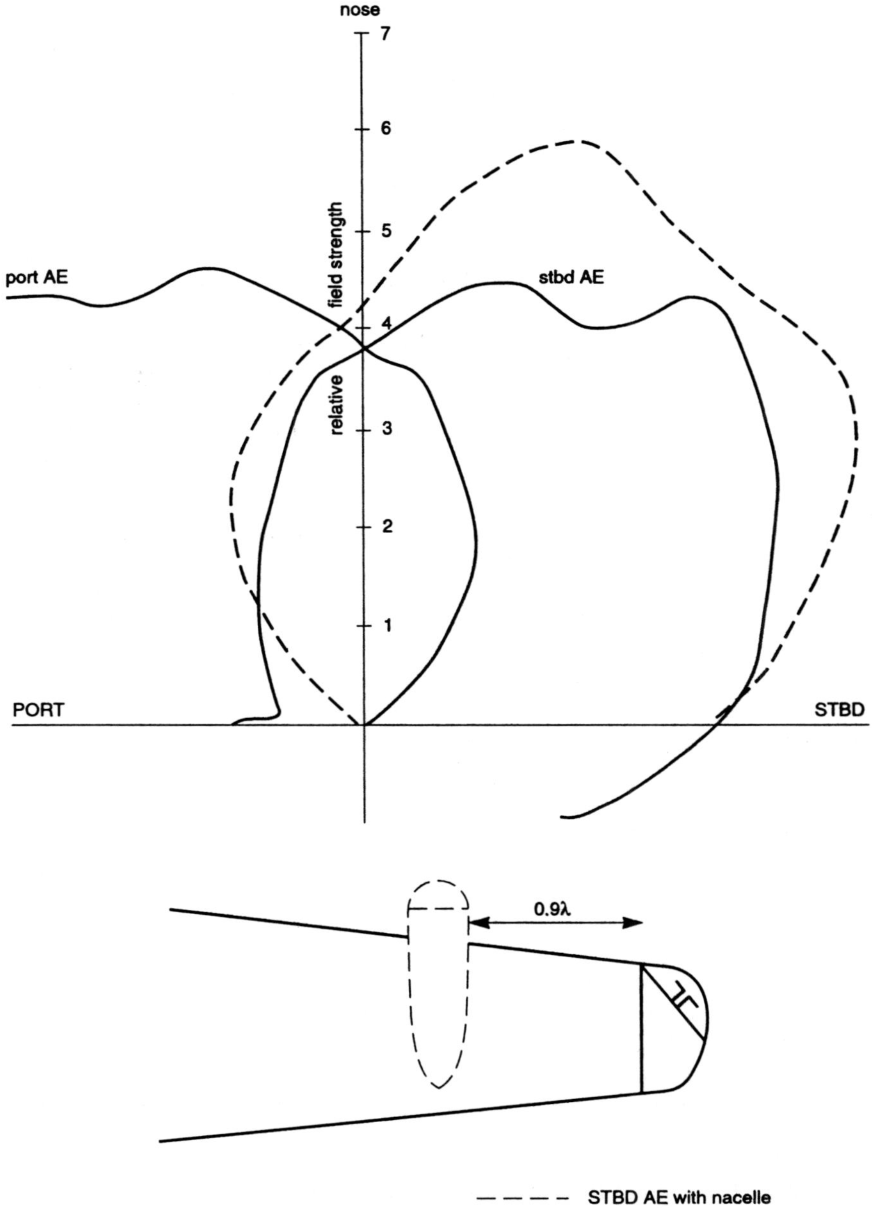

Figure 12.32 *Patterns of azimuth homing antennas showing effect of asymmetry*

(iv) Two orthogonal pairs of monopoles, each pair fed in antiphase and the two pairs fed in quadrature, will add on the vehicle axis and give a near-circular pattern in the roll-plane provided the vehicle diameter is not large.

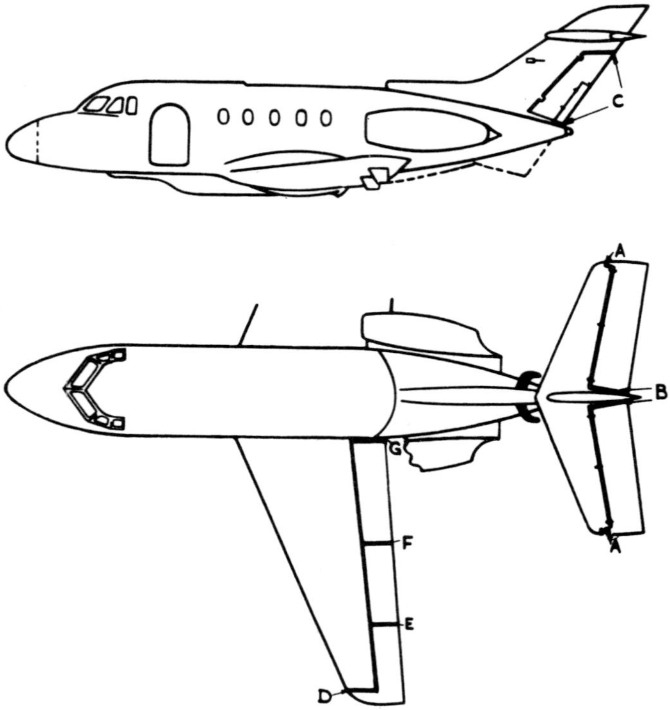

Figure 12.33 *Naturally occurring aircraft discontinuities*

A Tailplane-elevator
B Elevator-fin
C Fin-rudder
D Wing-aileron
E Aileron-flap
F Flap-flap
G Flap-fuelage

If the vehicle diameter is electrically large, more elements will be required to reduce the ripple. Pugh, Barker and Thomas [28] give figures for a number of circumferential slots round a cylinder, with a phase progression of one, two or three wavelengths around the array. Generally one cycle of phase progression will be satisfactory.

One difficult area to assess is the effect of vehicle efflux in rockets because of the different fuels. Solid fuels especially with metallic particles are believed to give the most rearward attenuation.

12.4.5 *Antennas for satellites*

As Pugh [29] wisely remarks 'Nearly every conceivable antenna could perform some useful function if it were practicable to place it on a spacecraft . . .'. Apart from severe acceleration and vibration loads in the launch phase which may last for a few minutes, the main problems for antenna designers are likely to be:

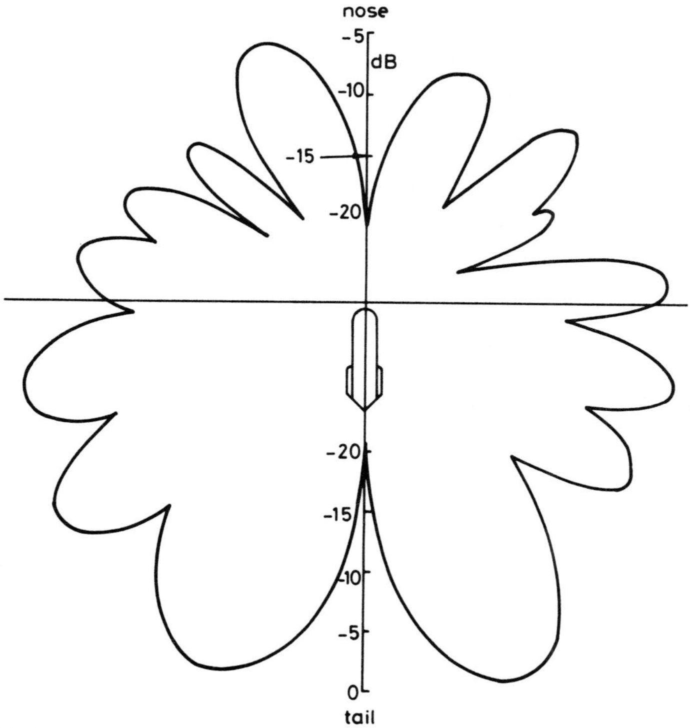

Figure 12.34 *Yaw plane pattern of nose spike antenna on 4λ slender missile*

(i) Very low pressure causing RF breakdown
(ii) Space constraints in the launch phase — antennas may have to be folded up under the launch fairing and only deployed after separation
(iii) Wide temperature variations.

For telemetry and telecommand (TT&C) systems isotropic coverage would be ideal because of the need to ensure two-way data transfer whatever the attitude of the satellite with respect to the ground station. Koob [21] introduced the term 'isotropy' as a measure of the overall coverage of an installed antenna. His coverage curves are a variant of that used by the author (Fig. 15.23) in that absolute gain levels are used. The area under the curve for gain levels up to 0 dBi is a measure of the isotropy. For an ideal source it is 100%, for an elementary dipole 90% and for a turnstile 90%. The difference is that at −3 dBi the turnstile provides 100% coverage, the dipole only 82%.

Three main antenna requirements will be addressed in this section. They are:

(i) TT&C antennas for 137 and 148 MHz bands
(ii) TT&C antennas for 2·2 GHz band
(iii) Higher gain antennas for 2·2 GHz.

12.4.5.1 *VHF antennas*

As Koob's paper suggests, the turnstile is a good solution for a near-isotropic antenna provided that the vehicle itself is not electrically large. Arrangements tend to follow one of the two schemes shown in Figs. 12.35 and 12.36. The four elements, usually $\lambda/4$ long, are fed with progressive 90° phase shifts. Studies of patterns of such antennas are reported by Albertsen *et al.* [1] and by TICRA [43]. During the bid stage for the GEOS satellite the author experimented with a canted turnstile on a boom above the satellite. Downward cover was found to be very poor but was much improved by adding $\lambda/4$ parasitic elements at 90° spacing around the lower circumference of the drum-shaped body.

Ancona and Do-Boi [2] used a rather different arrangement to provide toroidal coverage with good circular polarisation equatorially for the Symphonie satellite. The elements were four half-wave coaxial dipoles mounted normally to the flat surface at one end of the satellite. At their mid-point the antennas were bent parallel with the surface and curved so that the horizontal elements created an interrupted loop. They claimed that the use of half-wave rather than quarter-wave elements reduced the effect of the satellite body on radiation patterns.

A completely different solution was adopted by Lincoln Laboratory for two spin stabilised satellites about λ in diameter and $1{\cdot}3\lambda$ high. There is an equatorial band above and below which are solar panels. Gaps between the panels were used as vertical slot antennas, there being eight pairs. In one arrangement vertical full-wave dipoles were mounted over each collinear slot pair. In the other arrangement the vertical elements were pairs of half-wave elements. In the omnidirectional antenna all the slots were fed equally and in phase; similarly all the dipoles were equally fed in phase and combined with the slot feed with 90° phase shift. This resulted in an omnidirectional antenna, circularly polarised, with a gain of 3 dB.

Figure 12.35 *Turnstile with separate elements*

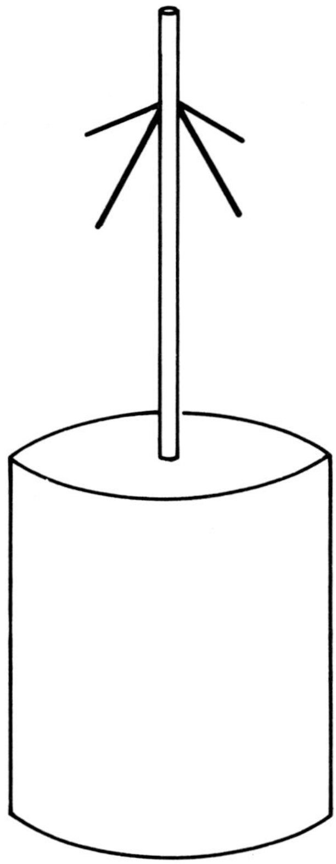

Figure 12.36 *Canted turnstile on satellite boom*

In the despun system two adjacent collinear pairs of slots and dipoles are fed with phasing such that a circularly polarised beam is directed towards Earth. The system is scanned in $22\frac{1}{2}°$ steps and provides a gain of approximately 10 dB. These results were reported by Rosenthal *et al.* [31].

12.4.5.2 *S-band* (2·2 *GHz*) *antennas*

The move to S-band meant that more complex antenna arrangements were necessary to achieve omnidirectional coverage since the satellites were still essentially of the same physical size. Scott and Soo Hoo [34] state that 'a null-free radiation pattern must contain all axial ratios of elliptical polarisation ranging from −1 (CP of one sense) through 0 (linear) to +1 (CP of the opposite sense)'. Koob [21] proposed an arrangement based on this principle using six radiators. These were placed at 90° spacing on the satellite body so that adjacent radiators were orthogonally polarised. If each antenna has a unidirectional pattern of 90° half power beamwidth the array pattern will be a circle in each of the principal planes. Koob claims a −3 dB coverage of 96% and an isotropy of 88% for such an array on a 10λ diameter body.

Another quasi-isotropic antenna designed by Galindo and Green [14] consists of a ring of crossed-slot radiators on a cylinder of about 1λ diameter. The slots are excited with a 360° phase progression round the ring. Results for the IME-D satellite are reported by Jensen and Pontoppidan [17].

12.4.5.3 *Higher gain antennas for* 2·2 *GHz*

In many instances the optimum solution is a circular belt array similar to that used by Rosenthal *et al.* [31] at VHF with electronic switching so that the beam is maintained in the required direction. An alternative scheme used on Helios is described by Tymann [38]. It consists of a vertical stack of eight cylinders around the antenna mast. Each cylinder has four half-wave circumferential slots around its equator. To reduce their length these are dumb-bell shaped. Each is fed by a three-wire transmission line balun the inner of which connects to a vertical transmission line. To obtain the correct phase the feed to successive rows of slots is taken alternately to the upper or lower edge of the slots. A gain of about 9 dB was achieved. Because of the very wide temperature range anticipated (+200°C to −100°C) no dielectrics were used, the three-wire baluns acting as the supports for the cylinders, a very neat arrangement.

12.5 References

1 ALBERTSEN, N.C. *et al.*: 'Computation of spacecraft antenna radiation patterns'. Laboratory of Electromagnetic Theory (LET), Technical University of Denmark (TUD), Lyngby, Report No R108, June 1972

2 ANCONA, C., and DO-BOI, H.: 'The Symphonie Satellite VHF circularly polarised antenna with toroidal pattern'. IEE Conf. Publ. 128, June 1975, pp. 148–153

3 ANCONA, C., and FROIDURE, P.: 'Systeme d'antennes à commutation realisant une couverture avion aux normes aerostat'. AGARD CP139, Nov. 1973, paper 18

4 ANDES, C.B.: 'Foreshortened center-fed VHF antenna'. ECOM report 02477-F, Dec. 1967

5 BLAND, R.G., and CLARKE, J.M.: 'A comparison of two L-band aircraft antennas for aeronautical satellite applications'. AGARD CP139, Nov. 1973, paper 21

6 BRAIN, D.J., and MARK, J.E.: 'The disc antenna. A possible L-band aircraft antenna'. IEE Conf. Publ. 95, March 1973, pp. 14–17

7 BRUECKMANN, H.: 'Improved wide-band VHF whip antenna', *IEEE Trans.*, 1966, **VC–15**, pp. 25–29

8 BURBERRY, R.A.: 'Aerials for Instrument Landing Systems' *in* 'Radio antennas for aircraft and aerospace vehicles'. Technivision Services, Maidenhead, England, Nov. 1967, pp. 51–66

9 BURNSIDE, W.D., MARHEFKA, R.J., and YU, C.L.: 'Roll plane analysis of on-aircraft antennas'. AGARD Conf. Publ. 139, Nov. 1973, paper 41

10 DAVIDSON, A.L., and TURNEY, W.J.: 'Mobile antenna gain in the multipath environment at 900 MHz', *IEEE Trans.*, 1977, **VT–26**, pp. 345–348

11 DEMMEL, F., and STARK, A.: 'Microprozessorgesteurte VHF-Fahrzeugantenne'. Antennen 82, Baden-Baden, Germany, 16–19 March 1982

12 DORIER, B.: 'The L-band aircraft antenna of the dioscures system (electronic scanning for satellite-aided navigation systems)'. IEE Conf. Publ. 77, June 1971, pp. 259–264

13 FRASER, W.M., and WILLIAMS, N.C.: 'An airborne phased array for use in an ATC satellite in L-band'. IEE Conf. Publ. 77, June 1971, pp. 118–126

14 GALINDO, V., and GREEN, K.: 'A near isotropic, circular polarised antenna for space vehicles', *IEEE Trans.*, 1965, **AP–13**, pp. 872–877

15 GRANGER, J.V.N., and BOLLJAHN, J.T.: 'Aircraft antennas', *Proc. IRE*, May 1955, pp. 533–550

16 HERSKIND, R.E.: 'A circular slot aperture with arbitrary polarisation for aerospace applications.' IEE Conf. Publ. 77, June 1971, pp. 265–275
17 JENSEN, F., and PONTOPPIDAN, K.: 'UHF radiation patterns of satellite antennas'. IEE Conf. Publ. 128, June 1975, pp. 41–46
18 JONES, I.L.: 'Movements of the phase centre of ILS airborne localiser aerials on a Varsity aircraft' *in* 'Radio antennas for aircraft and aerospace vehicles'. Technivision Services, Maidenhead, England, Nov. 1967, pp. 67–96
19 KING, H.E., and WONG, J.L.: 'A shallow ridged-cavity crossed slot antenna for the 240–400 MHz frequency range', *IEEE Trans.*, 1975, **AP–23**, pp. 687–689
20 KITCHEN, F.A.: 'The design of an omnidirectional aerial system for the frequency range 225–400 Mc/s', *Proc. IEE*, 1951, **98** Pt III, pp. 409–415
21 KOOB, K.: 'A near-isotropic antenna system for large aerospace vehicles'. IEE Conf. Publ. 77, June 1971, pp. 206–211
22 KUBINA, S.J.: 'Measurement and computer simulation of antennas on ships and aircraft for results of operational reliability'. AGARD Lecture Series 165, Oct. 1989, paper 3
23 KUNACHOWICZ, K.J.: 'Model testing of airborne VHF direction finding antenna system', IEE Conf. Publ. 128, June 1975, pp. 223–227
24 LAW, PRESTON, E.: 'Shipborne antennas' (Artech House Inc., Dedham, Mass.)
25 LONG, S.A., and SHEN, L.C.: 'A theoretical and experimental investigation of the circularly polarised elliptical printed circuit antenna'. IEE Conf. Publ. 195, April 1981, pp. 393–396
26 MAHONEY, J.: 'Fincap communication aerials'. IEE Conf. Publ. 77, June 1971, pp. 71–76
27 MAHONEY, J.: 'Upper L-band telemetry aerials for rockets and missiles'. AGARD CP139, Nov. 1973, paper 23
28 PUGH, B., BARKER, D.G., and THOMAS, D.C.: 'A sounding rocket omnidirectional antenna'. IEE Conf. Publ. 77, June 1971, pp. 1–6
29 PUGH, B.: 'Antennas in the space environment'. IEE Conf. Publ. 128, June 1975, pp. 21–28
30 RAHIM, T., GUY, J.R.F., and DAVIES, D.E.N.: 'A wideband UHF circular array'. IEE Conf. Publ. 195, April 1981, pp. 447–450
31 ROSENTHAL, M.L., DEVANE, M.E., and LAPAGE, B.F.: 'VHF antenna systems for spin-stabilised satellites', *IEEE Trans.*, 1969, **AP–17**, pp. 443–451
32 SANFORD, G.G., and KLEIN, L.: 'Development and test of a conformal microstrip airborne phased array for use with the ATS-6 satellite'. IEE Conf. Publ. 128, June 1975, pp. 115–122
33 SANFORD, G.G., and MUNSON, R.E.: 'Conformal VHF antenna for the Apollo-Soyuz test project'. IEE Conf. Publ. 128, June 1975, pp. 130–135
34 SCOTT, W.G., and SOO HOO, K.M.: 'A theorem on the polarisation of null-free antennas', IEEE Trans., 1966, **AP–14**, pp. 587–590
35 SHAFFER, E., and IKRATH, K.: 'Development of camouflaged vehicular VHF antennas'. US Army Electronics Command, Fort Monmouth, NJ, Rep ECOM–4261, Sept. 1974
36 SIDFORD, M.J.: 'A radiating element giving circularly polarised radiation over a large solid angle'. IEE Conf. Publ. 95, March 1973, pp. 18–25
37 SIDFORD, M.J.: 'Performance of an L-band aerosat antenna system for aircraft'. IEE Conf. Publ. 128, June 1975, pp. 123–129
38 TYMANN, G.: 'A new omnidirectional antenna for space application'. IEE Conf. Publ. 128, June 1975, pp. 142–147
39 WEBSTER, R.E.: '20–70 Mc Monopole antennas on ground-based vehicles'. *IRE Trans.*, 1957, **AP–5**, pp. 363–368
40 WERSTIUK, H.L., LAMBERT, J.D., MAYNARD, L.A., and CHINNICK, J.H.: 'UHF linear phased arrays for aeronautical satellite communications'. AGARD CP139, Nov. 1973, paper 20
41 WYATT, J.E.G.: 'The design of an omnidirectional UHF wrap-around antenna', *J. Royal Naval Scientific Service*, **29**, pp. 251–258
42 Proceedings of ECOM–ARO workshop on electrically small antennas, US Army Electronics Command, Fort Monmouth, NJ, Oct. 1976

43 'Study on VHF antenna for large satellites'. TICRA, Lyngby, Denmark. Report S–19–02, April 1974

The following papers by the author cover general aircraft antenna topics:

BURBERRY, R.A.: 'Aerial systems for aircraft', *J. Royal Aeronautical Society*, Feb. 1956, pp. 101–113

BURBERRY, R.A.: 'Progress in aircraft aerials', *Proc. IEE*, 1962, **109** Pt. B, pp. 431–444

KELLY, W.A., and BURBERRY, R.A.: 'A review of helicopter aerial problems'. IEE Conf. Publ. 77, June 1971, pp. 77–82

BURBERRY, R.A.: 'The rationalisation of aircraft antennas', IEE Conf. Publ. 128, June 1975, pp. 204–209

BURBERRY, R.A.: 'Accuracy of determination of aircraft antenna radiation patterns'. IEE Conf. Publ. 219, April 1953, pp. 97–100

BURBERRY, R.A.: 'Aircraft antennas', *Aeronautical J.* 1989, **93**, pp. 58–65

Chapter 13

Feed systems

13.1 Cabling

Coaxial transmission lines have been discussed in depth by many authors, notably by Blackband [3]. Among other points he considered is that of cable leakage, an increasingly important consideration in EMC studies. Also of importance in this area is Dummer and Blackband's [6] earlier study on wires and cables.

In many installations it is possible to connect antenna to radio equipment by a single coaxial cable chosen solely on its electrical characteristics. This is rarely the case in aircraft and aerospace vehicles where physical and mechanical constraints may be important. The types of cable may be limited by considerations of temperature range, humidity, pressure differential, resistance to fuels and hydraulic fluids and constrained by the need for flexibility and low weight.

In many types of vehicle it is operationally undesirable to have only one cable from antenna to radio. In a civil airliner there may be the need to take the cable through pressurised bulkheads and this may be best achieved if a sealed junction is incorporated in the bulkhead. Very long cables are tedious to install and very difficult to replace so junctions may be necessary for maintenance. In military aircraft the need to take the vehicle apart for transport may similarly necessitate cable breaks. It must also be appreciated that it is rarely possible to run cables in a direct line. Thus from an upper VHF antenna on a wide-bodied jet to the radio equipment in a direct line below it could well involve 10 m of cable. From the aircraft extremities considerable lengths will be required. In calculating the installed gain of any system, therefore, some estimate must be made for cable attenuation and mismatch effects. The type of assessment necessary is outlined below for a typical IFF installation operating in the 960–1220 MHz band.

As Chapter 12 showed, the best antenna arrangement for good coverage is given by a pair of antennas at or near the fuselage extremities. These are connected to an oscillating switch and thence to the radio equipment. Fig. 13.1 shows a typical arrangement: it is assumed that for ease of maintenance there is only a short cable on each antenna then a longer cable to the switch which is close to the radio equipment. Arbitrary values have been assigned to the VSWR and loss of each component:

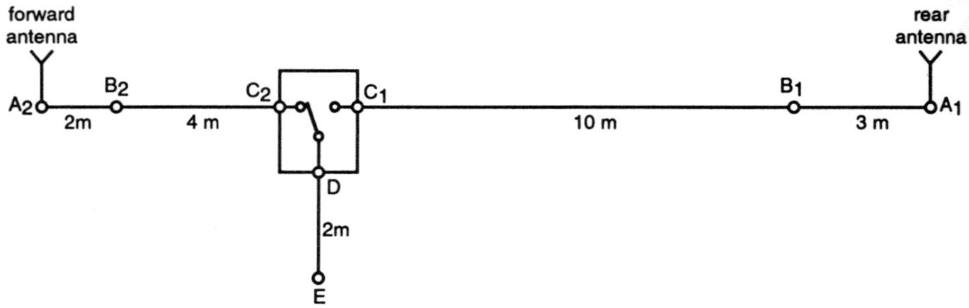

Figure 13.1 *Typical two-antenna installation in aircraft*

VSWR	Antenna	1·8	(5·1 dB)	
	Cable junction	1·1	(0·8 dB)	
	Switch	1·2	(1·6 dB)	
	Cable variation	1·08	(0·7 dB)	(±2 Ω in 50)
Attenuation	Cable		0·2 dB/m	
	Switch		0·5 dB	
	Cable junction		0·1 dB	

Because the cable lengths are arbitrary all the VSWRs are added: worst case. Fig. 13.2 shows the resulting VSWR and attenuation for each side of this arrangement. The attenuation has to be set against the antenna gain to give the true system performance; it may sometimes emerge that an apparently worse antenna coverage gives the better result because of reduced cable losses. The extreme instance of this general problem was in the specification of a lightweight IFF equipment for small helicopters. The general problems of siting on helicopters required antennas at the extremities, yet, such was the constraint on cable weight and attenuation that it was impossible to meet the specification for both antenna gain and cable attenuation on any of six helicopter types with which the author was concerned.

13.2 Baluns

For any balanced antenna or for a system of two separate elements as in Fig. 12.16 some form of balun is required to convert to a single coaxial output. In the simplest form this can take the shape of Fig. 2.7 — the 'quarter-wave can' or 'bazooka'. This maintains its balance over a limited frequency range for which the open-end impedance of the choke is high.

The twin line of Fig. 2.8*a* remains balanced so long as the reactance of the connecting line is small. It places a reactance across the twin terminals due to the short-circuited twin line. This reactance can be used to provide compensation as reported by Shnitkin and Levy [16] or Oltman [12], either by adjusting the length or the Z_0 of the twin line. For low VHF the quarter-wave line may be inconveniently long. It can be shortened by capacitance loading as in Fig. 13.3,

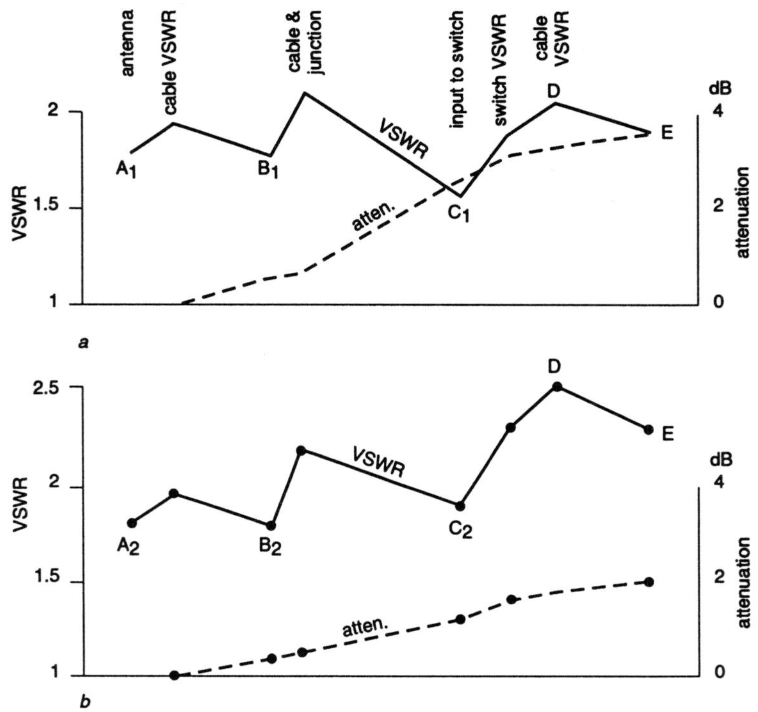

Figure 13.2 *VSWR and attenuation assessment*

a Rear system
VSWR at E = 1·926
Attenuation at E = 3·6 dB
b Forward system
VSWR at E = 2.32
Attenuation at E = 2.2 dB

which provides a much more rapid rate of change of reactance. The author has used baluns as short as 0·032λ in the 100 MHz band.

One objection to the conventional twin line of Fig. 2.8*a* is that it places a DC short circuit across the line and hence prevents insulation testing of the system. This can be obviated by the variant shown in Fig. 2.8*b*. The open-circuit line inside the second of the parallel lines is made a quarter-wavelength taking into account the velocity factor of the cable. It may be appropriate to enclose all these baluns in a metal case as in Fig. 2.8*c*: this alters the line Z_0 as shown.

An alternative physical arrangement is that of the collinear balun of Fig. 13.4 due to Marchand [10]. A DC open circuit version of this is of course possible on the same lines as Fig. 2.8*c*. There is also a capacitance-loaded version (Fig. 13.5) much used by the German Air Force in World War Two under the name Sperrtopf and known at least to the RAF as the 'milk bottle'. These collinear baluns are not suited for direct attachment to a dipole: for this the twin-line types are better suited as, properly designed, they can support the dipole arms. Further analysis of the Marchand balun has been done by Cloete [4,5]. Another

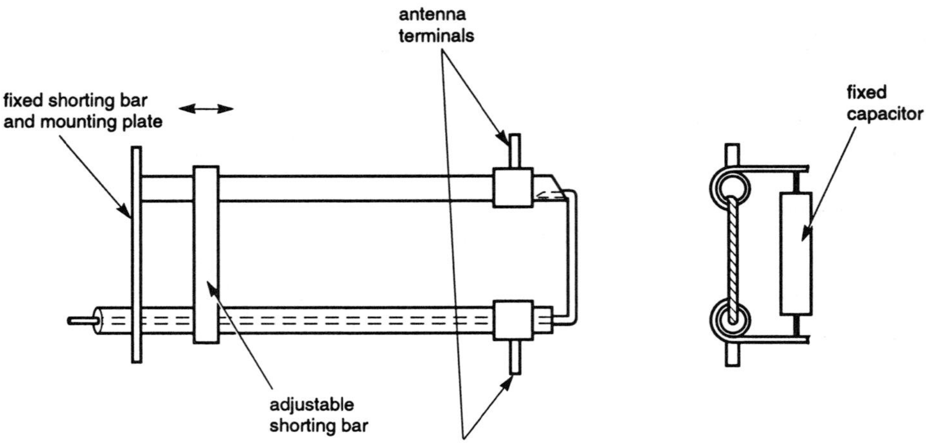

Figure 13.3 *Capacitance-loaded twin-line balun*

type which is suitable for frequencies above about 300 MHz is the split-tube balun of Fig. 2.9 also seen in a printed version in Fig. 2.10. The split tube is not mechanically strong if the tube diameter is small compared with the length of the split so is probably best used above 1 GHz. It can suffer from the effects of moisture across the slots so needs protection. If the inner line is surrounded by dielectric, which would prevent inward movement of the slotted tube and a consequent change in impedance, then the effective length of the slots will be altered and they will need to be shortened. A similar effect occurs if the slots are covered on the outside. The author experimented with polythene mouldings over a split-tube balun in order to seal it. Unfortunately polythene flowed into

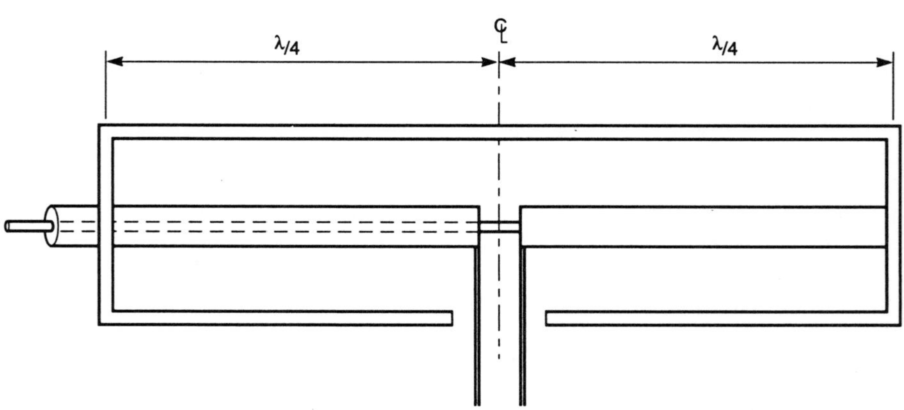

Figure 13.4 *Collinear balun*

Cross-section uniform but may be circular, square or rectangular

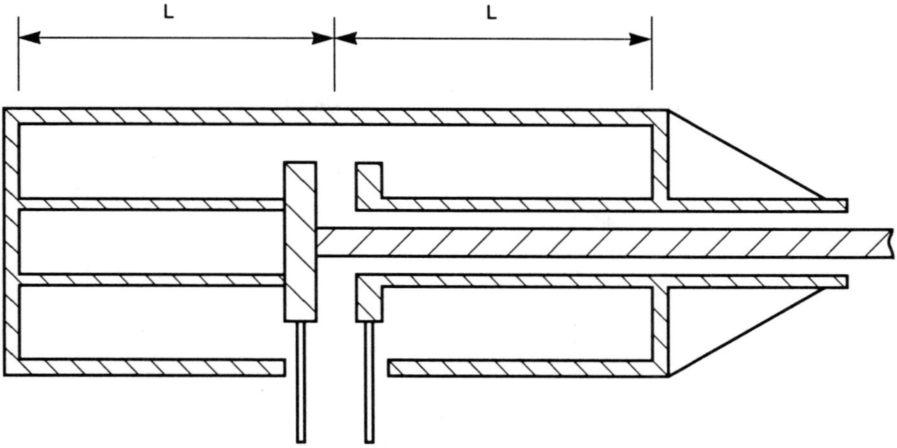

Figure 13.5 *Capacitance loaded collinear balun (Sperrtopf)*

the slots themselves producing variable effects on the antenna impedance. To avoid the transformation which occurs with a uniform diameter inner the latter is sometimes enlarged to keep Z_0 constant.

For low VHF or where a conventional balun is too large, a lumped circuit balun (Fig. 2.11) may be used. The author used such arrangements at 200 MHz during World War Two; today with better capacitors and with printed coils this type of balun should be capable of operation at much higher frequencies. A good description of the theory of the lattice network is given by Smith [17].

There have been innumerable papers on the design of wide band baluns. Among those of importance are papers by Roberts [15], O'Meara and Sydnor [13] and Duncan and Minerva [7]. For stripline applications, Bawer and Wolfe [2] and Jones and Shimizu [8] are worth consulting.

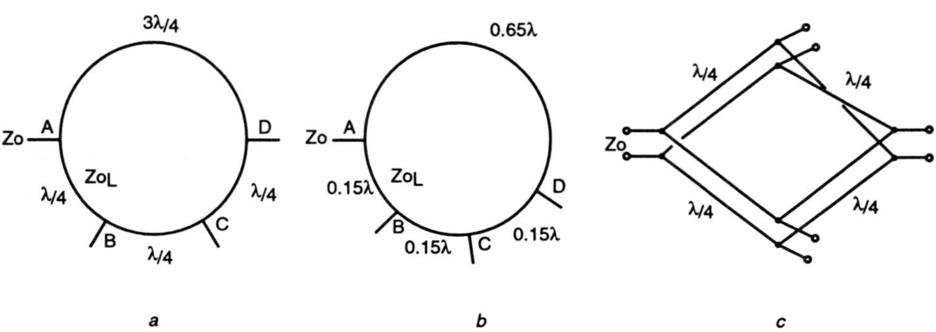

Figure 13.6 *Rat races*

a $Z_{0_L} = Z_0 \cdot \sqrt{2}$ } coaxial line
b $Z_{0_L} = Z_0$

c $Z_{0_L} = Z_0 \cdot \sqrt{2}$ twin line

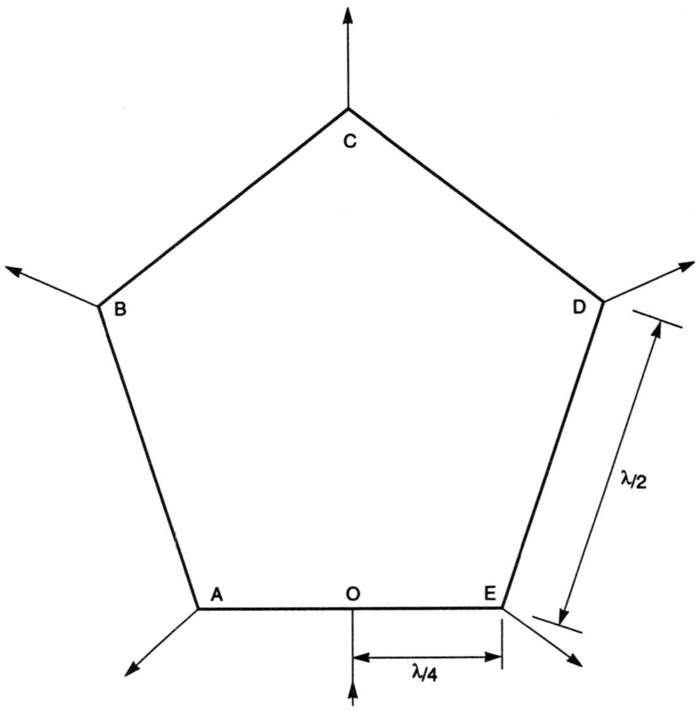

Figure 13.7 *Bagley polygon — five-sided version*
Impedance of outputs A–E $= Z_0$
Line impedance for match $= \dfrac{2Z_0}{\sqrt{N}}$
N = number of matched outputs

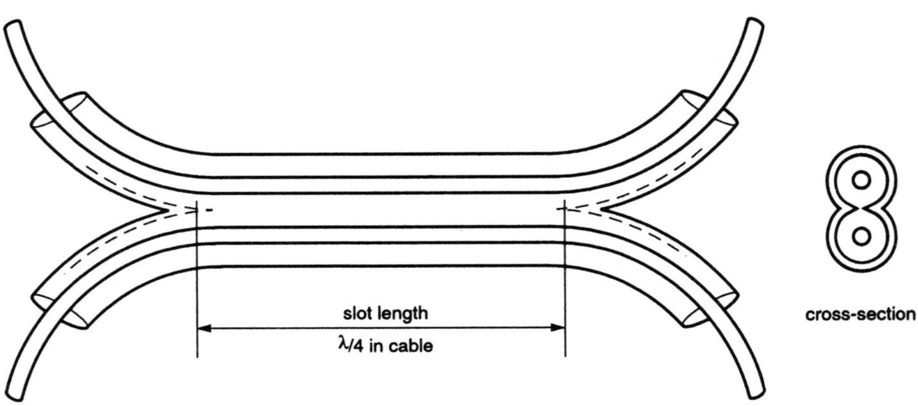

Figure 13.8 *Parallel line coupler with coaxial lines*

13.3 Power dividers

13.3.1 *Rat races*

Fig. 13.6 shows two types of rat race of which (*a*) is by far the better known. It consists of a loop of transmission line of circumference $3\lambda/2$. Connections into the loop are made at A, B, C and D which are spaced $\lambda/4$, $\lambda/4$, $\lambda/4$ and $3\lambda/4$, respectively. If a signal is fed in at C it is easily shown that equal amplitudes in phase will appear at B and D, there being no signal at A because the two routes from C differ by 180° and hence cancel.

Equally, a signal fed in at A appears at B and D with equal amplitude but with 180° phase difference. In this way the rat race can be used as a balun and it can be seen that for an input impedance of Z_0 at A the impedance across BD is $2Z_0$ so the unit also acts as an impedance transformer. For impedance matching the characteristic impedance of the line should be $Z_0\sqrt{2}$, i.e. 71 ohms if $Z_0 = 50$.

A further use for the rat race is to combine two signals to give sum and difference products. If signals are fed into the loop at B and D the output at C is their sum, at A their difference.

Physically the rat race can be constructed in coaxial cable, in twin line, or in stripline. If twin line is used the physical size can be reduced by inverting the $3\lambda/4$ line: a length of $\lambda/4$ only is then required, Fig. 13.6*c*. It is usual to

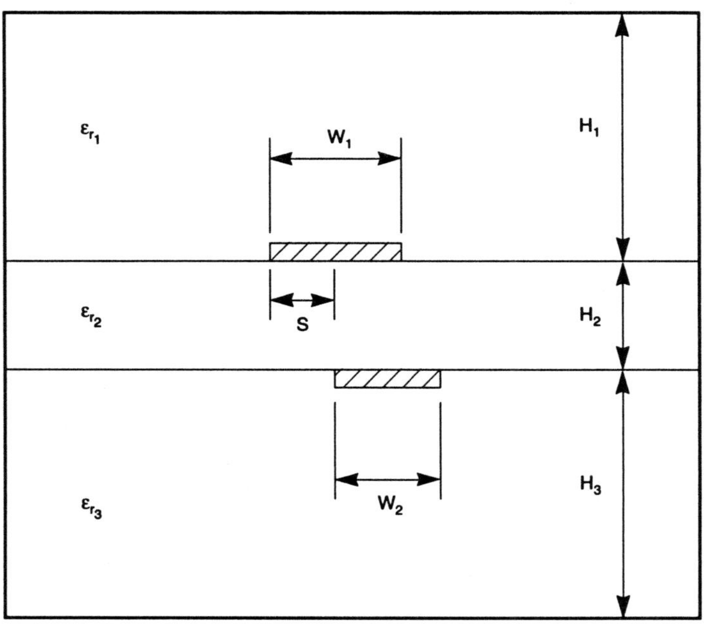

Reference 14

Figure 13.9 *Offset broadside-coupled strip lines*
Significant dimensions shown

terminate any unused terminals in Z_0. As can be seen from the above discussion the proportion of power dumped in these terminations should be zero and will remain small provided the line lengths do not deviate too much from the nominal dimensions. The physical shape of the loop is unimportant.

Fig. 13.6*b* shows another shortened form of rat race. This was widely used by the author's colleagues designing ILS equipment but its origins are unknown

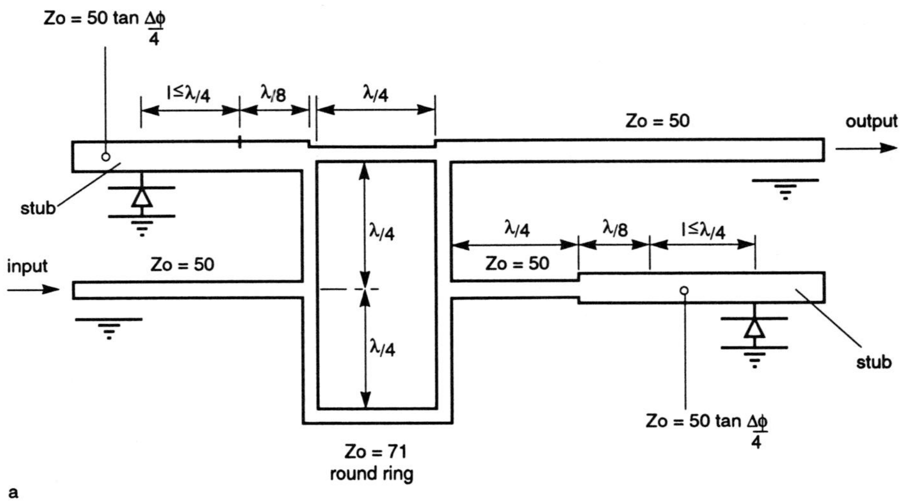

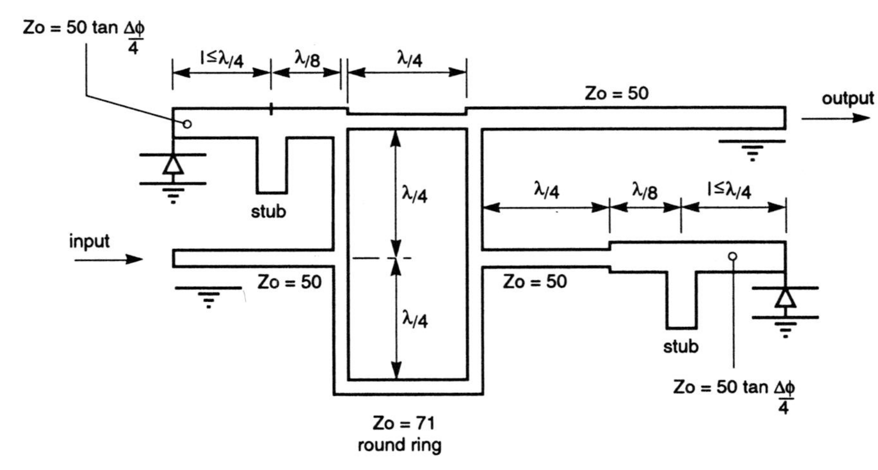

Figure 13.10 *p–i–n diode switched phase shifters*

a For phase shift $\Delta\phi \leqslant 135°$
b For phase shift $\Delta\phi \geqslant 135°$

and there does not appear to have been anything published on its performance. The loop length is only 1·1λ and the lines are of the same impedance as the input.

13.3.2 *The Bagley polygon*

This was devised by G. Bagley of the Royal Aircraft Establishment, Farnborough, and is described by Blackband [3]. It is a generalised form of rat race which can be used to produce *n* equal outputs or to combine *n* inputs (Fig. 13.7). The characteristic impedance $Z_0 = 2Z_0'/\sqrt{n}$ where *n* is the number of outputs each matched to Z_0'.

13.3.3 *The Wilkinson power-divider*

Another important *N*-way hybrid is due to Wilkinson [18].

13.3.4 *Parallel-line couplers*

Some of the earliest couplers were constructed of coaxial transmission lines in which the coupling was achieved by a slot between the outers of the coaxial lines as in Fig. 13.8. The outers of the lines were strapped together over the whole length of the coupler. Whilst this was effective it required the use of solid jacketed lines preferably with leaden outers which could be soldered together and careful machining was necessary to achieve the desired coupling factor.

Physically easier is the use of broadside-coupled strip transmission lines as described, for example, by Allen and Estes [1]. In a slightly different form offset lines are used as in Fig. 13.9. Details of their design are given by Levy [9]. One of these designs was used by Rehnmark and Lagerlof [4] in the band 136–150 MHz for satellite antennas, the length being accommodated by folding to form a meander pattern.

13.4 Phase shifters

The rat race described in Section 13.3.1 can be used as part of a variable phase shifter. Short-circuited lines of variable length are connected at B and D in Fig. 13.6*a*, their lengths being gauged so that they always differ by $\lambda/4$. If a signal is injected at C, the signal will appear at A with a phase shift dependent on the line lengths.

Fixed phase shifts of 90°, 135° and 180° can be obtained with lines using p–i–n-diode switching. Two such arrangements are shown in Fig. 13.10 due to Matthews and Markopoulos [11]. In this system the lines containing the switches have a characteristic impedance $Z_0 = 50 \tan \Delta\phi/4$ where $\Delta\phi$ is the desired phase shift.

13.5 References

1 ALLEN, J.L., and ESTES, M.F.: 'Broadside-coupled strips in a layered dielectric medium', *IEEE Trans.*, 1972, **MTT–20**, pp. 662–668

2 BAWER, R., and WOLFE, J.J.: 'A printed-circuit balun for use with spiral antennas', *IRE Trans.*, 1960, **MTT–8**, pp. 319–325

3 BLACKBAND, W.T.: 'Coaxial transmission lines and components' *in* 'Handbook of antenna design' (Peter Peregrinus Ltd., 1983) chap. 18
4 CLOETE, J.H.: 'Exact design of the Marchand balun', *Microwave J.*, May 1980, pp. 99–102
5 CLOETE, J.H.: 'Graphs of circuit elements for the Marchand balun', *Microwave J.*, May 1981, pp. 125–128
6 DUMMER, G.W.A., and BLACKBAND, W.T.: 'Wires and RF cables' (Pitman, London, 1961)
7 DUNCAN, J.W., and MINERVA, V.P.: '100:1 bandwidth balun transformer' *Proc. IRE*, 1960, **48**, pp. 156–164
8 JONES, E.M.T., and SHIMIZU, J.K.: 'A wide-band stripline balun', *IRE Trans.*, 1959, **MTT–7**, pp. 128–134
9 LEVY, R.: 'Directional couplers' *in* YOUNG, L. (Ed.): 'Advances in microwaves' Vol. 1 (Academic Press, NY, 1966) p. 121
10 MARCHAND, N.: 'Transmission line conversion', *Electronics*, 1944, **17**, pp. 142–145
11 MATTHEWS, P.A., and MARKOPOULOS, D.: 'A microstrip p-i-n diode-controlled L-band digital phase shifter'. IEE Conf. Publ. 77, June 1971, pp. 127–129
12 OLTMAN, G.: 'The compensated balun', *IEEE Trans.*, 1966, **MTT–14**, pp. 112–119
13 O'MEARA, T.R., and SYDNOR, R.L.: 'A very wide band balun transformer for VHF and UHF', *Proc. IRE*, 1958, **46**, pp. 1848–1860
14 REHNMARK, S., and LAGERLOF, R.: 'VHF antenna feeder power divider', *Microwave J.*, 1975, **18**, July, pp. 28C–D, 44A–D
15 ROBERTS, W.K.: 'A new wide-band balun', *Proc. IRE*, 1957, **45**, pp. 1628–1631
16 SHNITKIN, H., and LEVY, S.: 'Getting maximum bandwidth with dipole antennas', *Electronics*, 31 Aug. 1962, pp. 40–42
17 SMITH, R.A.: 'Aerials for metre and decimetre wavelengths' (Cambridge University Press, 1949) pp. 105–108
18 WILKINSON, E.J.: 'An N-way hybrid power divider', *IRE Trans.*, 1960, **MTT–8**, pp. 116–118
Microwave Engineers Handbook and Buyers Guide for 1967 and 1968 have some useful bibliographies as well as curves from some of the above References.

Chapter 14

Performance prediction

14.1 Introduction

Before undertaking any form of prediction of antenna performance it is essential to consider several aspects:

(*a*) Accuracy
(*b*) Time scale
(*c*) Cost

These are very much interrelated so some trade-offs may be possible.

14.1.1 *Accuracy*

The accuracy requirements will be determined by several factors:

(*a*) Ability to carry out accurate measurements on the final system
(*b*) Cost of building the system and rectifying any shortcomings
(*c*) Operational requirements of the system and the operating margins
(*d*) Extent to which the predictions will be used for system qualification and/or licensing.

A few examples may help to explain these factors:

(i) *Aircraft communications antenna*: Beyond the radio horizon the propagation loss increases much more rapidly than in the 'visual' region. There is therefore little advantage in achieving greater coverage gain than will enable good communication at the horizon. The accuracy required is therefore not high — perhaps 1 to 2 dB. There may be a case for higher gain if competition to be heard is important but this does not affect the order of accuracy either in gain or angular coverage. In general very narrow nulls are not important because not seen in operation.
(ii) *Spacecraft TT&C antenna*: Here the link budget can be accurately calculated and the coverage gain determined for a particular mission. What is important here is that the coverage gain should exceed a certain level: provided that this can be demonstrated it is not very important to know by how much it is exceeded.
(iii) *Point to point link antennas*: Because of crowded frequency bands there may be severe limitations on sidelobe levels and it then becomes necessary to achieve

sufficient accuracy to ensure that the sidelobe envelope has been met. In most instances it will be possible to measure the actual antenna with the required accuracy.
(iv) *Land vehicle antennas*: If these are operating in a multipath environment pattern shape is not very important. Maximum gain is what matters but it is probably cheaper to buy a scrap vehicle and measure antenna performance on it than attempt to predict it.

14.1.2 *Time scale*

Whilst the time scale for many projects may be measured in months and years rather than days, there are occasions when a rapid response is needed. One example would be the putting together of a proposal for a new aircraft or spacecraft on which a number of antennas were required. Some evidence that the coverage requirements could be met by the proposed installations would be essential, especially if some of the proposed solutions were unconventional or the vehicle shape did not allow comparison with other installations. In these circumstances a good first approximation which allows time for second thoughts would be the most useful attack: greater accuracy can follow when the contract has been won.

There may also be a need for a quick response in counter-measures operations whether against a military opponent or terrorists or criminals. This might well involve adding new radio systems and hence new antennas to an already well-covered vehicle. Any assessment would have to check that the performance of existing antenna installations was not impaired.

14.1.3 *Costs*

The costs of prediction may vary enormously. At one end of the range is the construction of very simple scale models for pattern measurement or the use of simplified mathematical modelling techniques, at the other the cost of a very complete model, e.g. of an aircraft, or the use of a full numerical computer programme. As an example of a very cheap model, the author used a model of a missile constructed of beer cans and oil cans of varying sizes, plywood, sheet metal and kitchen foil. For the use for which it was needed the model was accurate enough, cheap and quick to construct. A really accurate aircraft model would not only take several months to make but would cost £10 000 or more. On the other hand such a model, robustly constructed and well maintained, would last the development and operational life of the aircraft and, once made, would allow measurements to be made very rapidly and thus could be used when a rapid response was required. Bearing in mind the limitations of mathematical modelling in some areas this might well prove to be the best approach sometimes.

It will be clear from the above discussion that if a wide range of antenna problems have to be tackled there may well be a need for a flexible approach, using the best technique for the particular problem. There is a place both for scale modelling and mathematical modelling; the well-equipped antenna engineer will have both techniques at his disposal.

14.2 Impedance prediction

Examination of the literature will show that methods of impedance prediction are generally limited to antennas which are electrically thin, certainly much thinner than will normally be used in the VHF and UHF bands. Some of these methods are useful for very short antennas but, as can be seen from Chapter 9, their range of usefulness is somewhat limited. Most methods are not very good at predicting reactance and become excessively complex in handling feed point geometries. In most instances it will be more accurate to measure impedance rather than to calculate it.

There is a case for calculating the impedance of an antenna whose dimensions are such that its resistance is several orders smaller than its reactance. An example would be a very short monopole on an irregular structure. Its impedance cannot readily be measured with any accuracy so numerical methods will at least give some approximation. Wire grid modelling (method of moments) — see Section 14.3 — can give impedance but only if considerable care is taken in modelling the region around the antenna feed-point. The detail required for this is much greater than is needed for radiation pattern computation so may increase modelling costs. It is particularly important in wire grid modelling that the grid is selected correctly so that the field around the feed has the right form. The use of conformed field plotting is valuable here as a check on the grid. With electrically small antennas the loss resistance may outweigh the radiation resistance by an order or more. Although it is possible to incorporate loss resistance in the segmentation scheme this is a somewhat arbitrary process. Some of the advances in this method of modelling are discussed by Burke [3].

14.3 Numerical electromagnetics code — method of moments

This, usually abbreviated to NEC, is a very widely used computer modelling technique. It uses an electric field integral equation (EFIE) to model wires and wire-like objects and a magnetic field integral equation (MFIE) for surfaces. EFIE can also be used to model surfaces by treating them as grids of wires, hence the term wire grid modelling. This method has primarily been used for antennas on bodies which are not very large electrically because constraints on the grid size have meant that excessive computer run times were required for large objects, i.e. more than about λ cube.

Fig. 14.1 shows one side of a typical segmentation scheme, the grid covering the whole three dimensional surface of the vehicle. This is in fact a scheme for an HF wire antenna on a helicopter but it illustrates the principle. Guidelines to the grid dimensioning are summarised by Miller [7] in a very good survey of computational methods. The relevant data for wire grid modelling are given in Table 14.1.

An example of the use of NEC is shown in Fig. 14.2 in which the predicted and measured patterns for a low profile antenna on a military tracked vehicle are compared. The vehicle was $0{\cdot}8\lambda$ long, $0{\cdot}4\lambda$ wide and $0{\cdot}3\lambda$ high. Williams and Brammer [12] have used the method to examine the effects of imperfect ground

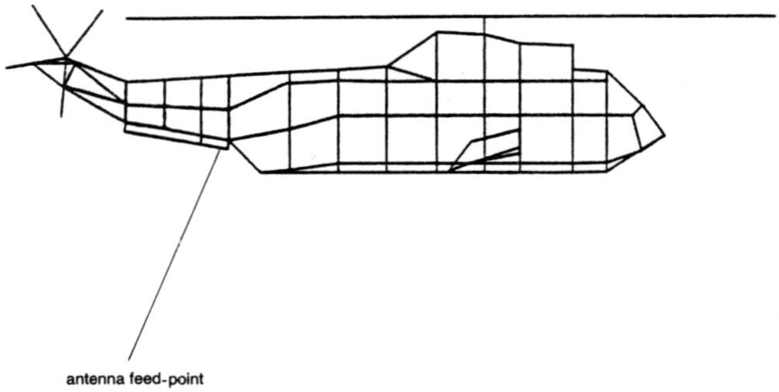

Figure 14.1 *Wire grid segmentation scheme*

on the azimuth patterns of a VHF antenna on a similar vehicle. It should be noted that MFIE does not appear to work very well on surfaces with sharp edges or non-continuous curvature. It seems to be better to use EFIE in these instances.

In spite of the continued development of NEC there are still some areas of difficulty. On a wire with a stepped radius, getting the correct charge distribution requires special care: a new programme was being developed in 1989. Electrically-small loops may also create problems.

One problem which persists is in validation. In many applications there are obvious indications such as negative resistance but this is not necessarily the case in some new applications where an approximate solution is not available. There is a natural tendency, when computed results do not agree with

Table 14.1 *Limitations on parameters used in wire grid modelling*

Parameter	Range	Reason
Wire length	$L>10d$	Neglect of end caps
Wire diameter d	$\lambda>\pi d$	Uniform current round circumference
Wire segment length		
max.	$l<\lambda/2\pi$	Constant current on segment
min.	$l>10^{-4}\lambda$	Seven place precision
	$l>10^{-8}\lambda$	Double precision
Wire radius a	$a<2l$	Thin-wire kernel
	$a<0{\cdot}5\,l$	Extended thin-wire kernel — straight wires only
Axial separation of parallel wires r	$r>3a$	
Wire mesh model of surface	$l<0{\cdot}1\,\lambda$	To reduce field leakage
Mesh size	$l\times l$	
Wire radius	$a=l/2\pi$	Wire area equal to surface area

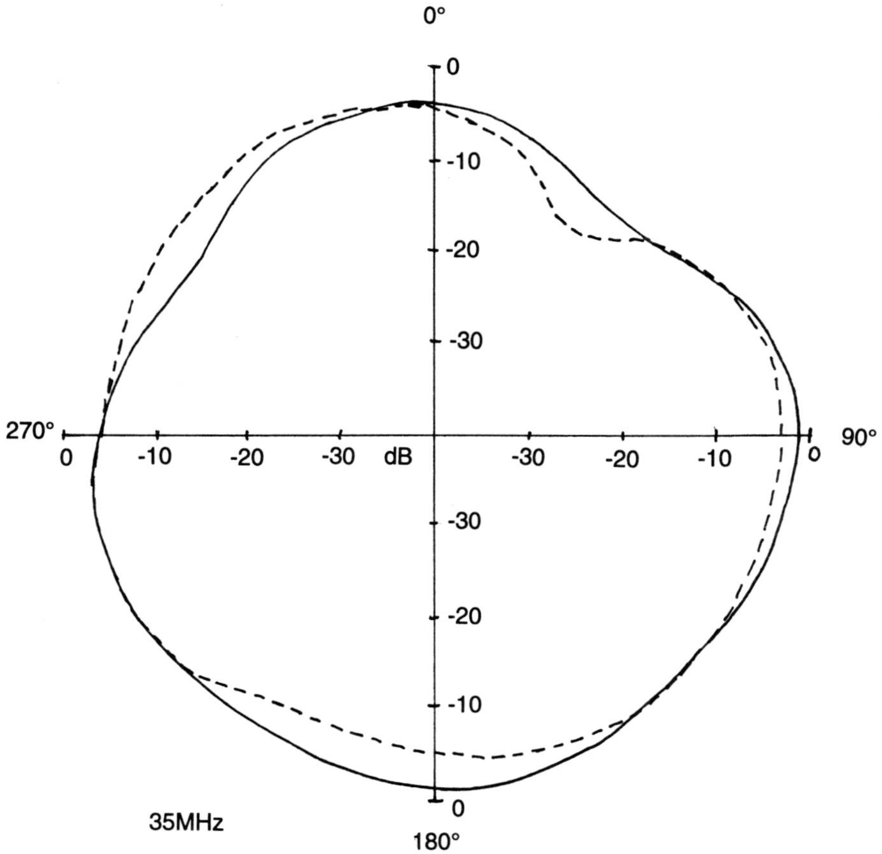

Figure 14.2 *Predicted and measured azimuth patterns on a tracked vehicle*

—— predicted
– – – measured
$\lambda/4$ monopole on nearside top front of 5 m long vehicle

measured ones, to adjust the programme or the segmentation scheme until good agreement is achieved. This does not improve confidence in mathematical modelling.

NEC is sponsored by US military development centres and not all the versions are available without authorisation from US Department of Defense.

14.4 Geometrical theory of diffraction (GTD)

This technique is best suited for large conducting surfaces. It uses a combination of geometrical optics plus diffraction from a variety of surfaces and edges; Fig. 14.3 shows some of the rays that may have to be considered in computing the field by this method. They include:

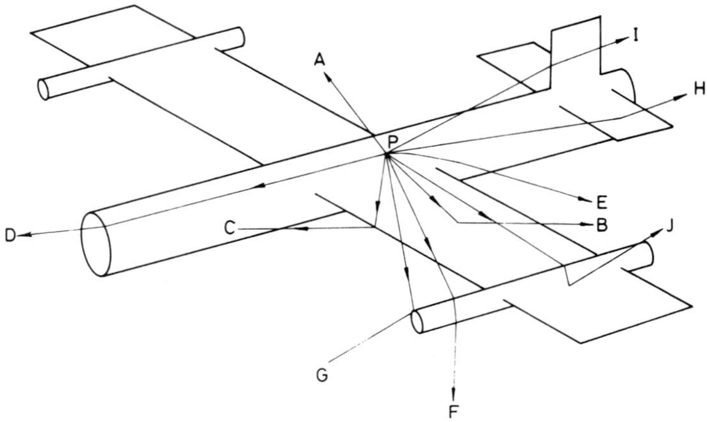

Figure 14.3 *Rays from an antenna at P used in pattern prediction by GTD*

A Direct
B Wing reflected
C Wing edge diffracted
D Nose diffracted
E Fuselage creeping wave
F Pod creeping wave
G Pod front diffracted
H Tail plane reflected
I Tail fin reflected
J Creeping wave and reflection

Direct rays
Reflected rays
Diffracted rays — single and multiple diffraction
Creeping waves — with and without reflection
Diffracted and reflected rays

The amplitude value of a diffracted ray is obtained by multiplying the field due to the ray incident on the diffraction surface by a diffraction coefficient appropriate to the surface. Most of these are available from solutions to the particular boundary value problem. Ray searching algorithms are used to determine the appropriate rays to be considered in any one direction. Some of these rays are illustrated and discussed by Molinet [8] whose paper summarises the development and recent advances in this technique.

Because it requires canonical solutions it is tempting to simplify the model of a scattering body, such as an aircraft, which may cause some lack of accuracy. Some of the problems are discussed by Pathak *et al.* [9]. As mentioned in this paper, one area where this technique offers the only sensible solution is in modelling antennas at frequencies above 1 GHz on large aircraft. Consider, for example, the installation of satellite communication or Navstar (GPS) antennas on a Boeing 747 at frequencies around 1500 MHz ($\lambda = 0{\cdot}2$ m). Assuming that a scale model aircraft can be built at 1/20 scale and the antennas can be modelled which is by no means simple and very expensive, the far field range will be of the order of 2500 m which is hardly practical. The enormous cost of scale

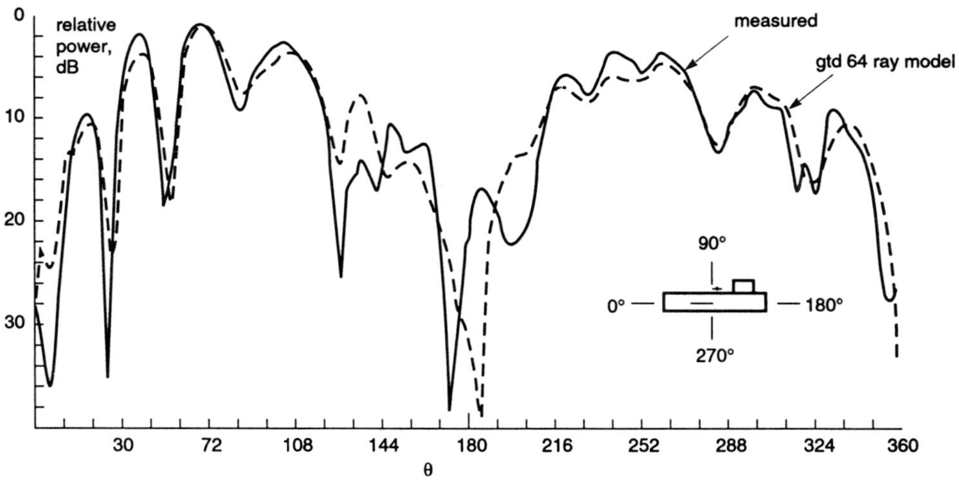

Figure 14.4 *Comparison of GTD and measured pitch plane patterns on a simplified aircraft model as Fig. 14.3*

modelling is here not justified: the aircraft is so large in terms of wavelength that some of the problems of longer wavelengths and smaller bodies no longer exist. Foster and Miller [6] considered ground planes of 5λ diameter (or 5λ side) to be about as small as could be handled. Some of their results were shown in Chapter 3.

GTD can still not deal very well with some aircraft problems:

- Dielectric structures such as radomes (Burnside and Pathek [4] have a solution for a thin dielectric slab)
- Resonant structures such as gaps between moving surfaces, pitot masts, fixed undercarriages especially skids
- Other antennas acting as parasites
- Antennas in the plane of fin- and wing-tips

Fig. 14.4 compares measured and computed patterns for an antenna on the simple model shown in Fig. 14.3. In general the largest variations in aircraft modelling occur along the aircraft axis probably because of difficulties in simulating mathematically the forward edge of the fuselage.

14.5 Simplified mathematical modelling

The need for rapid assessment of large numbers of aircraft installations for a new radio equipment led to the development by British Aerospace at Bristol of a system known as simplified mathematical modelling (SMM or SM^2 for short). This is based on Foster and Miller's programme of calculating elevation patterns of a monopole on a finite ground plane. It can be shown that the dominant features of these patterns are the direct ray and the diffracted ray in

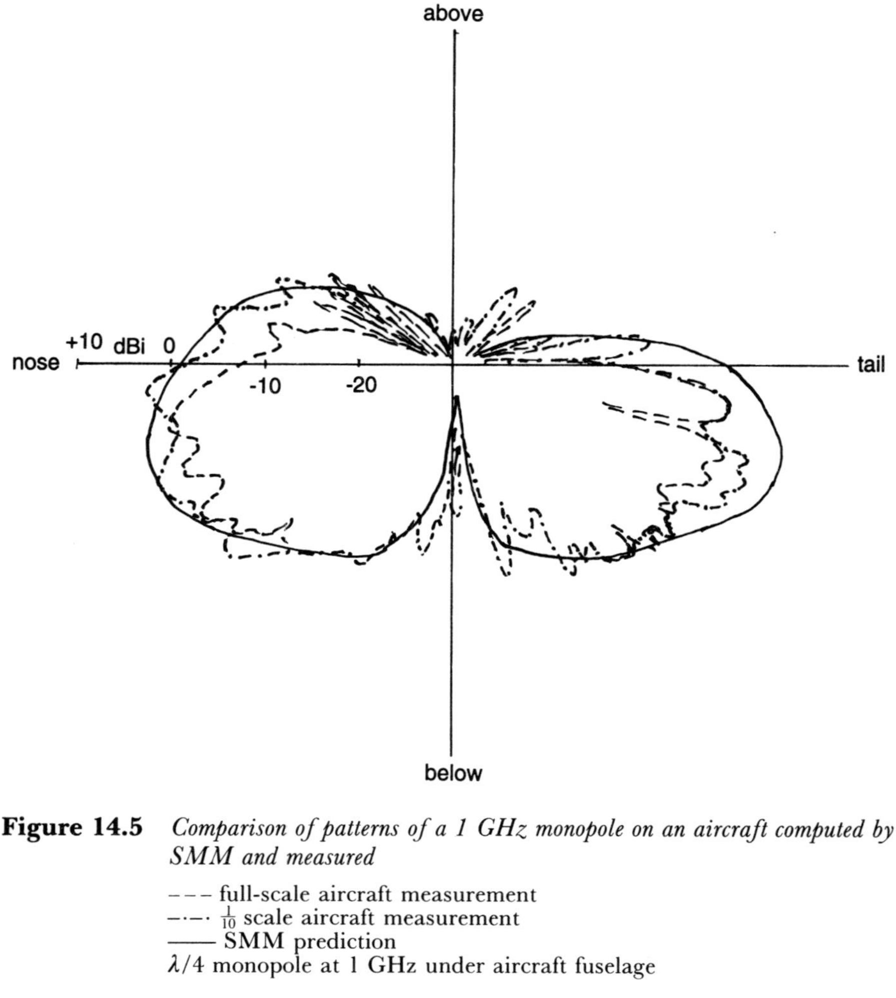

Figure 14.5 *Comparison of patterns of a 1 GHz monopole on an aircraft computed by SMM and measured*

--- full-scale aircraft measurement
-·-· $\frac{1}{10}$ scale aircraft measurement
—— SMM prediction
$\lambda/4$ monopole at 1 GHz under aircraft fuselage

Table 14.2 *Errors found in comparison of measurement methods*

Methods	Errors (dB)
Full scale/1/15th	±1
1/10th scale/SMM	±2 to ±5
GTD/full scale (at 0 dBi)	+0·75
(−25 dB nulls)	±5
GTD/1/8th scale (at 0 dBi)	±1
(−25 dB nulls)	±7

the direction of interest with no other significant contributions provided the edge is more than 2λ distant from the source antenna. It is therefore adequate to model the pattern in terms of the distance to the diffracting edge and by computing patterns for a series of edge distances. The elevation patterns of an antenna on an irregular surface can then be obtained simply by calling up the appropriate data from store.

An azimuth pattern can simply be obtained from the values at 0° elevation of a number of elevation patterns: the number required is a matter of judgment but is really determined by the periodicity of ripple which is a function of distance in wavelengths from the diffracting edge. Subsequent elaborations of the technique allow for reflections and for patterns on cylinders. Of course this method does not have the accuracy of a rigorous GTD but with it patterns for all the monopole antennas on an aircraft could be determined in one to two weeks — far faster than any other method. Only a three-view dimensional drawing of the aircraft is required. Fig. 14.5 compares patterns of a 1 GHz antenna (*a*) measured on a full scale aircraft elevated above ground, (*b*) measured at 1/10 scale and (*c*) modelled by SMM.

It is appropriate here to compare the accuracies achievable by various methods under the most favourable conditions. The author [2] and colleagues carried out an assessment of the following methods for aircraft antennas:

- Simple mathematical modelling
- Geometrical theory of diffraction

Table 14.3 *Summary of five measurement methods*

	SMM	GTD	Reduced scale	Full scale on ground	Flight
Surface accuracy	$\lambda/2$	$\lambda/10$	$\lambda/10$	NA	NA
Dielectrics modelled	No	No	Some	Yes	Yes
Dimensional accuracy	±5%	±2%	±2%	NA	NA
Elevation angles	±90°	±90°	±90°	Near 0°	Near 0°
Time per plane pattern	–	–	2–3 min	2–3 min	30–90 min
Errors (dB)					
Mean (Abs. gain)					
at +5 dBi	Envelope	–	±0·3(+0·8)	±0·4(+1·0)	–
at −5 dBi	Predicted	–	±0·35(±1·0)	±0·9(±2·5)	–
at −15 dBi	to ±1·5 dB	–	±0·55(±1·5)	Several dB	–
Mean (relative gain)					
at +5 dBi	Envelope	±1(±2)	±0·25(±0·5)	±0·3(±0·8)	±3
at −5 dBi	Predicted	±1·5(±3)	±0·3(±0·7)	±0·9(±1·85)	±3
at −15 dBi	to ±1·5 dB	±2·5(±5)	±0·5(±1·7)	Several dB	±3
Cost	Low	Medium	Medium	Medium to high	High

(*a*) Figures in brackets are worst case
(*b*) Cost of full scale measurement is moderate if aircraft is literally on ground, high if raised. Error limits apply to latter case.

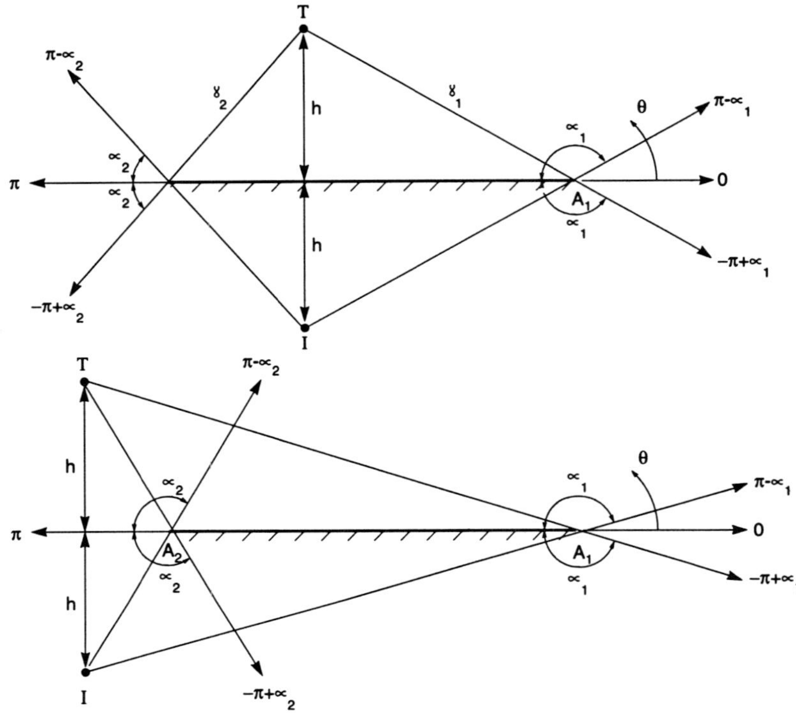

Figure 14.6 *Geometry of Booker's diffraction method*

- Reduced-scale modelling
- Ground measurements (full scale)
- Flight measurements

It is believed that the comments will apply to most linearly-polarised low gain omnidirectional antennas in the frequency range 30–1500 MHz.

The results are shown in Table 14.2 and 14.3.

Table 14.4 *Field due to ray theory*

Range of θ	Electric dipole	Magnetic dipole
$-\pi<\theta<-\pi+\alpha_2$	$\exp(ikh \sin \theta)$	$\exp(ikh \sin \theta)$
$-\pi+\alpha_2<\theta<-\pi+\alpha_1$	0	0
$-\pi+\alpha_1<\theta<\pi-\alpha_1$	$\exp(ikh \sin \theta)$	$\exp(ikh \sin \theta)$
$\pi-\alpha_1<\theta<\pi-\alpha_2$	$2i \sin (kh \sin \theta)$	$2 \cos (kh \sin \theta)$
$\pi-\alpha_2<\theta<+\pi$	$\exp(ikh \sin \theta)$	$\exp(ikh \sin \theta)$
$k=2\pi/\lambda$		

Table 14.5 *Diffracted rays*

Diffraction component	Electric dipole	Magnetic dipole	Field component
T_1	$\mp$	$\mp$	$\frac{1}{\sqrt{2}}\left\lvert \sin\frac{\theta-\alpha_1}{2}\right\rvert F\left\{\sqrt{\frac{8\gamma_1}{\lambda}}\left\lvert\cos\frac{\theta-\alpha_1}{2}\right\rvert\right\}$ $\exp\{ikh\sin\theta+\pi/4\}$ Ranges $-\pi+\alpha_1<\theta<\alpha_1$ $\alpha_1<\theta<\pi,\ -\pi<\theta<-\pi+\alpha_1$
T_2	$\pm$	$\pm$	$\frac{1}{\sqrt{2}}\left\lvert \sin\frac{\theta-\alpha_2}{2}\right\rvert F\left\{\sqrt{\frac{8\gamma_2}{\lambda}}\left\lvert\cos\frac{\theta-\alpha_2}{2}\right\rvert\right\}$ $\exp\{ikh\sin\theta+\pi/4\}$ Ranges $-\pi+\alpha_2<\theta<\alpha_2$ $\alpha_2<\theta<\pi,\ -\pi<\theta<-\pi+\alpha_2$
I_1	$\pm$	$\mp$	$\frac{1}{\sqrt{2}}\left\lvert \sin\frac{\theta+\alpha_1}{2}\right\rvert F\left\{\sqrt{\frac{8\gamma_1}{\lambda}}\left\lvert\cos\frac{\theta+\alpha_1}{2}\right\rvert\right\}$ $\exp(-ikh\sin\theta-\pi/4)$ Ranges $\pi-\alpha_1<\theta<\pi,\ -\pi<\theta<-\alpha_1$ $-\alpha_1<\theta<\pi-\alpha_1$
I_2	$\mp$	$\pm$	$\frac{1}{\sqrt{2}}\left\lvert \sin\frac{\theta+\alpha_2}{2}\right\rvert F\left\{\sqrt{\frac{8\gamma_2}{\lambda}}\left\lvert\cos\frac{\theta+\alpha_2}{2}\right\rvert\right\}$ $\exp(-ikh\sin\theta-\pi/4)$ Ranges $\pi-\alpha_2<\theta<\pi,\ -\pi<\theta<-\alpha_2$ $-\alpha_2<\theta<\pi-\alpha_2$

F is given by

$$F(u)=\tfrac{1}{2}(1-i)-\{C(u)-iS(u)\}$$

where C and S are tabulated Fresnel integrals. Where $\pm$ signs are given for electric and magnetic dipoles, the upper sign refers to the upper range(s) of θ, the lower sign to the lower range(s).

14.6 A simple diffraction method

The need during World War Two to evaluate the effect of reflecting sheets, including aeroplane wings, led Booker [1] to develop an approximate but nevertheless useful method for computing the radiation patterns of dipoles and loops adjacent to finite sheets. The geometry is shown in Fig. 14.6 where T is the source antenna and I its image in the sheet. For an elementary dipole whose axis is parallel to the edges of the sheet then the field for different angles of θ in the plane of the paper is given by ray theory by Table 14.4.

To this must be added waves T_1 and T_2 representing diffraction of waves from T at A_1 and A_2, as well as I_1 and I_2 representing diffraction of waves from I at A_1 and A_2. On the basis of simple Kirchhoff theory which here involves some approximations, these waves are as given in Table 14.5.

The patterns for an electric dipole parallel to the sheet and perpendicular to the diffracting edges may be obtained by multiplying the patterns obtained for the electric dipole parallel to the edges by cos θ. Similarly the patterns for an electric dipole perpendicular to the sheet are given approximately by multiplying by sin θ the patterns of a magnetic dipole with axis parallel to the edges.

These patterns are only approximate nevertheless they give a good appreciation of the situation. It will be noted that Booker [1] did not put a minimum distance to the edges. One advantage of this method is that it can be used without large computer programs. Several patterns are illustrated in Chapter 7.

14.7 Combined NEC–GTD programmes

Some problems such as fin-cap antennas on aircraft do not lend themselves to modelling solely by NEC or GTD. The latter cannot take account of the currents on the edges of the fin whilst the former which could handle it is usually ruled out by the problems of computer storage. Thiele and Newhouse [10] demonstrated the possibility of a hybrid technique in 1975. More recently, Burke [3] has reported the development of an addition to the NEC codes called NEC Hybrid. This combines NEC–3 and NEC–BSC (BSC = Basic Scattering Code, a general purpose modelling code based on GTD–UTD). Vendament and Buchmeyer [11] computed patterns for a horizontal dipole on an aircraft fin by both NEC–MOM and NEC–BSC and compared roll plane patterns with scale model results. Good agreement was found between all three methods in this plane although the BSC guidelines were 'broken and bent'.

Attempts by the author to model a fin-cap antenna for which scale model measurements were available were unsuccessful, the main problem being to determine how far down the fin the wire grid should extend. The size of the fin was such that a complete fin model would have required a very large number of segments. Ciccolella and Balma [5] have recently published a paper showing reasonable agreement between measured and calculated results. It required 999 segments to achieve this which added up to over 16 hours of CPU time on the computer available. It still appears cheaper to use scale models.

14.8 References

1 BOOKER, H.G.: 'Diffraction by aeroplane wings and aperiodic reflectors'. Telecommunications Research Establishment, England, March 1941

2 BURBERRY, R.A.: 'Accuracy of determination of aircraft antenna radiation patterns', IEE Conf. Publ. 219, April 1983, pp. 97–100

3 BURKE, G.J.: 'Recent advances to NEC: applications and validation'. AGARD Lecture Series 165, Oct. 1989, Reference 3

4 BURNSIDE, W.D., and PATHAK, P.H.: 'High frequency scattering by a thin dielectric slab'. IEE Conf. Publ. 195, April 1981, pp. 50–53

5 CICCOLELLA, A., and BALMA, M.: 'Radiation patterns on tail-cap antennas'. Conference on Electromagnetics in aerospace applications, Turin, Sept. 1989, pp. 9–12

6 FOSTER, P.R., and MILLER, T.: 'Radiation patterns of a quarter-wave monopole on a finite ground plane'. IEE Conf. Publ. 195, April 1981, pp. 451–455

7 MILLER, E.K.: 'A selective study of computational electromagnetics for antenna applications'. AGARD Lecture Series 165, Oct. 1989, Reference 2

8 MOLINET, F.A.: 'GTD/UTD: Brief history of successive development of theory and recent advances — applications to antennas on ships and aircraft'. AGARD Lecture Series 165, Oct. 1989, Reference 8
9 PATHAK, P.H., BURNSIDE, W.D., NAN WANG, and CHU, T.: 'Near and far field airborne antennas pattern analysis'. IEE Conf. Publ. 195, April 1981, pp. 247–252
10 THIELE, G.A., and NEWHOUSE, T.H.: 'A hybrid technique for combining moment methods with the geometrical theory of diffraction', *IEEE Trans.*, 1975, **AP–23**, pp. 62–69
11 VENDAMENT, C.H., and BUCHMEYER, S.K.: 'Comparison of pattern measurement and calculations for a VHF dipole on the empennage of an aircraft'. 3rd Annual review of progress ACES, 1987
12 WILLIAMS, D., and BRAMMER, D.J.: 'Moment method analysis of VHF antennas on vehicles on an imperfect ground'. IEE Conf. Publ. 195, April 1981, pp. 535–538

Chapter 15

Antenna measurements

Accurate measurements are essential for successful antenna design, in the development stage, in qualification, and in production testing. The proliferation of automatic test equipment for impedance measurement and for radiation pattern recording, while making faster measurements possible, has not necessarily improved the accuracy of measurement or reduced the likelihood of errors being made. In fact, when less complex equipment had to be used more care was taken in ensuring that the measurement conditions were appropriate and in taking the measurements. Like computers, modern automatic test equipment is equally good at producing large quantities of rubbish if inappropriately used.

15.1 Antenna specifications

To specify properly the antenna parameters that need to be measured and to set appropriate limits requires a complete knowledge of the system in which the antenna is intended to operate. Too many specifications indicate that a proper study has not been carried out. Even if the parameters are properly specified, there is still the problem of specifying how the measurements should be done. This is a particular problem when the antenna performance is specified on a vehicle. The International Electrotechnical Commission (IEC) has laboured for many years to produce specifications of methods of measurement which would be accepted world-wide and would therefore permit direct comparison of antennas produced in different countries. Some specifications have been issued but some of the more difficult vehicular problems have not been resolved. Some of these are discussed later in the chapter.

In any antenna design there will be three stages of measurement:

- During development
- In testing to prove compliance with specification
- In production testing

These may well all demand different measurements of the same parameters.

15.1.1 *Measurement parameters*

For all antennas the following parameters have to be considered:

- Impedance and VSWR
- Efficiency
- Radiation pattern coverage
- Gain

In addition, for transmitting antennas it is necessary to consider

- Power handling
- Voltage breakdown

For directional antennas more detailed information on radiation pattern may be necessary:

- Width of main beam
- Sidelobe level
- Front-to-back ratio
- Boresight error (difference in angle between physical alignment and electrical beam maximum)

In some systems polarisation purity may be important for reducing navigation errors or to minimise multipath fading.

In service, antennas may be subjected to a wide range of climatic conditions which may affect electrical performance. We are not here considering purely mechanical effects: some of these are discussed in the appropriate chapters on vehicular antennas. The conditions which have to be reckoned with are:

Vibration
Shock, e.g. satellite launch, gunfire
High temperature
Low temperature
Ice formation
Low pressure causing flashover and corona
High humidity
Rain

Because antennas are radiating elements they are susceptible to the presence of near-by objects. It is therefore not normally possible to put the antenna into a conventional environmental test chamber and to carry out electrical tests in situ. Some suggestions for simulating environmental conditions are given in Section 15.5. Environmental test specifications for radio equipment such as IEC 68 cannot be applied to antennas without modification and it will be up to each antenna designer to put forward a test programme which meets the spirit if not the letter of the tests demanded by approval authorities.

15.2 Impedance measurements

Impedance measurements are normally needed in the early stages of research and development but in subsequent stages it is often sufficient to measure VSWR since specifications are framed in terms of this parameter. It is not proposed to discuss impedance measuring equipment here but a few points should be noted. Network analysers are in widespread use and certainly facilitate rapid measurements but are not without drawbacks. In the first place they are often more susceptible to external EM fields than the equipment they displaced which did not have electronic circuitry to provide processing capability. Secondly, these equipments are not well suited to field measurements — their place is in the well-ordered environment of the laboratory. They are also not well suited to the accurate measurements of impedances well removed from 50 ohms. Small errors in phase and amplitude produce a much larger percentage error at the outside of a Smith chart than at the middle. For such impedances, which may well be encountered in initial investigations of a new antenna, there does not appear to be any substitute for RF bridges which are not based on 50 ohm coaxial systems. Unfortunately network analysers have made measurement so easy and so rapid that there is a tendency to forget that, like computers, they can produce rubbish results if inappropriately used.

The art of impedance measurement is to produce results which are both accurate and appropriate. Inaccuracy can arise from a number of sources:

(i) Positioning of the antenna so that energy is reflected back from near-by obstacles
(ii) Presence of harmonics in the frequency range
(iii) Leakage of energy into the receiving part of the measuring system
(iv) Presence of external sources radiating sufficient energy.

The influence of adjacent objects must be taken into serious consideration; one drawback to the adoption of less portable measuring equipment is a tendency to bring the antenna to the equipment instead of the reverse. This may result in attempting to measure impedance indoors where it should be obvious that the surroundings will influence the result. It is not necessarily the size of the room that matters but its reflectivity. If, for example, measurements are made in a well-sealed chamber with walls of high conductivity, most of the energy will be reflected back to the antenna and the perceived impedance will be wholly reactive except for loss resistance. This is precisely the situation in Wheeler's Box which is used to measure the efficiency of small antennas. It indicates the difficulty of impedance measurements in environmental test chambers which are often sealed structures with metal walls.

Large static impedance measuring equipment may represent one of the major adjacent objects. To separate equipment and antenna by long coaxial cables will reduce one problem but possibly at the expense of phase accuracy. Some equipments overcome this problem and the effect of cable attenuation by using a balancing cable of identical material to the measuring cable, i.e. from the same reel. This helps, but it must be remembered that flexible coaxial cables age so it is good policy to check measuring cables regularly and to replace them at any sign of deterioration.

A simple test for the effect of the surroundings is to move the antenna through a distance of at least $\lambda/4$ in all directions, including up and down, and to note the variation of VSWR. If this is within the acceptable limits for the particular measurement then the test site is satisfactory. Attempts have been made to provide a formula for calculating the allowable distance of reflecting objects for a desired accuracy, but none have proved universally applicable. Naturally, the simple test has to be modified to suit the use of the antenna. For one mounted on top of a land vehicle there is obviously no need to lift the vehicle $\lambda/4$ above the ground! Only the horizontal movement is necessary in this case. For large reflecting surfaces the formula

$$R = C_1 G e_\theta^2 \lambda$$

has been suggested where

R = distance to the surface

G = peak gain relative to a dipole
θ = angle between direction of peak gain and direction of the reflecting object
e_θ = field intensity at θ normalised to peak intensity

Typically $C_1 = 1{\cdot}2$ for 10% accuracy of VSWR
$= 2{\cdot}4$ for 5%
$= 4{\cdot}0$ for 3%

For highly directional antennas for frequencies above 1 GHz it may be desirable to erect screens faced with RF absorbent material (RAM) to reduce reflected energy. This becomes rather expensive at lower frequencies.

It should be noted that although metallic surfaces have the greatest reflectivity most other materials, if sufficiently thick, will reflect RF energy. Certainly wooden beams, trees, plastic pipes full of liquid, and of course the human body itself can give sufficient reflection if sufficiently close to the antenna under test. If a structure is needed to support an antenna at some height above ground to minimise ground reflections then care must be taken to keep the structure away from the high impedance parts of the antenna. Consider, for example, a VHF dipole raised above ground on a wooden or plastic pole. If possible the dipole arms should be horizontal, i.e. orthogonal to the pole rather than parallel to it. This also makes the dipole orthogonal to the measuring cable which is an advantage. However, the effect of the ground is greater on the impedance of a horizontal dipole than on a vertical one. Clearly the design of the whole experimental arrangement requires careful consideration of these factors. One very unsatisfactory arrangement is to mount a vertical dipole at the top of a vertical pole so that only one arm of the dipole is parallel to the pole. This causes unbalance and could create unwanted currents on the outside of the measuring cable.

The effect of high power external sources should not be overlooked. Many antennas possess higher passbands so the interference need not be within the intended operating band of the antenna. The characteristics of the measuring equipment will influence the effect as well. In some systems using directional couplers the coupler response may well increase with frequency thus emphasising the influence of external signals of higher frequencies. If complete anechoic chambers are inappropriate it could be necessary to move the set-up to an area of low external fields. The most probable sources of interference are VHF and UHF television transmitters and UHF radars.

15.2.1 *Ground planes and mock-ups*

In designing monopole-type antennas for mobile use flat, circular or square ground-planes are often used. These are cheap to construct, easy to move about, and provide some shielding of the measuring equipment from antenna radiation. Unfortunately they are often too small and a poor representation of the surface on which the antenna will be used. Storer [9] calculated the change of input impedance ΔZ of a base-fed monopole at the centre of a large circular ground compared with the impedance Z_0 on an infinite ground plane:

$$\Delta Z = Z - Z_0 = j\,\frac{60}{kd}\exp(-jkd)\left|k\int_0^h \frac{I(Z)}{I(0)}\,dZ\right|^2$$

where d = diameter of circular ground plane

h = height of monopole
$I(Z)$ = current distribution function of monopole
$I(0)$ = base current

Now for a quarter-wave monopole with sinusoidal current distribution,

$$\left| k \int_0^h \frac{I(Z)}{I(0)} dZ \right| = 1$$

hence

$$\Delta Z = j \frac{60}{kd} \exp(-jkd)$$

When d/λ is greater than 10, the change in either resistance or reactance is less than 1 ohm. But such large ground planes are impractical at low VHF and in any case are unlikely to represent correctly the effect of the structure on which the monopole is used in practice. Meier and Summers [7] produced experimental results for monopoles on circular ground planes down to 2λ diameter. They also showed that the variation on square ground planes of similar size was roughly half that of the same antenna on a circular ground plane. This is because, whilst the edge current on a circular sheet has the same phase around the sheet, the phase around the edge of a square sheet is not constant. There is, therefore, some practical advantage in using a square sheet for impedance measurement. This does not apply to radiation pattern measurement where the square sheet will create a distorted pattern.

Even sheets of 2λ diameter or 2λ side are uncomfortably large at, say, 30 MHz as well as being unrepresentative. Awadalla and Maclean [2] extended the calculations down to $0{\cdot}5\lambda$ diameter (Fig. 15.1). The rapid and large oscillations in resistance and reactance are obvious. Fig. 15.2 shows impedances of a quarter-wave monopole at three positions on the roof of a British 'family-size' saloon car. The size of the roof would be approximately λ long and $0{\cdot}7\lambda$ wide. Strong currents down the vertical members to the car body indicate the inadequacy of the roof as a ground plane at these frequencies.

If the antenna is intended for use on land vehicles a metal box with one open side will provide a much better simulation than will a flat sheet. If the open edges of the box are buried in the ground this will help to reduce still further the edge currents. For aircraft antennas, fuselage mounted, a metal cylinder of two or more metres diameter and one wavelength long should be representative of most aircraft bodies. The antenna should be mounted at mid-length. As an illustration of the difference between a flat sheet and a cylinder as a ground plane, consider the results obtained with a VHF bent sleeve monopole designed for the aircraft communications band 100–156 MHz. The design work was done on a fuselage of 2·4 m diameter ($0{\cdot}8\lambda$ at 100 MHz) and the specification figure of 2·5 VSWR was consistently met in production. Two Government agencies, who should have known better, chose to measure the VSWR on a square sheet of 1·8 m ($0{\cdot}6\lambda$) side and complained because the VSWR rose to 4 at the band-ends. When all parties measured the same antenna on the fuselage, albeit with different test equipments, all agreed that the specification had been met.

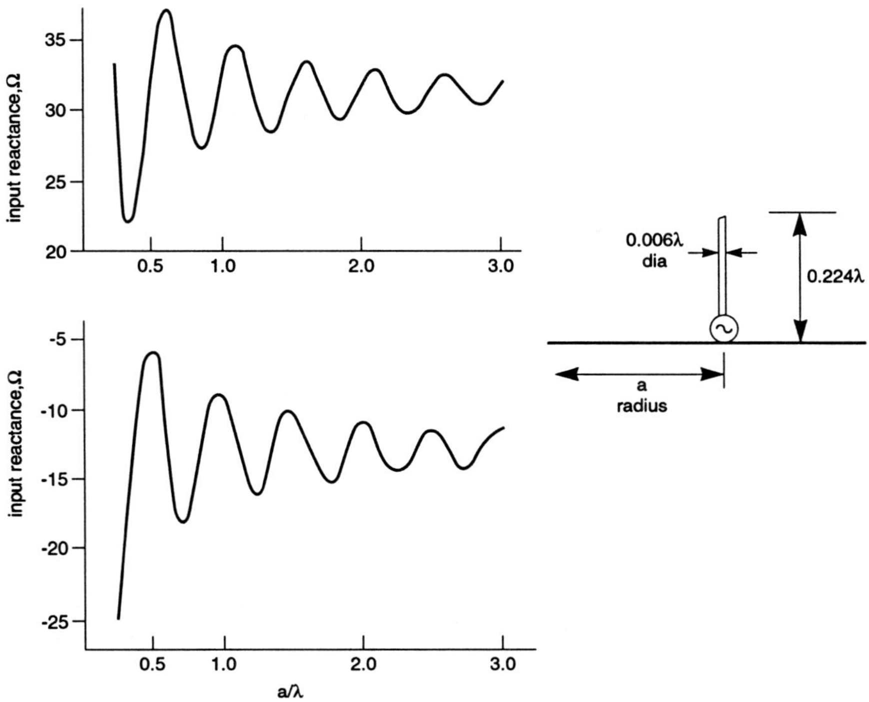

Figure 15.1 *Impedance of a λ/4 monopole on a circular ground plane*

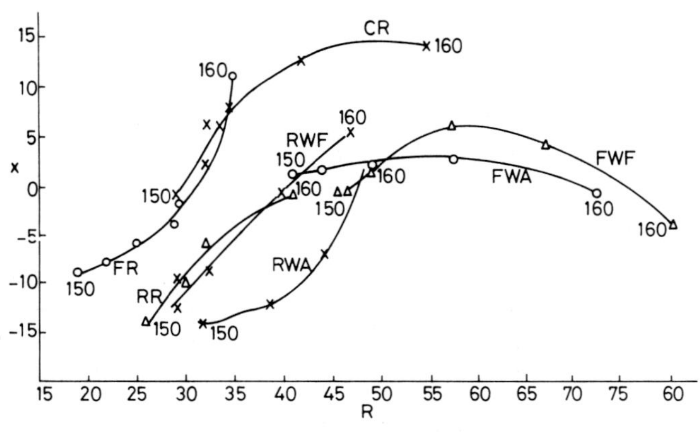

Figure 15.2 *Impedance of a λ/4 monopole on a car roof*

Antenna positions:
CR centre of roof
RR rear of roof
FR front of roof

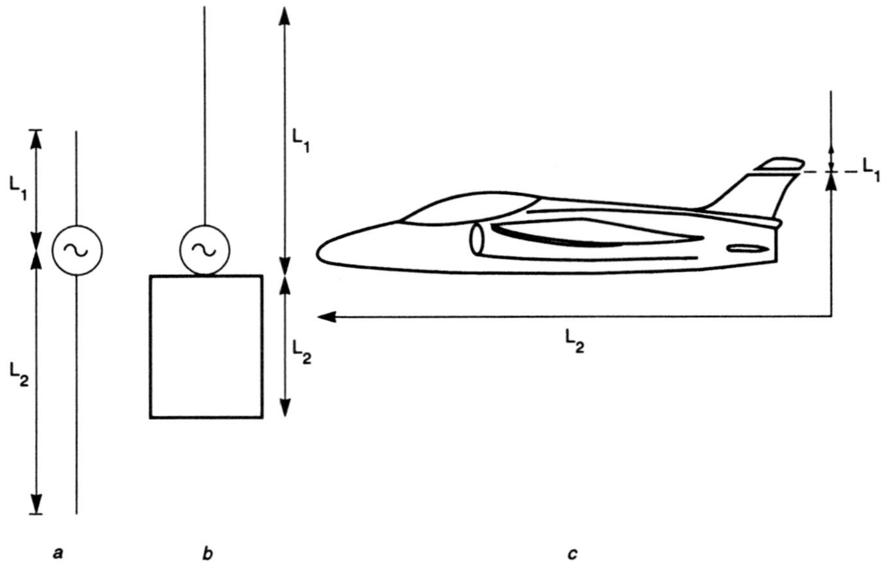

Figure 15.3 *Asymmetrically-fed dipoles*
a Diagrammatic
b Manpack
c Fin-cap antenna

If a monopole type antenna is mounted on the end of a conducting structure, without a separate counterpoise, then it forms with the structure an asymmetrically-fed dipole (Fig. 15.3). The mean value theorem of King [6] considers this dipole to consist of two monopoles of lengths L_1 and L_2 having impedances Z_1 and Z_2, respectively, where $2Z_1$ and $2Z_2$ are the impedances of the centre-fed dipoles of lengths $2L_1$ and $2L_2$. The approximate impedance of the asymmetric dipole is then $Z_1 + Z_2$. At first sight it would appear that the input impedance must be indeterminate as it depends on the length of the conducting structure. In most instances, however, this structure can be considered as a 'fat' dipole; examination of impedance curves for such dipoles, e.g. Brown and Woodward [3], shows that if the equivalent diameter of the structure is greater than $0{\cdot}05\lambda$, then the impedance will not vary significantly if its length is greater than $0{\cdot}5\lambda$. This result has great practical significance: it implies that for, say, an antenna at the top of an aircraft fin it is only important to make a mock-up $0{\cdot}5\lambda$ long at the lowest frequency of interest. If the structure is non-circular in cross-section an equivalent diameter can be determined; see for example Jasik [5] p. 3.6. Some of these equivalents are shown in Fig. 15.4.

15.2.2 *VSWR measurements*

Since specifications usually quote VSWR rather than impedance, this type of measurement is preferable for qualification, production and *in situ* testing. One great advantage is that the measuring device itself is small and can therefore be

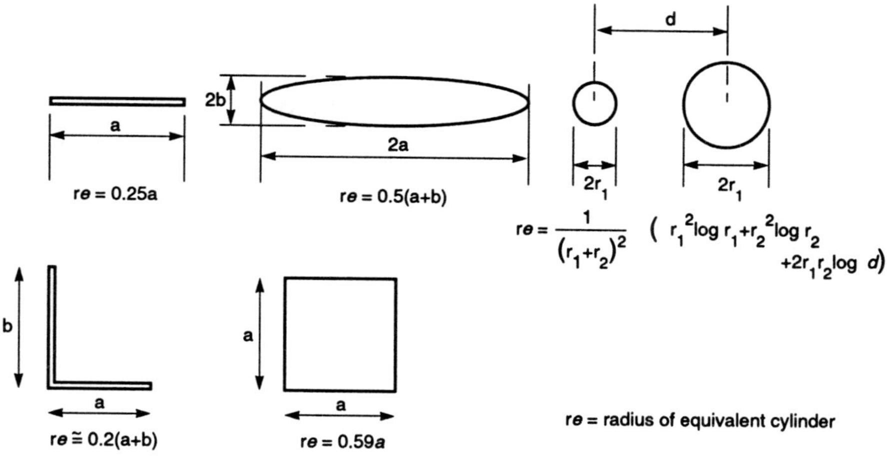

Figure 15.4 *Equivalent diameters of non-circular cross-sections*

attached to the feeder cable in places which would be inaccessible to, say, a network analyser. This is particularly useful in *in situ* testing on vehicles, particularly aircraft in which the radio equipment itself may be mounted in odd corners.

VSWR bridges were used in the 1940s and were literally Wheatstone bridges carefully designed to operate reasonably accurately up to several hundred MHz. They use a series of resistive loads of known mismatch to calibrate the system. These have improved over the years so that satisfactory equipment is available up to 1 GHz. Some VSWR test sets are completely self-contained, consisting of a range of signal sources, a bridge, and a calibrated indicator. As they are battery operated they can be used in the field which is a valuable facility. These equipments give single frequency readings but can be set to any frequency within the range 30–400 MHz and also include coverage of the 1 GHz airborne band. VSWR bridges are generally not suitable for permanent monitoring: for this a system using double directional couplers is required. Using these with a swept frequency source and an *X–Y* recorder the VSWR over a desired frequency band can be plotted out directly. This system can be left permanently on-line if required.

Either system can be used at a single frequency with a tunable antenna. Fig. 15.5 shows VSWR curves for a switched-tuned antenna operating in five sub-bands. Each tuning circuit needs to be set to a chosen frequency to ensure correct overlap between the sub-bands.

VSWR measurements can be used in routine testing of installations particularly in aircraft. Some international airlines and military air forces have adopted regular testing using VSWR test sets. The initial figures for any new antenna installation are recorded and if subsequent readings differ the system is investigated. It should be noted that an improvement in VSWR over a period of time is not an occasion for celebration but is a warning probably of increased attenuation in the RF cables. In an aircraft there are likely to be several cables in series between the antenna and the radio equipment. Using a VSWR test set

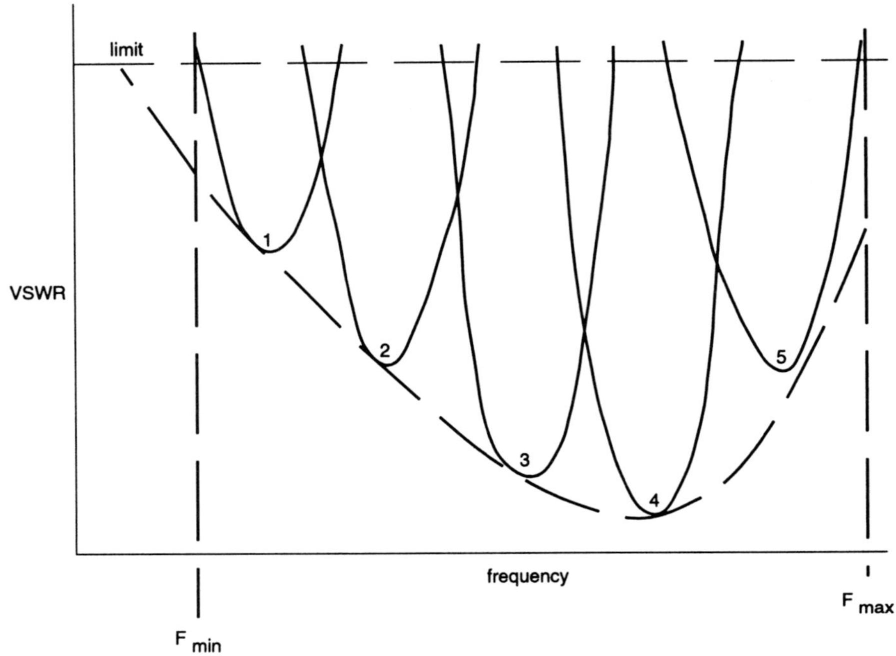

Figure 15.5 *VSWR curves for five-band antenna*
Typical VSWR frequency curves for five-band aerial

and a series of calibrated loads it is possible to measure *in situ* the attenuation of each cable provided that each end is accessible. The measurement is based on the following relationship for the system shown in Fig. 15.6:

$$R = \frac{(S+1)|(S-1) + \exp 2n}{(S+1)|(S-1) - \exp 2n}$$

where R is the VSWR at the far end of the lossy line

S = measured VSWR
n = attenuation, nepers = $0{\cdot}115\ \alpha l$
α = line attenuation, dB/m
l = line length, metres

This equation can be manipulated to yield

$$\exp 2n = \frac{S+1}{S-1}\,\frac{R-1}{R+1}$$

of which both R and S are known if the line is terminated in a load of known VSWR.

Figs. 15.7 and 15.8 show the relationship between true and measured VSWR for a range of cable attenuation. The effect of even a small amount of

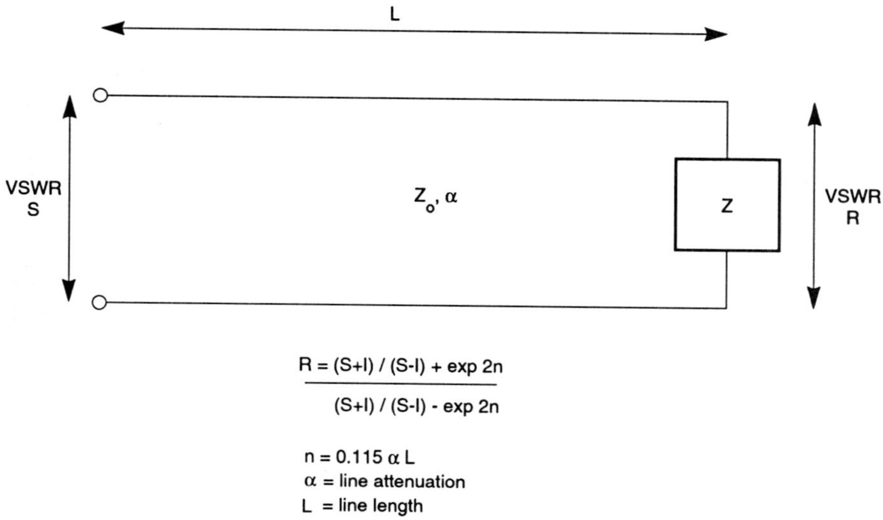

Figure 15.6 *Perceived VSWR on a lossy line*

attenuation in reducing a large VSWR significantly should be noted. Unless the measuring equipment automatically compensates for cable attenuation any measurements will have to be corrected to allow for it.

15.3 Radiation pattern measurements

All test ranges have to satisfy a number of basic conditions if they are to provide accurate results. This will often mean that a range is suitable only for a particular class of measurements or for a specific frequency range. These conditions are considered below with reference to a typical outdoor test range.

15.3.1 *Outdoor test ranges*

Probably the most common form of test site is the ground reflection range shown in Fig. 15.9. Most outdoor ranges are derived from this so the mathematics will be discussed below. Assuming initially that reflections other than from the ground can be neglected, energy from the source antenna arrives at the antenna under test by two paths, direct and by reflection at the ground. The ground is assumed to be smooth, plane and uniform.

With the dimensions shown the path lengths are

$$R_1 = \sqrt{[D^2 + (h_2 - h_1)^2]}$$
$$R_2 = \sqrt{[D^2 + (h_2 + h_1)^2]}$$

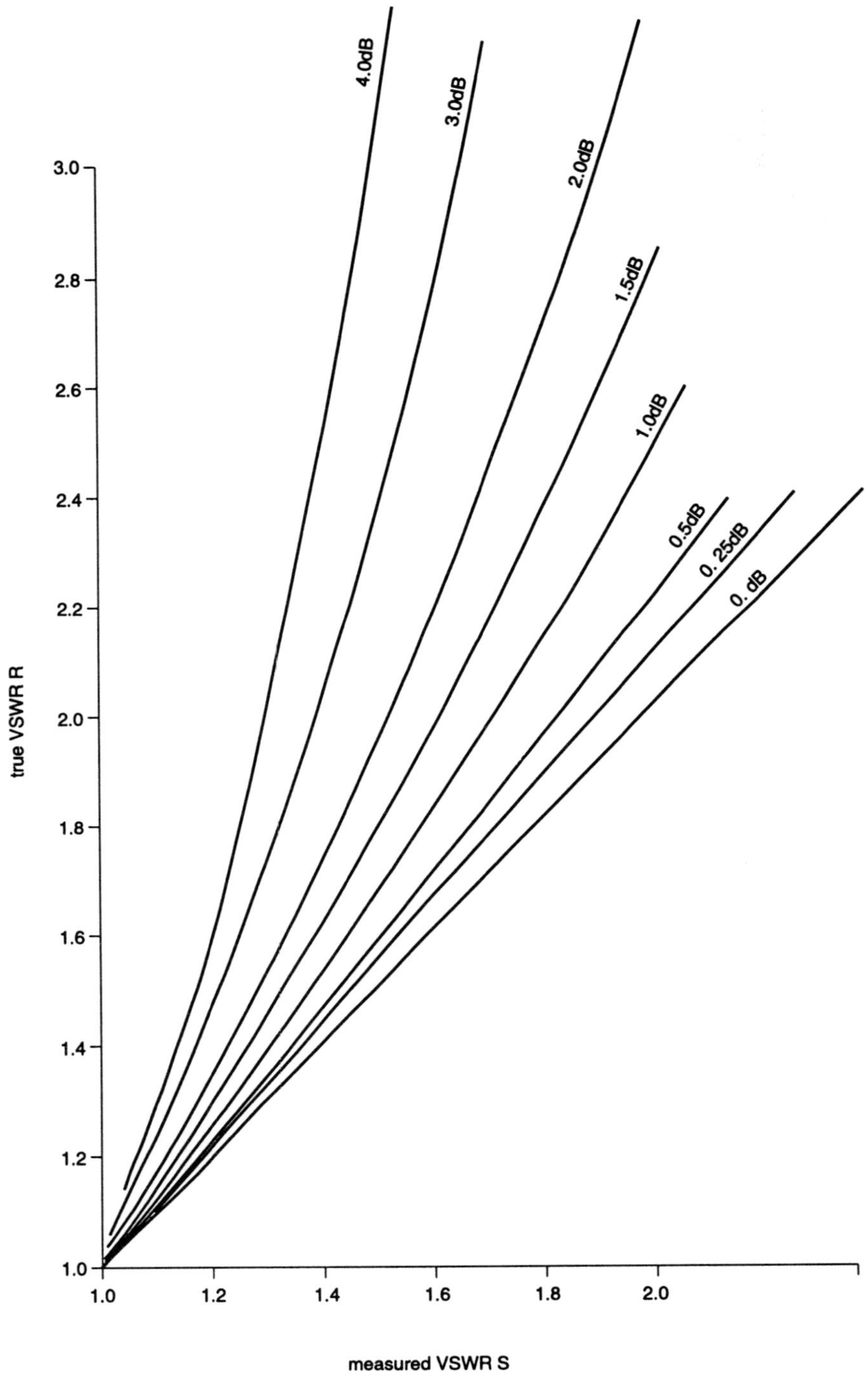

Figure 15.7 *VSWR on a lossy line, for small VSWR*

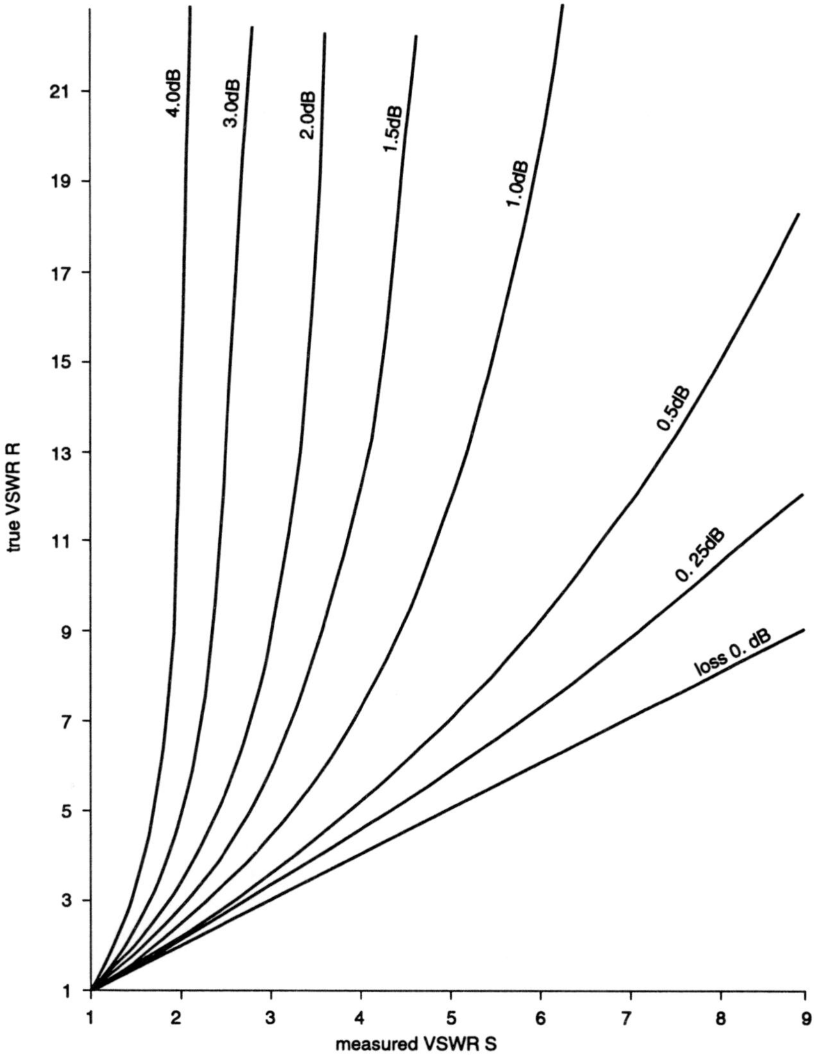

Figure 15.8 *VSWR on a lossy line, for large VSWR*

If $D \gg h_1$ and h_2 these can be expanded to

$$R_1 = D\left[1 + \frac{(h_2 - h_1)^2}{2D^2}\right]$$

$$R_2 = D\left[1 + \frac{(h_2 + h_1)^2}{2D^2}\right]$$

Then the path difference, $R_2 - R_1$, becomes

$$\frac{2h_1h_2}{D}$$

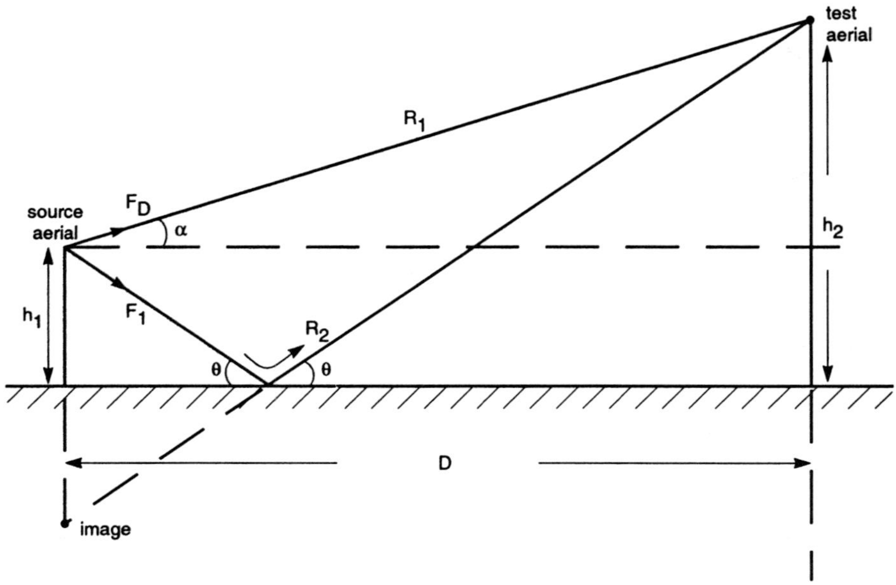

Figure 15.9 *Ground reflection range*

$$\text{The phase difference} = \frac{2\pi}{\lambda}\,\frac{2h_1h_2}{D} = \delta.$$

The ground ray is reflected with an amplitude coefficient ρ and phase ϕ which are functions of polarisation, ground permittivity, frequency and grazing angle θ. This analysis assumes plane waves and a flat earth which are reasonable approximations for test ranges. The source antenna has relative directivity factors F_D and F_I and a peak gain G_1. If E_0 is the free space field strength at the receiver due to a doublet, then the field due to the source antenna becomes

$$E = \sqrt{G_1}E_0\sqrt{(F_D^2 + \rho^2F_I^2 - 2F_DF_I\rho\cos\Omega)}$$

where $\Omega = \delta + \phi - \pi$

Typical field strength curves for a test site on wet ground are shown in Fig. 15.10. If the surface of the site is considered to be homogeneous soil, it can be treated as a dielectric material of relative permeability unity and relative permeability $\varepsilon = \varepsilon_r + j\varepsilon_i$ where $\varepsilon_i = g/\omega\varepsilon_0$ and g is conductivity in mho/m and $\varepsilon_0 = 1/(36\pi \times 10^9)$ farad/m. The reflection coefficient ρ is then given by

$$\rho_H = \frac{\sin\theta - \sqrt{(\varepsilon - \cos^2\theta)}}{\sin\theta + \sqrt{(\varepsilon - \cos^2\theta)}}$$

$$\rho_V = \frac{\varepsilon\sin\theta - \sqrt{(\varepsilon - \cos^2\theta)}}{\varepsilon\sin\theta + \sqrt{(\varepsilon - \cos^2\theta)}}$$

where ρ_H and ρ_V are the complex reflection coefficients for horizontal and vertical polarisation respectively.

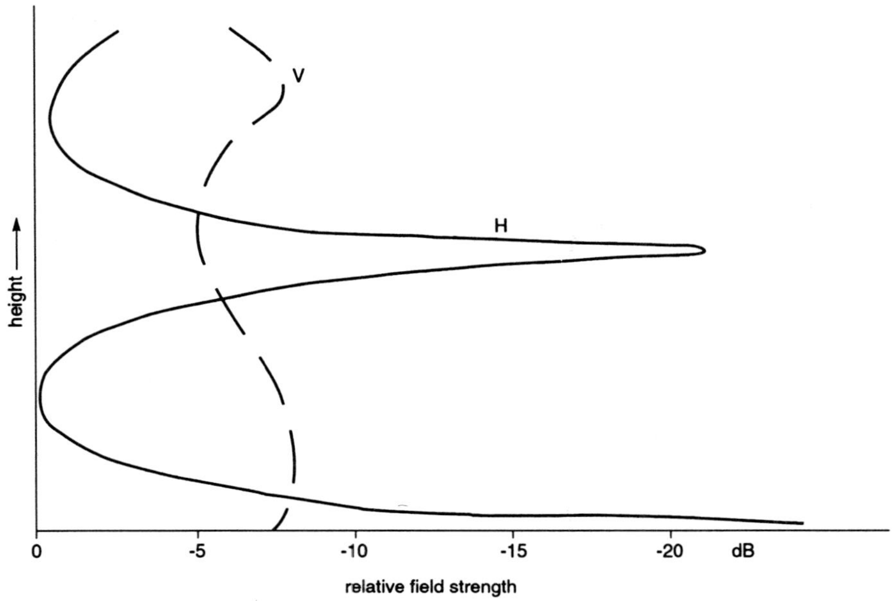

Figure 15.10 *Variation of field on a ground reflection range*

It should be obvious that the largest reflection occurs from a surface with high conductivity, i.e. metal or water. It is therefore an advantage to have a test range on soil of poor conductivity such as sand. Stone chippings of about 30 mm stones size are even better as they allow water to seep away rapidly. There may be appreciable wave penetration into soil so the depth of the water table may be significant as Ford [4] shows in an analysis of the antenna test range at the European Space Research and Technology Centre in Holland. If there is likely to be marked variation in the water table at a given site it may be necessary to use a metal surface in order to obtain consistent results. The relative costs of this or a sufficient height of chippings above ground need to be weighed for each site. In any case, snow on the ground will upset the conditions.

To give some idea of the relative dimensions of the permittivity components the following typical figures should be noted:

	ε_r	g(mho/m)
Desert	3	0·011
Average soil	15	0·028
Marsh	30	0·11
Salt water	80	5

Any real antenna occupies a finite space and the aim must be to illuminate this space as uniformly as possible. If the antenna under test is to be rotated about a vertical axis (to give a horizontal radiation pattern) then the space is a vertical cylinder of radius equal to the greatest excursion of the antenna and height equal to the effective height of the antenna. It should be standard practice to probe this volume at sufficient frequencies to cover the band over which tests are to be done.

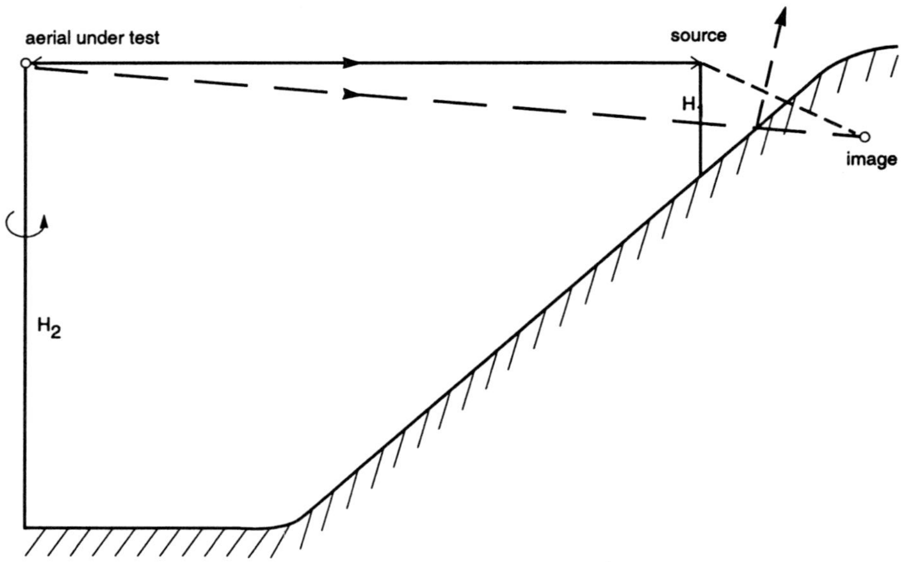

Figure 15.11 *Slant range using natural slope of ground*

Ground reflections may be minimised by using a source antenna of high directivity in the vertical plane and by using as high and as short a range as the general criteria, discussed below, will permit or physical considerations will allow. In some areas it may be possible to use a slant range such as Fig. 15.11. This arrangement has the advantage of requiring rotation about a vertical axis whereas in a true slant range with the source antenna on the ground, a tilted axis of rotation is required. In either case a large structure may be required to elevate the antenna under test. One method used by the Royal Aircraft Establishment in England to elevate quite large aircraft models was a 12 m fibreglass tube 0·6 m in diameter which could sink down into a vertical shaft in the ground to bring the antenna to ground level. This may have some advantage over a pivoted pole which can result in some severe mechanical design problems.

The ultimate slant range is the vertical range (Fig. 15.12) which avoids ground reflections at the expense of mechanical limitations. Its main use is for measurement on aerospace vehicles using models of HF and low VHF antennas. Thus a 1/60 scale model of an HF antenna on an aircraft would operate in the range 120–1800 MHz. The main problem with the vertical range is to avoid reflections from the vertical support structure.

An alternative method of reducing ground reflections is by the use of diffraction fences. Fig. 15.13 shows a typical arrangement. Note that secondary fences F_2 and F_3 have been positioned to minimise ground reflections which would re-radiate from F_1.

So far we have only considered the effects of ground reflection in determining range performance. Reflections from structures in the horizontal plane may also affect the uniformity of the field. One method of pinpointing the main reflectors is to rotate about a vertical axis an antenna of high horizontal directivity and to

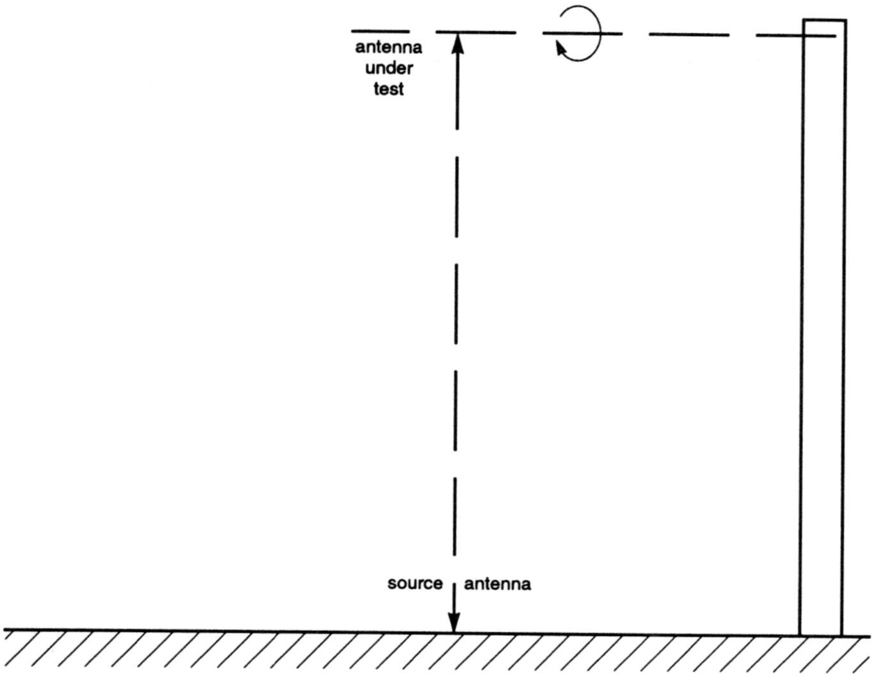

Figure 15.12 *Vertical range*

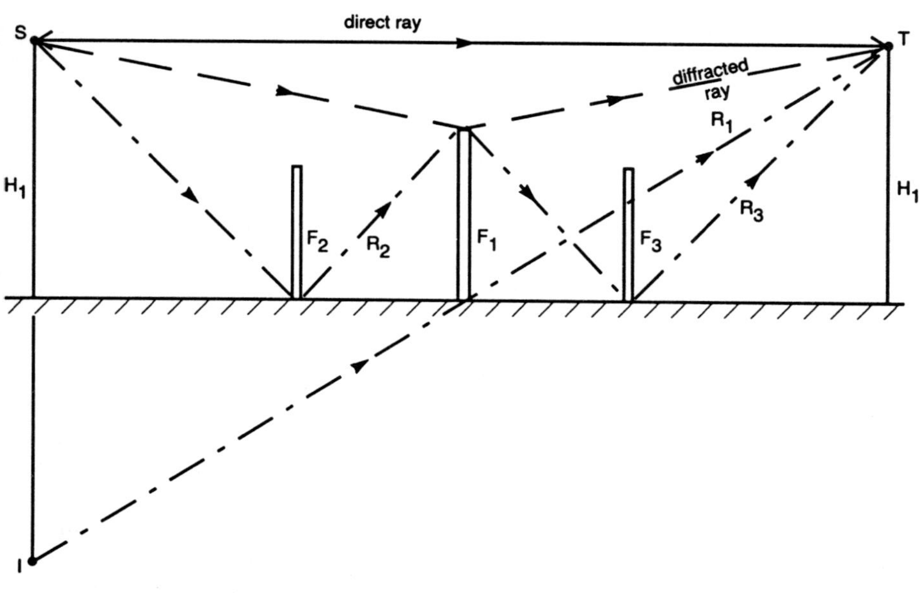

Figure 15.13 *Ground range with diffraction fences*

note the position of asymmetric side lobes in the recorded pattern. To test for possible asymmetry in the antenna itself, this should be inverted and the pattern repeated. If an unavoidable scatterer is found the range should be oriented so that the scatterer is behind the source antenna thus minimising its illumination.

15.3.1.1 *Range criteria*

The illumination of the test aperture needs to be uniform in both phase and amplitude. Consider first the phase constraints with reference to Fig. 15.14. The difference in phase between the centre and the extremities of the aperture is ΔR given by the equation

$$(R+\Delta R)^2=\left(\frac{D}{2}\right)^2+R^2$$

which can be expanded, and when $D \ll R$, gives

$$R=D^2/8R$$

hence the phase variation $=\dfrac{2\pi}{\lambda}\dfrac{D^2}{8R}$

For most antennas, a variation of $\pi/8$ is acceptable giving the well-known

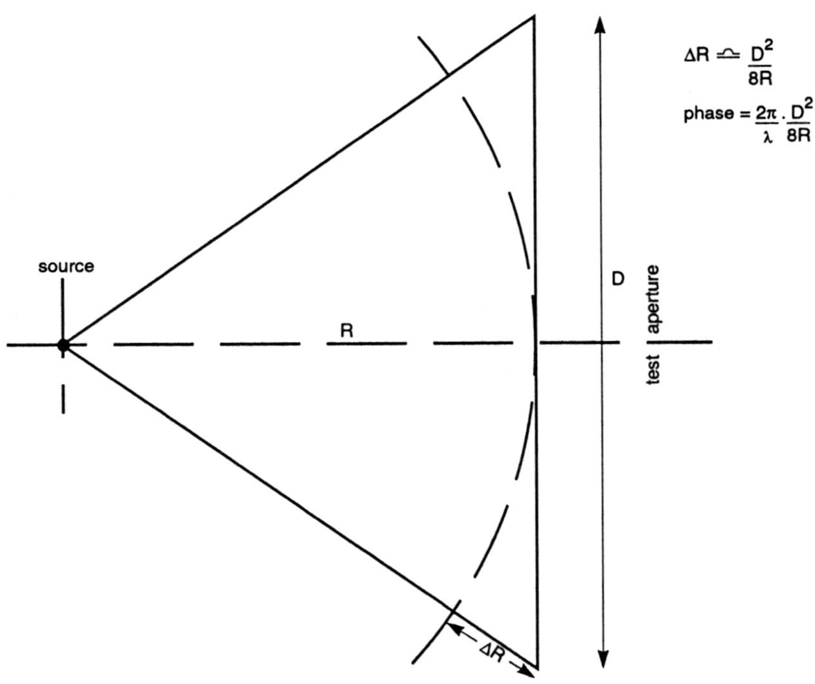

Figure 15.14 *Phase variation over aperture*

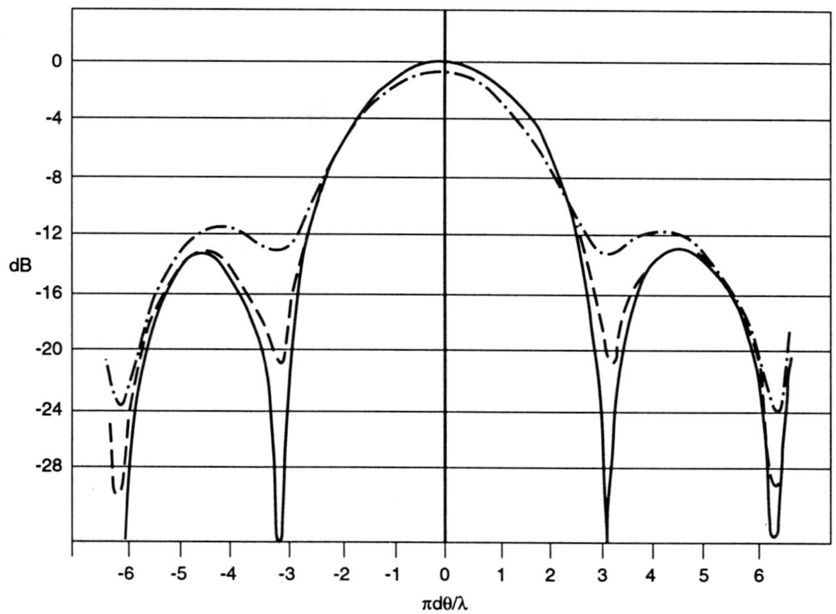

Figure 15.15 *Effect of phase errors on a typical pattern*
—— $\delta = 0°$
– – – $\delta = 22{\cdot}5°$
–·–· $\delta = 52{\cdot}6°$

criterion

$$R \geqslant 2D^2/\lambda$$

The effect of errors in phase is generally to fill in the minima between lobes without greatly affecting the size of main lobe or side lobes. This is shown in Fig. 15.15. Except where it is important to know the depths of the minima accurately, this range definition will be sufficient. It can be seen that for large antennas at very short wavelengths the minimum range can become appreciable.

Whilst the aperture of an array or a dish antenna is obvious, this is not the case when the antenna is mounted on a vehicle. Consider for example an antenna operating at 1600 MHz on an aircraft whose major dimensions are 60 m. With a wavelength of just 0·18 m the $2D^2/\lambda$ distance would be $2 \times 30 \times 30/0{\cdot}18 = 10\,000$ m which is clearly impractical for anything but in-flight measurements. Using a 1/15 scale model would reduce this to 667 m which is still impossibly large. This problem was recognised many years ago by Sichak and Nail [8] who carried out measurements on scale models at different distances and demonstrated that azimuth patterns of a vertical monopole on top of the fuselage only differed by a slight blurring of the nulls even down to distances of the order of 0·1 D^2/λ. A few moments consideration will show that this is not surprising as the contributions due to, say, the wings make negligible effect in the forward direction.

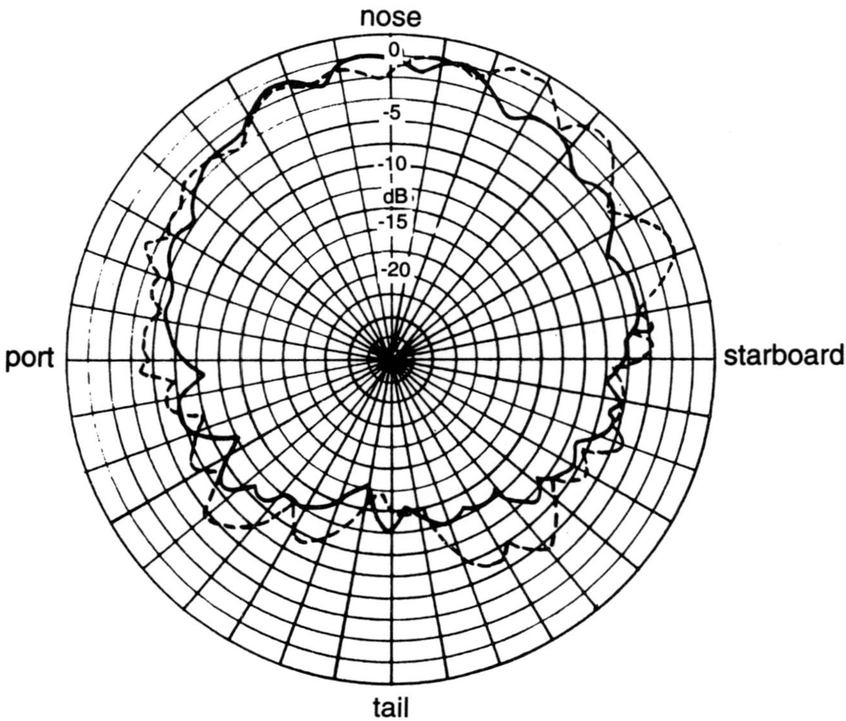

Figure 15.16 *Comparison of measurements on full-scale and reduced-scale aircraft models*

Azimuth radiation pattern of ILS localiser aerial in nose radome
Full scale frequency 112 MHz
—— Full size aircraft measured on ground
--- $\frac{1}{5}$th scale model

Fig. 15.16 compares measurements of an Instrument Landing System (ILS) localiser antenna on a real aircraft on the ground and on a 1/15 scale model. Because the antenna was in the aircraft nose the whole length of the aircraft is taken as D. In the full scale measurements the distance was 0·123 D^2/λ and in the scale model case 1·206 D^2/λ.

A further range limitation is given by the need to reduce the induction field to a sufficiently low figure. The ratio between the first and second terms (the largest two) in the field components may be chosen as parameter. In the case of an electric doublet the first two terms are

$$E_\theta = 60\pi I\left[\frac{j}{\lambda r} + \frac{1}{kr^2}\right] \sin\theta \, dl \, \exp\{jk(ct - r)\}$$

where dl is the length of the element.
The ratio ρ is therefore $1/kr$

Typically this ratio between the induction field and the radiation field is chosen to be 36 dB giving $r=10\lambda$. This ensures that in radiation measurements the coupling between the antennas can be regarded as negligible compared with the radiated field. Clearly the penalty in measuring at $r=5\lambda$ is unlikely to be serious in many situations.

15.3.1.2 *Effect of amplitude variation in illumination*
Given the range requirements of the previous section it is now possible to determine the appropriate radiation pattern of the source antenna to give proper illumination of the antenna under test. If the swept aperture of this antenna has a width D and height H then the source antenna should produce near constant illumination over ± angles of $\tan^{-1}D/2R$ in width and $\tan^{-1} H/2R$ in height. This is the situation in free space measurement and will be modified by ground reflection as we saw earlier.

The effect of amplitude errors is most pronounced on the lower levels in the radiation pattern. A variation of 1 dB in amplitude could result in errors of +3·9 dB to −7·3 dB at 20 dB below the main lobe maximum whilst 0·5 dB variation produces errors of +2·2 dB to −3 dB. Clearly, accurate measurement of low side-lobe levels requires a high degree of uniformity in field illumination.

15.3.2 *Indoor test ranges*

Outdoor test ranges have many advantages in that lack of space is rarely a problem and the cost of setting up a range will be much lower than an indoor facility. There are, however, a number of disadvantages which may make an indoor range attractive:

- Effects of rain, snow and long periods of sunshine on ground reflection characteristics
- Effects of wind on stability of measurement antennas and need to design structures for high wind loads
- Interference from local transmitters
- Limitations on frequency use on the test site
- High cost of avoiding ground reflections for antennas normally operating in space, i.e. aerospace systems.

For these reasons anechoic chambers may be favoured for 'free-space' measurements. There is little point in using them for measurements of antennas on land vehicles or on ships for which a ground range is more appropriate.

The anechoic chamber is a room in which energy incident on the walls is either absorbed or, more rarely, directed away from the area occupied by the antenna under test. The chamber does not have to be rectangular like Fig. 15.17 but may consist of a rectangular working section joined to a tapered section near the apex of which the source antenna is mounted. This minimises reflections from the walls besides reducing the amount of RAM (RF absorbent material) needed for lining. Fig. 15.18 shows a typical tapered chamber. Such chambers are most useful for frequencies above 1 GHz and even at this frequency there may be difficulty in matching the radiation pattern of the source antenna to the shape of the chamber to produce uniform illumination at the antenna under test.

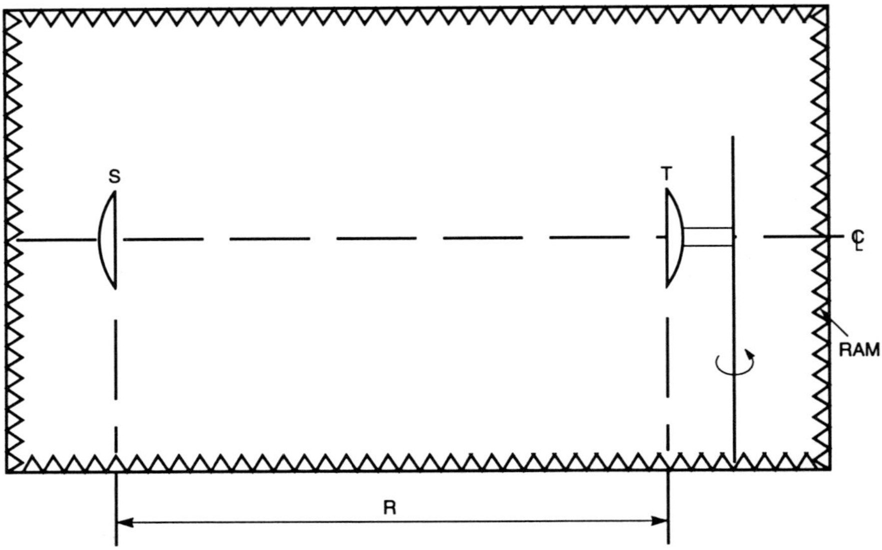

Figure 15.17 *Basic anechoic chamber*

It is usual to describe the quality of an anechoic chamber by the size of Quiet Zone in which the signal variation is below some specified level; the size of this zone will be decreased by reflections from the walls. It is, therefore, not

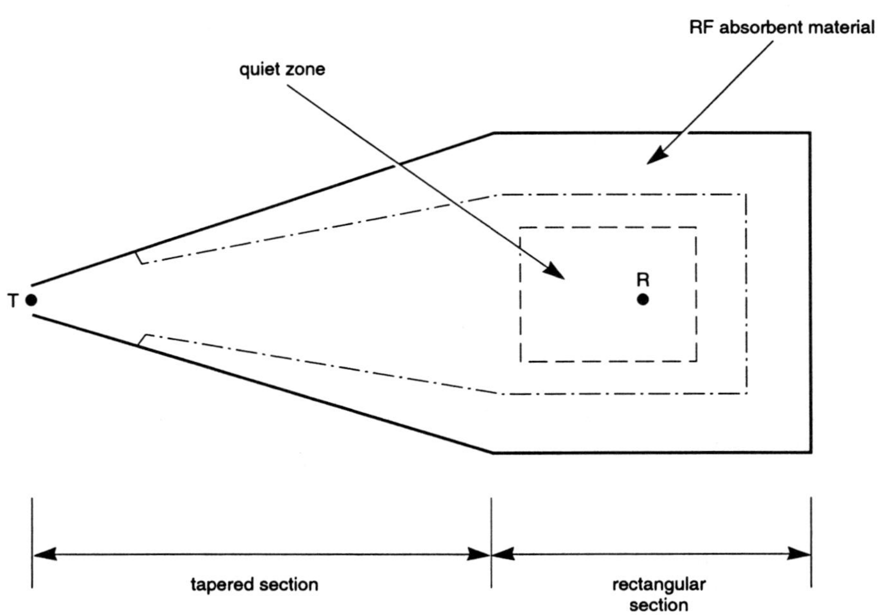

Figure 15.18 *Tapered chamber*

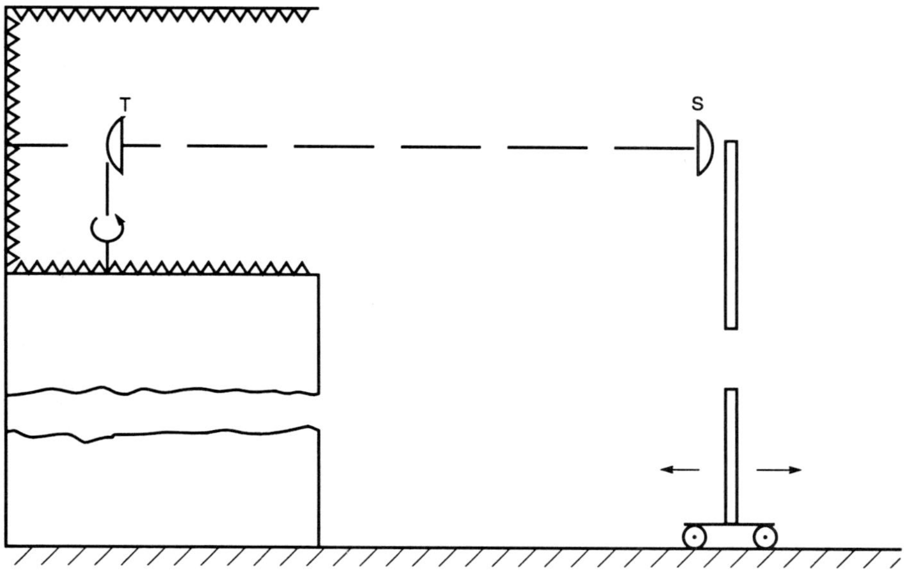

Figure 15.19 *Partially open chamber*

sufficient to line the chamber with RAM: the radiation patterns of the source antenna must be carefully chosen, too. On the one hand the test antenna must be uniformly illuminated (a taper of 0·25 dB is usually acceptable) whilst on the other hand illumination of the side walls needs to be minimised. The mid-point of the walls, floor and ceiling between the two antennas is likely to be the most critical area in a rectangular chamber. In one large rectangular chamber used over a very wide frequency range poor uniformity of illumination at some frequencies was found to be due to the regular spaced pyramids of RAM round the chamber mid-point. A random arrangement of RAM in the appropriate areas gave considerable improvement as did the use of source antennas with a null directed at the mid-point. This meant that the source antenna had to have a near-constant radiation pattern, at least as far as the critical null was concerned, and hence the use of very broadband source antennas was undesirable. In any case with a constant-sized test object such as a scale-model aircraft the illuminating beamwidth needs to be reasonably constant.

The same rules on uniformity of field apply as for outdoor ranges so the use an anechoic chamber is limited both by reflection from the walls and the length available. Absorption by a given size of RAM decreases with frequency so that, depending on the type of antenna to be measured, a lower frequency limit can be set. The most stringent requirements apply to directional antennas where the depths of null are important; a given chamber can be used for lower frequencies for near-omnidirectional patterns. Very few chambers are suitable for frequencies much below 1 GHz for the cost of a large building and suitable lining material is prohibitive. One method of overcoming the range and frequency limitations is by the use of a partially open chamber (Fig. 15.19); this is a rectangular chamber with one wall removed, the others lined with RAM. It has to be mounted high above the ground. The source antenna is mounted on a tower whose distance can, ideally, be adjusted to suit the range requirements, Alia and Rispoli [1].

Other methods of overcoming the range limitations are by the use of a compact range or by near-field probing. Such methods are most appropriate for large microwave antennas and will not be discussed here. One other method for extending the use of a chamber is by using scale models.

15.3.2.1 *Scale modelling*

A reduced scale model of an antenna will have the same characteristics as its full scale version providing that the operating wavelength is reduced in the same proportion. Thus a 1/10 scale model of a 100 MHz antenna can be tested at 1 GHz. Since both the aperture and the wavelength are reduced in the same ratio the far field criterion, $R=2D^2/\lambda$, is reduced by the scaling factor.

In theory the conductivity of the antenna materials and of any structure on which it is mounted should be increased by the scaling factor; if the conductivity is already high this requirement can be neglected since the current flow over these surfaces will be scarcely altered. If, however, some parts of the system under test are of poorly conductive material, scaling of the conductivity is essential if significant changes in current distribution are to be avoided. Particular problems arise with materials such as carbon-fibre composites (CFC) whose conductivity depends on many factors including lay-up of the fibres, resin and notably frequency. Above about 300 MHz full scale most CFC structures behave in an EM field as if they are metals. This has been demonstrated both for ground planes and reflectors. Below this frequency region there is likely to be some divergence in the performance of different composites and some testing of materials will be necessary to determine how to model structures of such materials. This will be particularly difficult where scale models are being used to measure the performance of HF antennas on composite aircraft. The changes in conductivity ought to mean that different models are required for different parts of the frequency range.

In any model of a vehicle with transparencies, e.g. a motor car or an aircraft, the transparencies themselves can be neglected in the model. It may, however, be necessary to model in some detail the structure behind the transparencies. The extreme example of this is a helicopter with a largely transparent nose: it was found necessary in one instance to model the interior of the cockpit in some detail, particularly the edges of any metal sheets such as the floor. Fig. 15.20 shows a typical helicopter with transparencies, glass-fibre radome and panels. Some of the latter could well be made of CFC in future as in fact the whole tail might be.

The treatment of radomes in a scale model aircraft has always been difficult and in most instances the radome has been ignored. Some pattern measurements have been made with a small blade antenna adjacent to a 10 GHz radome (Fig. 15.21); they indicate that the radome can largely be neglected for frequencies below 3 GHz. Above this frequency considerable lobing was apparent and to date there is not a good mathematical model. Full scale measurements at these frequencies seem the only safe method of determining the effects.

Producing an adequate scale model of a vehicle such as an aircraft or a satellite requires an intimate knowledge of the structure. Any discontinuities in

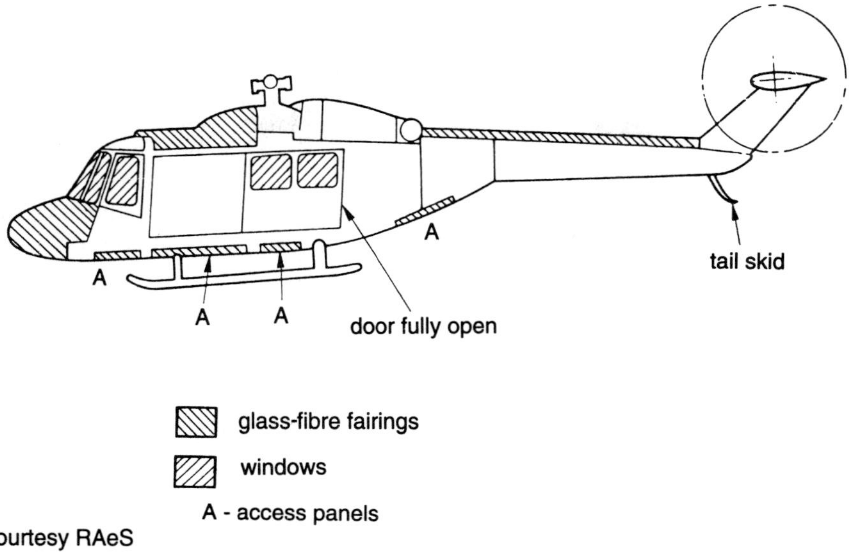

Figure 15.20 *Structure of typical helicopter*

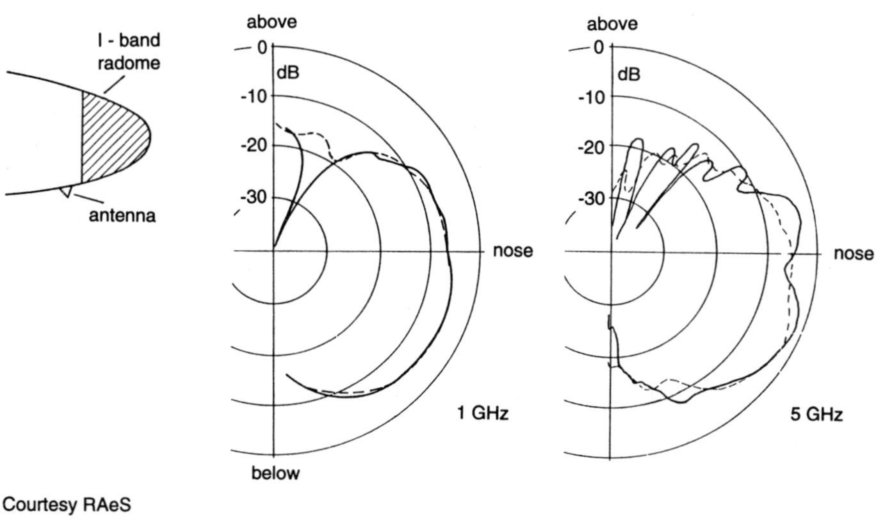

Figure 15.21 *Effect of radome on radiation pattern of adjacent antenna*

—— with radome
– – – without radome

metal structures must be modelled since they can have significant effect on radiation patterns. The gaps between fixed and moving surfaces on wings and tail are obvious examples but it is not always easy to find where these are electrically bridged. In many aircraft, particularly the smaller military ones, the wings may not be attached to the fuselage at all at the edges but only in the central region. Similarly in small helicopters the tail may only be connected to the front fuselage by a few bolts with no skin joint at all. To get a representative model which can be used to model antennas over a wide frequency range considerable detective work is necessary: in all probability no one man in an aircraft design office knows all the answers.

When the scaled antenna is to be mounted on a vehicle or other structure it is first necessary to ensure that it fully represents the full-scale item. This is most easily done by measuring the performance of both under the same conditions. Once the antenna has been installed it is difficult to separate effects due to the antenna from those of the vehicle in measured radiation patterns. The ability to scale the antenna successfully is likely to be the major limitation in using large scaling factors. An upper scaled frequency limit of 40 GHz is probably as far as one can go without the cost of making the scale antenna becoming excessive.

Scale modelling is very largely used for measurements of the performance of antennas on aerospace vehicles where the shape of the vehicle dominates the radiation patterns. A modern military aircraft may have as many as 40 antennas and a civil airliner 30. Although mathematical modelling has made great strides there are still a number of areas of uncertainty for which scale modelling may be the only solution. A good model will last the in-service life of an aircraft and will permit rapid assessment of the effects of any changes during that period.

15.3.3 *Displaying radiation pattern measurements*

Once upon a time, all radiation pattern measurements resulted in piles of single plane plots either in cartesian or polar form. As more and more information is being demanded this arrangement, although adequate for many purposes, no longer satisfies the ultimate user who has neither the time to wade through piles of individual plots nor perhaps the expertise to interpret them. If this user is a system engineer then he may only be interested in seeing where the antenna fails to give the required coverage. This implies that the information given to him shall be referenced to some gain standard; radiation patterns without a gain reference are of little use to anyone. There has sometimes been a tendency to look at the uniformity of pattern rather than the gain level. This can be neatly illustrated in Fig. 15.22 which shows azimuth radiation patterns of two ILS antennas. The pedestal-mounted one clearly has the more uniform pattern but at all points its gain is below that of the fin-mounted antenna. In service the latter has always been preferred.

One method of comparing patterns of disparate shape is the percentage coverage plot shown in Fig. 15.23. This shows the proportion of coverage in a single plane or over a limited sector or over the complete sphere which exceeds some chosen gain level. In the Figure both antennas A and B have the same mean gain but A has the higher gain over the greater percentage of the coverage. If these were azimuth radiation patterns of communications antennas

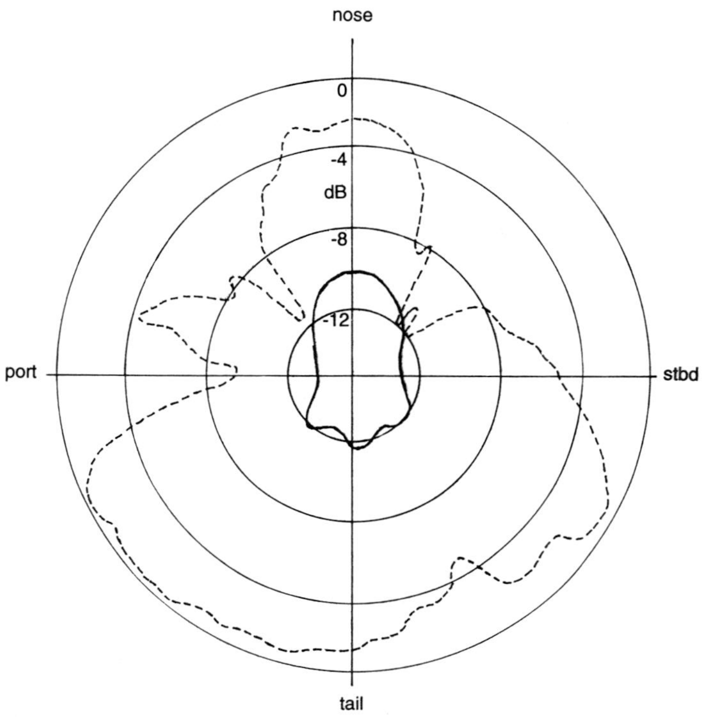

Figure 15.22 *Pattern of two ILS localiser antennas*
—— antenna 0·11λ above fuselage
- - - antenna on tail-fin

on an aircraft or a land vehicle, where omnidirectional cover is required, then A would be preferred. This method of assessment has proved very valuable in the mobile antenna field. The data can easily be obtained from cartesian or polar plots by measuring the percentage of the coverage angle for each of a series of gain levels.

When the cover required extends over a range of elevation angles as well as in azimuth the results of single pattern plots can be incorporated on a single graph showing gain levels over the required coverage. Although this can be done by hand it is a tedious process and can be done with less pain if the recording equipment can take digitised samples. These can then be printed out directly. These samples can also be used to determine antenna directivity as described in the following section. One can move a step further by programming the controlling software only to print out those samples which fall below the required gain level. This removal of redundant information makes it so much easier for the user, who may not be an antenna engineer, to interpret the results.

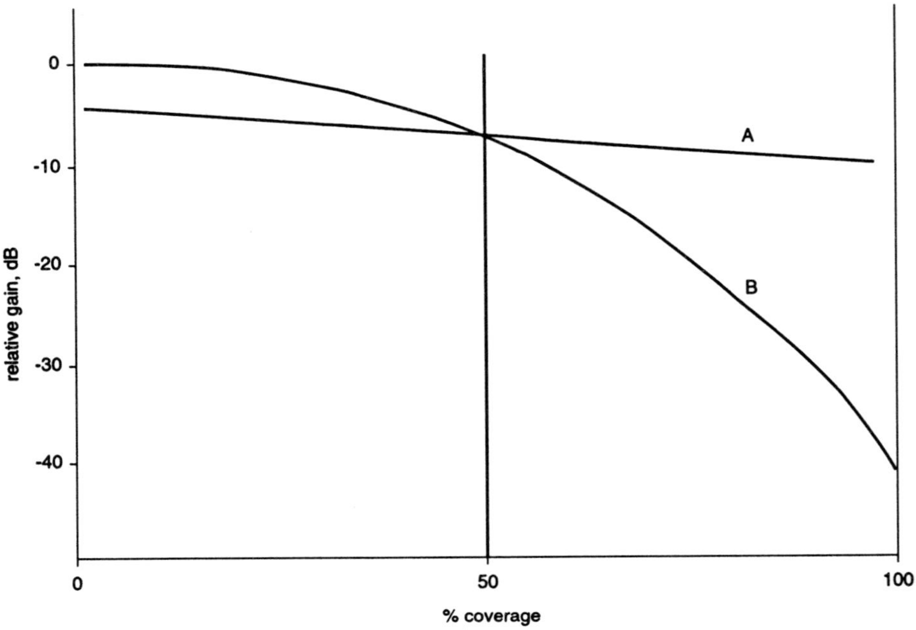

Figure 15.23 *Percentage coverage plot*

15.4 Gain measurements

15.4.1 *Definitions of gain and directivity*

15.4.1.1 *Directivity*

Directivity is defined as $D = \dfrac{\text{maximum radiation intensity}}{\text{average radiation intensity}}$

$$= \frac{4\,\pi\ (\text{maximum radiation intensity})}{\text{total power radiated}}$$

The radiation intensity pattern U can be expressed as

$$U = U_a F(\theta,\phi)$$

where U_a is a constant and θ and ϕ are elevation and azimuth angles respectively.

Then $D = \dfrac{4\pi F(\theta,\,\phi)_{max}}{\iint F(\theta,\,\phi)\; d\Omega}$

If the field pattern is used instead of the power density and normalised to the peak level,

$$D = \frac{4\pi}{\iint F^2(\theta,\,\phi)\,\sin\theta\; d\theta\; d\phi}$$

where $F(\theta,\phi)$ is the normalised field pattern.

It must be noted that $F(\theta,\phi)$ is the total field at any point. Considering the field at any point to consist of two orthogonal components F_1 and F_2 then

$$D=\frac{4\pi}{\int_0^{2\pi}\int_0^{\pi}[F_1^2(\theta,\phi)+F_2^2(\theta,\phi)]\sin\theta\, d\theta\, d\phi}$$

To obtain D from measured patterns, two polarisations must be measured and a series of patterns taken over a complete sphere. Values of $F(\theta,\phi)$ have then to be extracted: the simplest method is by sampling at regular intervals in each plane. The number of sampling points to give a desired accuracy is a function of the antenna aperture. As an example an aircraft with 30 m span has been chosen. For 1% integration accuracy 52 data points in 360° are needed at 30 MHz, 5184 points at 3 GHz.

Directivity is sometimes known as 'pattern gain'.

It will be noted that the efficiency of the antenna does not enter into the definition of directivity. It can also be regarded as the comparison between a lossless test antenna and a hypothetical, lossless isotropic radiator. The relationship between gain G_0 and directivity is simply

$$G_0=aD$$

where a = efficiency factor ($\leqslant 1$)

15.4.1.2 *Gain*

Gain is defined as the ratio

$$G=\frac{\text{maximum radiation intensity (test antenna)}}{\text{maximum radiation intensity (reference antenna)}}$$

for the same input power to each radiator.

The gain of the reference antenna is known with reference to an isotropic antenna, possibly by relating to a secondary standard; hence

$$G_0=bG$$

where b is the gain of the reference antenna referred to isotropic ($b>1$).

A half-wave dipole is often used as a secondary standard, the gain of a lossless dipole with respect to isotropic being 2·15 dB. With careful construction dipoles with a loss of less than 0·05 dB may be made.

G_0 is sometimes referred to as 'absolute gain'. When the gain is referred to a half-wave dipole it is written as G(dBd), whilst absolute gain is written as G(dBi).

15.4.2 *Direct measurement of gain*

In principle, direct measurements of gain by substitution are simple; in practice they are beset with traps for the unwary. The most difficult antennas to measure accurately are those with the broadest radiation patterns at the lowest frequencies.

On an antenna test range a known power is radiated from a source antenna and received by the antenna under test, the peak signal level being noted S_1. This antenna is then replaced by the reference antenna and the signal level again noted S_2. The gain is then the ratio of these levels expressed as a ratio of powers: $G = 20 \log S_1/S_2$ dB. To avoid problems of non-linearity in the detection system, the transmitted power can be adjusted to produce the same received signal, the gain then being $G = 10 \log P_2/P_1$.

This measurement is simple if carried out in free space conditions. With most VHF and UHF antennas this will be far from the case and ground reflections will normally be present unless a vertical test range is available. Even this will have limitations in range and antenna size. If we assume that there will be ground reflections it is clear that unless the elevation patterns of the antenna under test and the reference antenna are identical the measured gain will be in error. If the height and distance of the transmitting and receiving antennas are chosen to make the direct and reflected fields in phase, then the apparent gain over the reference antenna is

$$G' = G \frac{(F_D^T + \rho F_I^T)^2}{(F_D^R + \rho F_I^R)^2}$$

where F_D and F_I are the relative directivity factors of direct and reflected waves and the superscripts T and R refer to the test antenna and reference antenna respectively.

Two possible methods of overcoming this problem are suggested. The first is to make the measurements with the antennas as close to the ground as is practical without affecting their impedance and at as large a distance as possible. By reducing the grazing angle in this way there is less chance that the elevation patterns of the two antennas will differ significantly. The second method is to make measurements at a series of distances by moving either the receiving or transmitting antennas. The distance moved has to be such that the field strength passes through a number of maxima and minima. If the field strengths are plotted against distance for each antenna it should be possible to determine an asymptotic value of the relative gain of the antenna under test.

15.4.2.1 *Reference antennas*

The absolute gain of horn antennas can be determined with a high degree of accuracy and such reference antennas are often used in determining the gain of high gain microwave antennas. But horns are not suitable as standards for measurement of gains under 10–15 dBi and this is the region for most VHF and UHF antennas. Some small antennas may indeed have significantly negative gains.

The classical free space standard is the thin half-wave dipole quoted as having an absolute gain of 2·15 dBi. For slightly higher gains the US National Bureau of Standards has developed a double dipole on a ground screen (Fig. 15.24). The gain of about 8 dB is known within 0·1 dB for frequencies above 148 MHz. It can readily be scaled for frequencies up to 1 GHz without loss of accuracy. The International Electrotechnical Commission (IEC) have designed a series of monopoles for use on standard ground planes in the range

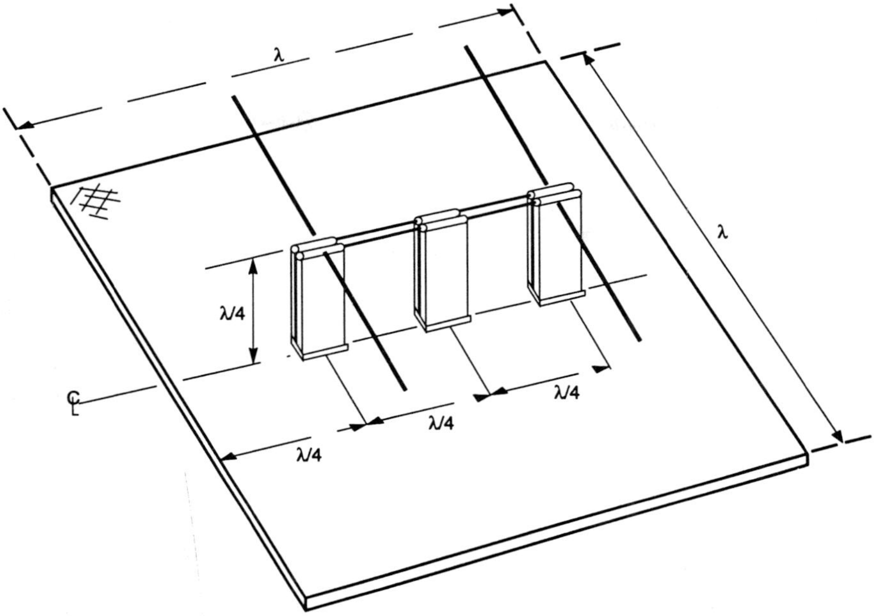

Figure 15.24 *Double dipole gain standard*

30–1000 MHz. These are not absolute standards but are for comparison only with antennas mounted on the same ground plane. Both antennas are dimensioned in IEC 489 [11]. Below 100 MHz the accurate measurement of gain becomes increasingly difficult especially when the antenna is mounted on a vehicle of irregular shape. The best that can usually be achieved is comparison with some single antenna, such as a resonant monopole, similarly mounted.

15.4.2.2 *Measurement of efficiency*
Efficiency is defined as the ratio $R_A/R_A + R_p$ where R_A is the radiation resistance and R_p is the loss resistance. A closed metal container will reflect all the energy radiated by an antenna inside it, except for that absorbed by the walls. If these are highly conductive the loss is negligible and the measured resistance of the antenna becomes R_p. Measuring the antenna impedance in free space gives $R_A + R_p$ and this, with the measurement in the box, gives both parameters. This method was proposed by H.A. Wheeler [10] and is usually known as the Wheeler method. It is particularly useful for antennas on a ground plane as a box with one open side can readily be attached to the ground plane (Fig. 15.25). Good contact with the ground plane is of paramount importance.

One disadvantage with this method becomes apparent with highly efficient antennas: where R_p is small very large mismatches have to be measured. For antennas of low efficiency and small size the method is simple and accurate. Typically the box should have sides of about $\lambda/3$, smaller boxes producing larger reactive components.

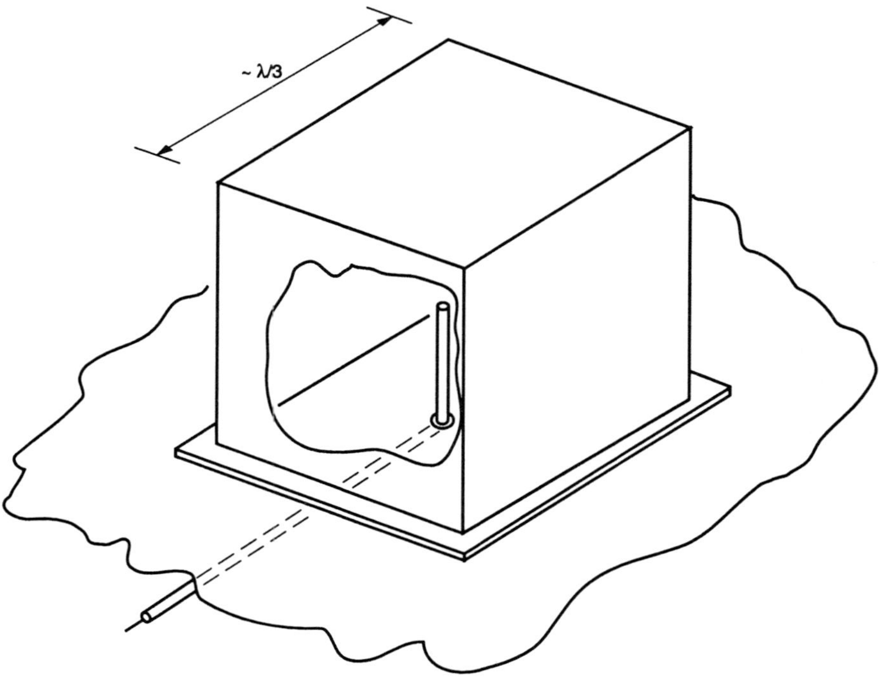

Figure 15.25 *The Wheeler box*

15.5 Measurement under environmental conditions

All antennas, unless they are used indoors, are subject to a number of environmental forces to a greater or lesser degree. At one end of the scale are domestic television antennas subject to wind force and vibration, heat, cold, rain and perhaps icing. At the other end are antennas for high flying, high speed aircraft which have to withstand very high aerodynamic loads, shock, extremes of heat and cold, driving rain, humidity, icing, low pressure, corrosion from aircraft fluids, salt and fungus.

For many of the tests the antenna has to be enclosed in special chambers in which no RF testing is possible although it may be practical to measure insulation resistance if the design of the antenna permits this. Usually it will be adequate to measure the VSWR under defined standard conditions at room temperature and pressure before and after each cycle of environmental testing. The order in which tests are carried out is important and will be laid down in the particular specifications the antenna is intended to meet. Thus heat and cold may affect the mechanical strength of the antenna and should therefore precede any vibration testing. Similarly they may affect antenna sealing and should precede humidity and driving rain tests.

Driving rain is one of the few tests in which VSWR testing can be carried out during the test. The antenna should be mounted as for standard conditions and subjected to water sprayed from equipment which is sufficiently far from the antenna not to modify the VSWR. Because this is an accelerated test the 'rainfall' rate will be much higher than would occur in practice and it is important to note how long the antenna takes after spraying has ceased to return to normal. Where possible the antenna should be mounted so that surface water on, for example, the ground plane does not submerge the antenna.

When the antenna contains fixed tuning elements, particularly capacitors, or active elements it is essential to carry out VSWR testing during hot and cold cycles. This can be done by enclosing the antenna in a non-conducting box with polythene or other suitable plastic walls and blowing in hot or cold air to satisfy the required temperature conditions. If changes in VSWR do occur during testing it is again essential to check that the VSWR returns to normal under standard conditions.

Power handling and voltage breakdown at reduced pressures can be a problem if the antenna is designed for very high altitudes encountered in aircraft or space craft. The most effective method of testing is to connect the antenna to a source capable of supplying the required power with in-line dual directional couplers to monitor the VSWR. The antenna must be enclosed in a plastic pressure vessel which can be evacuated to the required low pressure. Any breakdown will be obvious by rapid changes to the VSWR. Because of expense the pressure vessel may not be very large and may modify slightly the VSWR. In this case the VSWR at normal pressure should be taken as the standard.

The build-up of ice on antennas can affect the VSWR to some extent. It may also modify the radiation pattern of directional antennas, the effects being most noticeable at UHF. Experiments have shown that with UHF Yagi antennas the direction of peak gain can actually be turned through 180° due to ice build-up on directors. The shape of ice build-up is governed by wind speed and tests may be necessary in icing tunnels to check the shape occurring under operational conditions. There is no standard test for ice to date; there is some evidence that its effect on Yagi antennas can be simulated by fitting GRP tubes around the elements but more work needs to be done on this. It would be possible to ice-up an antenna artificially and then check its performance in a non-conductive chamber kept at −5°C. If this were mounted on a turntable it would even be possible to measure radiation patterns.

15.6 References

1 ALIA, F., and RISPOLI, F.: 'A new anechoic and shielded chamber used in either semi-open or closed configuration'. IEE Conf. Publ. 219, 1983, pp. 229–233

2 AWADALLA, K.H., and MACLEAN, T.S.M.: 'Input impedance of a monopole antenna at the centre of a finite ground plane', *IEEE Trans.*, 1978, **AP–26**, pp. 244–248

3 BROWN, G.H., and WOODWARD, O.M.: 'Experimentally-determined impedance characteristics of cylindrical antennas', *Proc. IRE*, Apr. 1945, pp. 257–262

4 FORD, E.T.: 'The ground-reflection mode used on an antenna test site for VHF and UHF measurements' in 'Aerospace antennas'. IEE Conf. Publ. 77, June 1971, pp. 89–97

5 JASIK, H. (Ed.): 'Antenna engineering handbook' (McGraw Hill Book Co., 1936) p. 3.6

6 KING, R.: 'Asymmetrically-driven antennas and the sleeve dipole'. *Proc. IRE*, 1950, **38**, pp. 1154–1164
7 MEIER, A.S., and SUMMERS, W.P.: 'Measured impedance of vertical antennas over finite ground planes', *Proc. IRE*, June 1949, pp. 609–616
8 SICHAK, W., and NAIL, J.J.: 'UHF omnidirectional antenna systems for large aircraft', *IEEE Trans.*, 1954, **AP–2**, pp. 6–15
9 STORER, J.E.: 'The impedance of an antenna over a large circular screen', *J. Appl. Phys.*, 1951, **12**, p. 1058
10 WHEELER, H.A.: 'Fundamental limitations of small antennas', *Proc. IRE*, Dec. 1947, pp. 1477–1484
11 'Methods of measurement of antennas in the mobile service'. IEC 489 Part 2, International Electrotechnical Commission, Geneva

Appendix 1
Calculation of loss resistance

Although the loss resistance per unit length of an antenna remains constant, the power loss will vary if the current is not constant. This is the case, for example, when the antenna is open-ended since the current is zero at the open end. This is strictly true only for an infinitely thin element but is sufficiently true for antennas of large length/diameter. An equivalent total loss resistance at the base of the antenna is needed to add to the base radiation resistance.

Consider a monopole of length l above a perfectly conducting ground plane. If $I(z)$ is the current at height z above the base of the antenna and the loss per unit length is R_L,

$$I_0^2 R_1 = R_L \int_0^l I^2(z)\, dz$$

where I_0 is the base current and R_1 the equivalent loss resistance.

In the general case the current is sinusoidal and

$$I(z) = I_0 \frac{\sin k(l-z)}{\sin kl}$$

which gives $R_1 = \dfrac{R_L l}{2 \sin^2 kl}\left[1 - \dfrac{\sin 2kl}{2kl}\right]$

If the antenna is electrically short, $l/\lambda < 1/30$, then the current distribution may be considered as linearly decreasing with length:

$$I(z) = I_0\left(\frac{l-z}{l}\right) \qquad \text{from which } R_1 = \tfrac{1}{3} R_L l$$

These values are for a single element above ground and must be doubled for a dipole.

Index